Erratum

ASTRONOMY AND ASTROPHYSICS LIBRARY
Volker Schönfelder (Ed.)
The Universe in Gamma Rays
ISBN 3-540-67874-3

Dear Reader,
We regret, that Figure 11.2. on page 277 in this book was reproduced incorrectly. The correct figure is shown below.

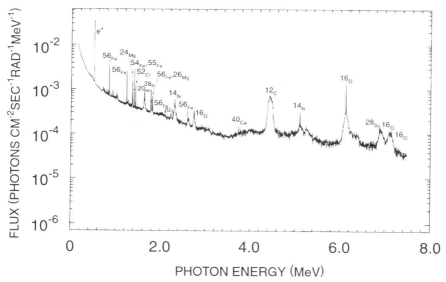

Fig. 11.2. Gamma-ray line spectrum expected (under certain assumptions, see Sect. 11.1) from nuclear reactions in interstellar space towards the general direction of the Galactic Center, superimposed on a bremsstrahlung continuum component. (Reproduced with permission from Ramaty et al. 1979.) The total flux in each line is obtained by integrating over the entire line profile

© Springer-Verlag Berlin Heidelberg 2002

ASTRONOMY AND ASTROPHYSICS LIBRARY

Series Editors: I. Appenzeller, Heidelberg, Germany
G. Börner, Garching, Germany
M. Harwit, Washington, DC, USA
R. Kippenhahn, Göttingen, Germany
J. Lequeux, Paris, France
P. A. Strittmatter, Tucson, AZ, USA
V. Trimble, College Park, MD, and Irvine, CA, USA

Springer
*Berlin
Heidelberg
New York
Barcelona
Hong Kong
London
Milan
Paris
Singapore
Tokyo*

Physics and Astronomy ONLINE LIBRARY

http://www.springer.de/phys/

Volker Schönfelder (Ed.)

The Universe in Gamma Rays

With 141 Figures Including 28 Color Figures

Springer

Prof. Volker Schönfelder (Ed.)
Max-Planck-Institut für extraterrestrische Physik
Giessenbachstrasse
85748 Garching
Germany

Cover picture: All-sky map obtained by COMPTEL on the Compton Gamma-Ray Observatory in the light of the 1.809 MeV gamma-ray line from radioactive ^{26}AL.

Library of Congress Cataloging-in-Publication Data applied for.

Die Deutsche Bibliothek - CIP-Einheitsaufnahme
The universe in gamma rays : with 4 tables / Volker Schönfelder (ed.). - Berlin ; Heidelberg ; New York ; Barcelona ; Hong Kong ; London ; Milan ; Paris ; Singapore ; Tokyo : Springer, 2001 (Astronomy and astrophysics library) (Physics and astronomy online library) ISBN 3-540-67874-3

ISSN 0941-7834
ISBN 3-540-67874-3 Springer-Verlag Berlin Heidelberg New York

This work is subject to copyright. All rights are reserved, whether the whole or part of the material is concerned, specifically the rights of translation, reprinting, reuse of illustrations, recitation, broadcasting, reproduction on microfilm or in any other way, and storage in data banks. Duplication of this publication or parts thereof is permitted only under the provisions of the German Copyright Law of September 9, 1965, in its current version, and permission for use must always be obtained from Springer-Verlag. Violations are liable for prosecution under the German Copyright Law.

Springer-Verlag Berlin Heidelberg New York
a member of BertelsmannSpringer Science+Business Media GmbH

http://www.springer.de

© Springer-Verlag Berlin Heidelberg 2001
Printed in Germany

The use of general descriptive names, registered names, trademarks, etc. in this publication does not imply, even in the absence of a specific statement, that such names are exempt from the relevant protective laws and regulations and therefore free for general use.

Typesetting by the authors using a Springer TEX macro package.
Final typesetting and figure processing: LE-TEX Jelonek, Schmidt & Vöckler GbR, 04229 Leipzig.
Cover design: *design & production* GmbH, Heidelberg

Printed on acid-free paper   SPIN: 10678431   55/3141/mf - 5 4 3 2 1 0

Preface

This is an ideal time to write a book about the Universe as seen in γ-rays. During the last decade, γ-ray astronomy has experienced a period of fruition. The first-ever complete sky surveys have been produced by the Compton Gamma Ray Observatory, with a large number of exciting discoveries; the more than 25-year-old puzzle of the origin of γ-ray bursts is on its way to being solved; and ground-based γ-ray astronomy at TeV energies has achieved its first break-through. We have learned that the γ-ray Universe is violent, energetic and variable. The recent results in this field have attracted scientists from all branches of astronomy and astrophysics. Gamma-ray astronomy has now become an integrated part of astronomy.

This book summarizes recent achievements. The anticipated readers are students, graduates, astronomers and astrophysicists in general, and experts working in the field. The book was written by the team of γ-ray astronomers at the Max-Planck-Institut für extraterrestrische Physik in Garching, Germany. By means of many working sessions we have tried to produce a coherent description of the entire field of γ-ray astronomy. The book was written in 1998 and early 1999, and reflects the state of the research field up to that date.

The distinction between X-ray and γ-ray astronomy is not well defined (the classical definition of X-radiation being produced in atomic shells and γ-radiation in atomic nuclei does not make sense in astronomy). In this book, we consider γ-ray astronomy to begin at photon energies around 500 keV. Sometimes lower energy photons are considered as well, if this turns out to be important.

Following the introduction, which is devoted to the history of γ-ray astronomy, the book starts with two chapters describing the processes of cosmic γ-ray production and absorption and the basic principles of instrumentation used in γ-ray astronomy. The main part of the book (starting with Chap. 4) deals with the astronomical results and their interpretation. Those readers who are not interested in the instrumentation and/or the production processes, may skip Chaps. 2 and 3. The collection of references and the subject index at the end of the book may be a helpful starting point for readers who want to go deeper into this new research field.

Many friends and colleagues have given us their support in improving the book. We want to thank particularly Drs. Peter von Ballmoos, Bernd Aschenbach, Edward L. Chupp, Dieter Hartmann, Felix Aharonian, Werner Becker, Markus Böttcher, Karen Brazier, Andreas von Kienlin, Stefan Plüschke, John Mattox, Bonnard Teegarden, David J. Thomson, and Jörn Wilms, who all made valuable suggestions on early drafts of the manuscript. We are also indebted to Mrs. Helga Haber for her supporting role in fulfilling the many editorial tasks and to Dr. Robert Georgii, who took great care of all technical editorial matters.

As editor of this book, I would like to take this opportunity to express my special gratitude to my wife, Bärbel, who has provided essential support and encouragement to me throughout my entire professional life.

Garching, May 2001 *Volker Schönfelder*

Contents

1 Introduction
Volker Schönfelder 1
References 7

2 Gamma-Ray Production and Absorption Processes
Roland Diehl 9
2.1 Introduction to Astronomy with Gamma-Rays 9
2.2 Production Processes 12
    2.2.1 Charged Particles in Strong Electric or Magnetic Fields 13
    2.2.2 Inverse Compton Scattering 16
    2.2.3 Nuclear Transitions 17
    2.2.4 Decays and Annihilation 18
    2.2.5 Charged particles bound in strong magnetic fields 19
2.3 Gamma-Ray Interactions, Absorption Processes 20
2.4 Cosmic Gamma-Ray Sources 21
2.5 The Fate of Gamma-Rays Between Source and Detector 23
References 25

3 Instruments
Giselher Lichti and Robert Georgii 27
3.1 Detection of Gamma-Rays from Space 27
    3.1.1 The Opaqueness of the Atmosphere 27
    3.1.2 The Low Cosmic Gamma-Ray Fluxes 27
    3.1.3 The Background of Gamma-Ray Telescopes 28
3.2 Gamma-Ray Detection Techniques 31
    3.2.1 Scintillation Techniques 31
    3.2.2 Solid-State Detectors 35
    3.2.3 Spark Chambers 37
    3.2.4 Cherenkov Detectors 38
    3.2.5 Drift Chambers 39
3.3 Gamma-Ray Telescopes 39
    3.3.1 The Chopper Technique 41
    3.3.2 The On/Off Technique 41
    3.3.3 Compton Telescopes 50

3.3.4 Pair-Tracking Telescopes 56
 3.3.5 Air-Shower Detection Techniques 60
 3.3.6 Gamma-Ray Burst Instruments 65
3.4 Specific Calibration and Data-Analysis Problems 66
3.5 Current Developments and Future Instrumentation 68
 3.5.1 Advanced Compton Telescopes 68
 3.5.2 Advanced Pair-Production Telescopes 71
 3.5.3 The Gamma-Ray Lens 71
References 73

4 Summary of the Gamma-Ray Sky
Volker Schönfelder 77

4.1 Introduction 77
4.2 All-Sky Maps in the MeV–GeV Region 78
4.3 The Sub-MeV Domain 80
4.4 The TeV Gamma-Ray Sky 86
4.5 Gamma-Ray Bursts 87
References 89

5 The Sun as a Gamma-Ray Source
Erich Rieger and Gerhard Rank 91

5.1 Introduction 91
5.2 Production of Gamma-Rays and Neutrons 94
 5.2.1 Flare Scenario 94
 5.2.2 Conditions for Gamma-Ray Production
  in the Solar Flare Environment 95
5.3 The Solar Flare Gamma-Ray Spectrum 100
 5.3.1 The Observed Gamma-Ray Spectrum 100
 5.3.2 Spectral Analysis 103
5.4 Temporal History of Gamma-Ray Emission During Flares 111
 5.4.1 Evolution of Flares: Impulsive and Gradual Emission 111
 5.4.2 Timing of Electron and Ion Interactions 113
 5.4.3 Extended Gamma-Ray Emission 114
5.5 Acceleration Mechanisms 116
 5.5.1 Theoretical Considerations of Particle Acceleration 116
 5.5.2 Acceleration Models for Solar Flares 119
5.6 Conclusions 121
 5.6.1 Our Current Understanding of the Flare Phenomenon 121
 5.6.2 Expectations for the Future 122
References 123

6 Gamma-Ray Pulsars
Gottfried Kanbach 127

6.1 Introduction 127
    6.1.1 Radio Pulsars: Discovery of Rotating Neutron Stars 127
    6.1.2 The Formation of Neutron Stars
          and Their Basic Properties 128
    6.1.3 Expectation of High-Energy Phenomena
          Around Young Pulsars................................. 131
    6.1.4 Early History of the Observation of Gamma-Ray Pulsars... 134
6.2 Current Observational Results of High-Energy Emission
    from Pulsars 135
    6.2.1 Detections of High-Energy Pulsars 135
    6.2.2 Gamma-Ray Searches and Limits for Other Radio Pulsars.. 138
    6.2.3 Light Curves of Gamma-Ray Pulsars 140
    6.2.4 Multi-Wavelength Light Curves 141
    6.2.5 Spectra of Electromagnetic Radiation from Pulsars........ 143
    6.2.6 Phase-Dependent Gamma-Ray Spectra of Pulsars 147
    6.2.7 Pulsar Luminosities and Gamma-Ray Efficiencies 147
    6.2.8 Search for Identification of Other Gamma-Ray Sources 149
6.3 Theory of Pulsar High-Energy Emissions 150
    6.3.1 Acceleration Zones and Production of Relativistic Particles. 150
    6.3.2 Radiation and Propagation Processes.................... 151
    6.3.3 Field Line Geometries, Beaming and Light Curves 153
    6.3.4 Spectra, Luminosities and Efficiencies 154
References 156

7 X-Ray Binaries as Gamma-Ray Sources
Werner Collmar 159

7.1 The Nature of X-Ray Binaries............................... 159
7.2 High-Energy Emission from Black-Hole Systems 161
    7.2.1 Time Variability 164
    7.2.2 Energy Spectra 168
    7.2.3 Interpretation 171
7.3 High-Energy Emission from Neutron-Star Binaries
    and Cataclysmic Variables 175
7.4 Peculiar Cases 177
    7.4.1 Cygnus X-3 177
    7.4.2 Centaurus X-3.. 178
7.5 TeV Observations of X-Ray Binaries 179
7.6 Summary and Conclusions................................... 181
References 183

8 Continuum Gamma Ray Emission from Supernova Remnants
Anatoli Iyudin and Gottfried Kanbach 185

8.1 Introduction 185
    8.1.1 Shell-type SNRs 188
    8.1.2 Crab-like SNRs 188
    8.1.3 Composite SNRs....................................... 189
8.2 Continuum Gamma Rays from SNR 191
8.3 SNRs and Particle Acceleration 197
8.4 Discussion and Outlook 200
8.5 Conclusions.. 204
References 204

9 Diffuse Galactic Continuum Gamma-Rays
Andrew W. Strong and Igor V. Moskalenko 207

9.1 Introduction 207
9.2 Theoretical Background 207
    9.2.1 Basic Processes of Gamma-Ray Production................ 207
    9.2.2 Cosmic-Rays 210
    9.2.3 Galactic Structure 213
9.3 Gamma-Ray Observations 215
9.4 Interpretation 217
    9.4.1 Longitude and Latitude Distributions.................... 218
    9.4.2 Spectra of the Inner Galaxy 219
    9.4.3 Implications for Cosmic Rays 221
    9.4.4 Multi-component Fitting Approach...................... 223
    9.4.5 Implications for ISM: H_2/CO Calibration 225
    9.4.6 Local Clouds 227
    9.4.7 Contribution of Unresolved Sources to 'Diffuse' Emission ... 227
    9.4.8 High Latitudes 227
9.5 Global Gamma-Ray Properties of Our Galaxy................... 228
9.6 Extragalactic Objects 229
    9.6.1 The Large and Small Magellanic Clouds 229
    9.6.2 Other Galaxies 230
References 230

10 Nucleosynthesis
Roland Diehl 233

10.1 Introduction 233
10.2 Nucleosynthesis Processes 236
10.3 Sites of Nucleosynthesis 242
    10.3.1 Hydrostatic Nuclear Burning 242
    10.3.2 Explosive Nucleosynthesis............................. 244
    10.3.3 Other Nucleosynthesis Sites 252

10.4 Gamma-Ray Lessons on Supernovae 254
    10.4.1 Individual Nucleosynthesis Sources 254
    10.4.2 Integrated Nucleosynthesis 262
10.5 Summary and Perspective 271
References 272

11 Nuclear Interaction Gamma-Ray Lines
Volker Schönfelder and Andrew W. Strong 275

11.1 Introduction 275
11.2 Nuclear Interaction Lines from the Inner Galaxy 278
11.3 Nuclear Interaction Lines from Accretion Processes 279
11.4 The Observations of the Orion Complex 280
References 282

12 Gamma-Ray Emission of Active Galaxies
Werner Collmar 285

12.1 The Nature of Active Galaxies 285
    12.1.1 Differences Between Normal and Active Galaxies 285
    12.1.2 The Source of Energy for Active Galaxies 286
    12.1.3 Unification Schemes for Active Galaxies 287
12.2 Gamma-Ray Blazars 291
    12.2.1 Observational Status on AGN before CGRO 291
    12.2.2 CGRO Source Detections 291
    12.2.3 Observational Properties 293
    12.2.4 Broadband Emission of CGRO-Detected Blazars 296
    12.2.5 TeV Emission from Blazars 298
    12.2.6 Interpretation 300
    12.2.7 Summary and Open Questions 306
12.3 Seyfert and Radio Galaxies 307
    12.3.1 Seyfert Galaxies 307
    12.3.2 Centaurus A 312
    12.3.3 Summary 313
12.4 Summary and Conclusions 314
References 316

13 Unidentified Gamma-Ray Sources
Olaf Reimer 319

13.1 Objective 319
    13.1.1 Gamma-Ray Source Detection History 319
    13.1.2 Compton Gamma-Ray Observatory Era 320
13.2 The Observations 321
13.3 Correlation Studies 328
    13.3.1 Correlations with Source Populations 328
    13.3.2 Prominent Individual Unidentified Sources 332

14 The Extragalactic Gamma-Ray Background
Georg Weidenspointner and Martin Varendorff............ 339

14.1 Introduction 339
14.2 History 343
14.3 Current Results 346
14.4 Possible Origins 350
    14.4.1 Unresolved Point Sources 351
    14.4.2 Truly Diffuse Sources 357
14.5 Conclusions and Prospects 360
References 364

15 Gamma-Ray Bursts
Martin Varendorff.. 367

15.1 Introduction 367
    15.1.1 The History of Gamma-Ray Burst Measurements 367
    15.1.2 Soft Gamma-Ray Repeater 371
    15.1.3 Bursts from Earth 372
15.2 Properties of Gamma-Ray Bursts 372
    15.2.1 Time Characteristics of Gamma-Ray Bursts 373
    15.2.2 Spectral Characteristics of Gamma-Ray Bursts 376
    15.2.3 Spatial Distribution of Gamma-Ray Bursts 379
15.3 Counterparts of Gamma-Ray Bursts 380
    15.3.1 Search for Counterparts in Other Wavelengths 380
    15.3.2 A New Approach with the Beppo-Sax Satellite and First Success 381
15.4 Models for Gamma-Ray Bursts 385
    15.4.1 The Distance to the Gamma-Ray Burst Sources 385
    15.4.2 Cosmic Fireball: The Evolution of the Blast Wave ... 387
    15.4.3 The Central Engine 390
15.5 Prospects 393
References 395

Index 397

13.4 Conclusion 336
References 337

List of Contributors

V. Schönfelder
Max-Planck Institut
für extraterrestrische Physik
Postfach 1312
85741 Garching, Germany
vos@mpe.mpg.de

W. Collmar
Max-Planck Institut
für extraterrestrische Physik
Postfach 1312
85741 Garching, Germany
wec@mpe.mpg.de

R. Diehl
Max-Planck Institut
für extraterrestrische Physik
Postfach 1312
85741 Garching, Germany
rod@mpe.mpg.de

R. Georgii
Max-Planck Institut
für extraterrestrische Physik
Postfach 1312
85741 Garching, Germany
rog@mpe.mpg.de

A. Iyudin
Max-Planck Institut
für extraterrestrische Physik
Postfach 1312
85741 Garching, Germany
ani@mpe.mpg.de

G. Kanbach
Max-Planck Institut
für extraterrestrische Physik
Postfach 1312
85741 Garching, Germany
gok@mpe.mpg.de

G. Lichti
Max-Planck Institut
für extraterrestrische Physik
Postfach 1312
85741 Garching, Germany
grl@mpe.mpg.de

I. Moskalenko
presently at:
NASA Goddard
Space Flight Center, Code 660
National Research Council
Senior Research Associate
Greenbelt, MD 20771, USA

D. V. Skobeltsyn Institute
of Nuclear Physics
Moscow State University
Moscow 119899, Russia
imos@galprop.gsfc.nasa.gov

G. Rank
Max-Planck Institut
für extraterrestrische Physik
Postfach 1312
85741 Garching, Germany
gzr@mpe.mpg.de

O. Reimer
presently at:
NASA Goddard
Space Flight Center, Code 661
National Research Council
Resident Research Associate
Greenbelt, MD 20771, USA
olr@egret.gsfc.nasa.gov

E. Rieger
Max-Planck Institut
für extraterrestrische Physik
Postfach 1312
85741 Garching, Germany
SCHAFI24@hotmail.com

A. Strong
Max-Planck Institut
für extraterrestrische Physik
Postfach 1312
85741 Garching, Germany
aws@mpe.mpg.de

M. Varendorff
Max-Planck Institut
für extraterrestrische Physik
Postfach 1312
85741 Garching, Germany
martin@Varendorff.de

G. Weidenspointner
presently at:
NASA Goddard
Space Flight Center, Code 661
Greenbelt MD 20771, USA

Universities
Space Research Association
7501 Forbes Blvd. 206
Seabrook, MD 20706-2253, USA
ggw@tgrosf.gsfc.nasa.gov

1 Introduction

Volker Schönfelder

Astronomy is now performed over the entire range of the electromagnetic spectrum, from radio to γ-ray energies. From the so-called "New Astronomies", which are performed outside the optical window, we learned during the second half of the 20th century that each spectral range provides specific information which cannot be obtained by other means. Gamma radiation represents the most energetic part of the electromagnetic spectrum (see Fig. 1.1). Therefore it is natural that it provides information about the most energetic processes and phenomena in the Universe.

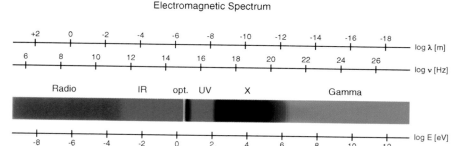

Fig. 1.1. The electromagnetic spectrum, from radio to γ-ray energies. The electromagnetic radiation can be characterized either by its photon energy (measured in eV) or by its frequency (measured in Hz) or by its wavelength (measured in m)

The energy band of γ-ray astronomy extends over more than seven orders of magnitude – from typically several 100 keV (i.e., from 500 keV) to more than 1 TeV – which is similar to the band from radio to optical astronomy. It is thus not surprising that a wide variety of different objects and phenomena can be studied.

The entire field of γ-ray astronomy can be separated into two broad domains. The first one is the domain of "spaceborne γ-ray astronomy", which ranges from 500 keV to about 100 GeV. Gamma rays in this energy range cannot penetrate the Earth's atmosphere without being absorbed and scattered. These γ-rays can only be detected from space with satellite experiments or with high-altitude balloon experiments. The other domain is that of "ground-based γ-ray astronomy", which begins at energies above several 100 GeV. At these energies electromagnetic cascades are developed by the γ-rays in the Earth's atmosphere, producing Cherenkov light in the direction of the photon trajectory. The Cherenkov light is detected by optical telescopes from the ground.

In both domains, great successes with spectacular discoveries have been achieved during the last decade. The field of γ-ray astronomy is currently experiencing a golden age. In spaceborne γ-ray astronomy the first complete systematic sky surveys have been accomplished between 1 MeV and > 1 GeV by NASA's Compton Observatory with its two telescopes COMPTEL and EGRET. From these surveys and from measurements by other telescopes like OSSE and BATSE on the Compton Observatory and SIGMA on the Russian mission GRANAT, we have learned that the sky is rich in high-energy phenomena.

Specifically, the most compact and energetic objects emit γ-rays – neutron stars, stellar and massive black holes, supernova explosions/remnants, and cosmic rays, via their interactions with matter and fields. In addition, it appears that most of the γ-ray sky is continuously changing. With γ-rays we see the most violent parts of the Universe.

In the field of ground-based γ-ray astronomy the real breakthrough came in the late 1980s after instrumental techniques had been developed which could separate the Cherenkov light associated with γ-rays from the light produced by high-energy cosmic ray interactions in the Earth's atmosphere. Since then the detection of a few TeV γ-ray sources has been well established.

Gamma-ray astronomy has become an integral part of astronomy and astrophysics. It has been recognized that objects exist in the Universe, such as radio pulsars, quasars or the famous γ-ray burst sources, which have their peak luminosities at γ-ray energies. It is practically impossible to understand these objects without knowing their γ-ray properties. The large number of spectacular discoveries in γ-ray astronomy will find the attention of astrophysicists in many areas for a long time to come.

It took tremendous effort lasting over several decades to reach the present status. Initial attempts to place upper limits on the fraction of γ-rays in the primary cosmic radiation were performed with balloon and rocket experiments in the 1940s and early 1950s (e.g., Hulsizer and Rossi 1944; Perlow and Kissinger 1951; Rest, Reiffel and Stone 1951). The history of γ-ray astronomy really started in the 1950s with predictions: Hayakawa (1952) predicted the diffuse γ-ray emission following the decay of π°-mesons from cosmic-ray interstellar matter interactions, and Hutchinson (1952) predicted the γ-ray emission from cosmic-ray bremsstrahlung. The papers by Burbidge, Burbidge, Fowler and Hoyle (1957) and Morrison (1958) raised hopes that it might be relatively easy to discover cosmic γ-rays from special celestial objects, in line or continuum emission respectively. Especially after the Morrison paper numerous efforts with balloon experiments were made to discover the first cosmic γ-rays. We now know that Morrison's predictions were too optimistic, by several orders of magnitude, and no instruments at that time had enough sensitivity, and all had too much background. A review of the early history of γ-ray astronomy is given by Greisen (1966).

Measurements with early instruments were often limited by statistics or systematic uncertainties. These difficulties have led to some spurious and – as we now know – often wrong results.

Though much of the pioneering work in the area of instrument development for γ-ray astronomy was achieved with balloon instruments, nearly all the important discoveries were made from satellites. The major milestones in observational γ-ray astronomy are summarized in the timeline shown in Fig. 1.2.

The first reliable detections of cosmic γ-rays from space and from the Earth's atmosphere were those made by Explorer 11 in 1961 (Kraushaar et al. 1965) and by OSO-III in 1968 (Kraushaar et al. 1972) from the Milky Way above 100 MeV.

The cosmic γ-ray burst phenomenon was discovered at about the same time (in 1967) by the network of Vela satellites of the U.S. Department of Defense, which were designed to monitor for evidence of clandestine nuclear tests after the Nuclear Test Ban Treaty of 1963. The discovery was not made public for several years; in 1973 (Klebesadel et al. 1973) it was publicized as a new astronomical phenomenon, whose origin remained a puzzle for another 25 years.

In the 1960s and early 1970s, several satellites were launched to explore the cislunar space up to 400 000 km from Earth; they contained omnidirectional sensitive scintillation γ-ray detectors operating typically from 100 keV to 30 MeV [Ranger 3 and 5 (Metzger et al. 1964), Apollo 15 (Trombka et al. 1973)]. From these measurements a diffuse cosmic background component was derived whose energy spectrum showed a bump at photon energies of a few MeV. The origin of the bump was the subject of quite a number of theoretical γ-ray astrophysics papers in the early 1970s. We now know that this spectral feature was actually an artifact caused by instrumental background radiation.

In August 1972 the first direct evidence for specific γ-ray lines associated with solar flares was obtained (Chupp et al. 1973). The emission observed by OSO-VII showed strong annihilation (511 keV) and neutron capture lines (at 2.23 MeV), but also nuclear interaction lines at 4.4 and 6.1 MeV from excited carbon and oxygen nuclei with weaker fluxes. Pure continuum emission from solar flares up to 3,7 MeV had been detected already earlier – in May 1967 – by ERS-18 (Gruber et al. 1973).

A major step forward in Galactic γ-ray astronomy at high energies was taken with the two satellite missions SAS-II and COS-B (>30 and >70 MeV, respectively), which were launched in 1972 and 1975, and both of which contained spark chamber experiments. Whereas the OSO-III mission still suffered from low statistics (in total only 621 cosmic γ-rays were detected), SAS-II

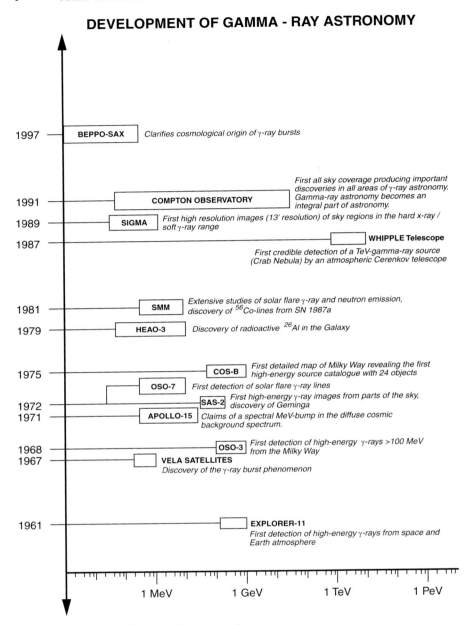

Fig. 1.2. Timeline of the development of γ-ray astronomy

and COS-B provided the first accurate maps of the Milky Way. The measured longitude and latitude profile resulted in a lively discussion on a possible gradient of cosmic rays in the Galaxy and on the question of a galactic or extragalactic origin of the bulk of cosmic rays observed near Earth (Kniffen and Fichtel et al. 1981; Mayer-Hasselwander et al. 1982). Apart from the study of the diffuse galactic γ-ray emission the two missions also provided the first real source detections. The strongest source features were found from the Crab and Vela pulsars and from a source called Geminga, which remained unidentified for another twenty years, and was finally identified as a pulsar as well (Halpern and Holt 1992). The COS-B source catalog contained in total 25 objects (Swanenburg et al. 1981), of which only one (3C 273) was extragalactic.

In the field of cosmic γ-ray line spectroscopy a milestone was achieved by the Germanium spectrometer on board HEAO-C, with the first detection of a nucleosynthetic line: the 1.809 MeV line from radioactive ^{26}Al, which was discovered from the general direction of the Galactic Center region (Mahoney et al. 1984). The same spectrometer also provided an accurate measurement of the line profile of the 511 keV line from the Galactic Center region with a continuum tail towards lower energies from three-photon positronium decay (Riegler et al. 1985). The first convincing detection of this line had been made by the Bell–Sandia balloon group a few years earlier (Leventhal et al. 1978).

The scintillation γ-ray spectrometer on the Solar-Maximum Mission, which was launched in 1981 and stayed in orbit for nine years, provided a wealth of information on solar flare γ-ray emission in lines and continuum up to 10 MeV (Chupp 1984). In addition, it also made important studies of cosmic γ-ray lines, e.g., of the 511-keV annihilation line (Share et al. 1988) and the 1.809-MeV ^{26}Al line (Share et al. 1985), and in particular, it impressively succeeded in detecting nucleosynthetic lines from ^{56}Co decay in SN 1987a in the Large Magellanic Cloud (Leising and Share 1990).

In the 1990s γ-ray astronomy matured. This "golden epoch" started with the launch of the French telescope SIGMA on board the Russian GRANAT mission in December 1989. SIGMA allowed imaging in the transition region between X-ray and γ-ray astronomy (mainly around 100 keV) with an unprecedented angular resolution of the order of 10 arc minutes. It mainly observed the Galactic Center region, where it detected about 30 sources representing a great variety of objects (see Vargas et al. (1997)). Examples are the so-called X-ray novae, which are believed to be accreting binaries with a stellar black hole and which produce variable X/γ-ray emission with huge outbursts, and the so-called "Galactic microquasars", which are able to produce strong features around 511 keV and which show a double-sided jet structure emanating from a compact cone at radio wavelengths. Most of the SIGMA sources turned out to be highly variable.

The wealth of exciting discoveries from the Compton Observatory, which was in operation from April 1991 to June 2000, established the role of γ-ray astronomy as an important branch of astronomy and astrophysics, in general (Gehrels et al. 1993). The presently known celestial γ-ray sources cover a large variety of different objects, such as the Sun, isolated spin-down pulsars, accreting binaries with stellar neutron stars and black holes, supernovae, supernova remnants, the interstellar medium, normal galaxies, radio galaxies, Seyfert galaxies, quasars, and the famous γ-ray burst sources. About 300 celestial γ-ray sources (excluding the more than 2000 γ-ray burst sources) are currently known, of which two thirds are still unidentified.

The long-awaited breakthrough in the field of γ-ray bursts came in 1997, when the Italian/Dutch X-ray satellite Beppo-Sax succeeded in observing X-ray afterglows from a few burst sources (Costa et al. 1997). The subsequent observations of these objects at optical – and in a few cases also radio – wavelengths clearly established the extragalactic origin of the γ-ray burst sources (Kouveliotou 1997). Another milestone was reached on January 23, 1999, when bright optical emission was detected for the first time from a burst (GRB 990113), while it was still in progress (Akerlof et al. 1999). The energy released from these objects in γ-rays is of the order of 10^{52} to several 10^{54} erg (assuming isotropic emission); these values are comparable to the entire rest-mass of the Sun. This enormous amount of energy might be released during the coalescence of compact objects like neutron stars or black holes or during the collapse of a massive star leading to the formation of a hyper-accreting black hole.

The recent exciting discoveries in ground-based γ-ray astronomy around 1 TeV, made by using atmospheric Cherenkov imaging, have their origin in the mid-1980s, when Hillas (1985) developed a technique to discriminate γ-ray-initiated air showers from those generated by protons. Crucial for the difference between these showers is the shape of the Cherenkov image. This technique resulted in the first credible detection of a TeV source, namely the Crab Nebula, at these energies (Weekes et al. 1989). With this detection, a long-lasting period of spurious reports of source detections of TeV-radiating objects ended. Since then, the number of well-established TeV sources is constantly increasing; definite detections have been made of the Crab Nebula, PSR 1706-44 (unpulsed), and the active galaxies Markarian 421 and Markarian 501 (Weekes et al. 1997).

Gamma-ray astronomy has opened a new and fascinating window in astronomy. The large number of exciting and important results have stimulated new theories and plans for new γ-ray astronomy missions, which will increase our understanding of the Universe even further. In order to reach the present status, tremendous difficulties had to be overcome. The fact that X-ray astronomy developed much faster than γ-ray astronomy is mainly due to three things: first, X-ray fluxes from celestial objects are orders of magnitude higher than γ-ray fluxes; second, conversion lengths of X-rays in detector materials are of the order of milligrams per cm^{-2}, whereas those of γ-rays are $10\,g\,cm^{-2}$ or more – hence massive and large-area detectors are an absolute must for γ-ray observations; third, focussing is not practically feasible at γ-ray energies above 500 keV, which is why the angular resolution achieved with presently existing telescopes lags far behind that achieved in X-ray astronomy. In spite of these difficulties, the results obtained so far in γ-ray astronomy are fascinating and manifest the tremendous effort made in the past.

References

Akerlof, C., Balsano, R., Barthelmy, S. et al., 1999, Nature **398**, 400
Burbidge, E.M., Burbidge, G.R., Fowler, W.A., Hoyle, F., 1957, Prog. Theor. Phys. **29**, 457
Chupp, E.L., 1984, Annu. Rev. Astron. Astrophys. **22**, 359
Chupp, E.L., Forrest, D.J., Higbie, P.R. et al., 1973, Nature **241**, 333
Costa, E., Frontera, F., Heise, J. et al., 1997, Nature **387**, 783
Gehrels, N., Fichtel, C.E., Fishman, G.J., Kurfess, J.D., Schönfelder, V., 1993, Sci. Am. Dec 93, 38
Greisen, K., 1966, in: *Perspective of Modern Physics*, ed. by R.E. Marshak, John Wiley & Sons, New York, p. 355
Gruber, D.E., Peterson, L.E., Vette, J.I., 1973, ed. by Ramaty, R., Stone, E.C., NASA-SP-342, 147
Halpern, J.-P., Holt, S.S., 1992, Nature **357**, 222
Hayakawa, S., 1952, Prog. Theor. Phys. **8**, 571
Hillas, A.M., 1985, Proc. of 13th Int. Cosmic Ray Conf. (Denver), ed. by Chasson, R.L., Vol. **3**, 445
Hulsizer, R.I., Rossi, B., 1944, Phys. Rev. **73**, 1402
Hutchinson, G.W., 1952, Philos. Mag. **43**, 847
Klebesadel, R.W., Strong, I.B., Olson, R.A., 1973, Astrophys. J. **182**, L85
Kniffen, D.A., Fichtel, C.E., 1981, Astrophys. J. **250**, 389
Kouveliotou, C., 1997, Sci **277**, 1257
Kraushaar, W.L., Clark, G.W., Garmire, G.P., Helmken, H. et al., 1965, Astrophys. J. **141**, 845
Kraushaar, W.L., Clark, G.W., Garmire, G.P. et al., 1972, Astrophys. J. **177**, 341
Leising, M.D., Share, G.H., 1990, Astrophys. J. **357**, 638
Leventhal, M., MacCallum, C.J., Stang, P., 1978, Astrophys. J. **225**, L11
Mahoney, W.A., Ling, J.C., Wheaton, W.A., Jacobson, A.S., 1984, Astrophys. J. **286**, 578
Mayer-Hasselwander et al., 1982, Astron. Astrophys. **105**, 164
Metzger, A.E., Anderson, E.C., Van Dilla, M.A., Arnold, J.R., 1964, Nature **204**, 766
Morrison, P., 1958, Nuo Cim **7**, 858
Perlow, G.J., Kissinger, C.W., 1951, Phys. Rev. **81**, 552
Rest, F.G., Reiffel, L., Stone, C.A., 1951, Phys. Rev. **81**, 894
Riegler, G.R., Ling, J.C., Mahoney, W.A. et al., 1985, Astrophys. J. **294**, L13 and correction in Astrophys. J. **305**, L33 (1986)
Share, G.H., Kinzer, R.L., Kurfess, J.D. et al., 1985, Astrophys. J. **292**, L61
Share, G.H., Kinzer, R.L., Kurfess, J.D. et al., 1988, Astrophys. J. **326**, 717
Swanenburg, B.N., Bennett, K., Bignami, G.F. et al., 1981, Astrophys. J. Letter **243**, L69

Trombka, J.I., Metzger, A.E., Arnold, J.R. et al., 1973, Astrophys. J. **181**, 737
Vargas, M., Paul, J., Goldwurm, A. et al., 1997, Proc. of 2nd INTEGRAL Workshop (St. Malo), ed. by Winkler, C., Courvoisier, T.J.-L., Durouchoux, P., ESA-SP-382, 129
Weekes, T.C., Cawley, R.F., Feagan, D.J. et al., 1989, Astrophys. J. **342**, 379
Weekes, T.C., Aharonian, F., Fegan, D.J., Kifune, T., 1997, Proc. of 4th Compton Symposium (Williamsburg), ed. by Dermer, C.D., Strickman, M.S., Kurfess, J.D., AIP Conf. Proc. **410**, 361

2 Gamma-Ray Production and Absorption Processes

Roland Diehl

2.1 Introduction to Astronomy with Gamma-Rays

How do γ-rays compare with other types of radiation? 'Radiation' in common language describes 'energy packages' which travel on straight paths. 'Electromagnetic radiation' is characterized by variations of electric and magnetic fields in space and time. Another type of 'radiation' are 'cosmic rays', very energetic particles discovered early in the 20th century in the upper atmosphere of the Earth and known to pervade interstellar space. These particles are called 'cosmic radiation' because with their high energies they propagate at the speed of light and in certain aspects behave like photons of similar energies, γ-rays.

In order to understand the nature of γ-rays in comparison to other electromagnetic radiation, we remind the reader about 'wave–particle' dualism: We may usefully describe electromagnetic radiation *either* in terms of a propagating *particle* with energy[1] $E = \hbar\omega$ *or* in terms of *waves* with wavelength[2] $\lambda = c/\nu$ with characteristic diffraction and interference patterns. The more appropriate description depends on the energy of the radiation, more precisely on the relation between the wavelength of the radiation and the structural dimensions of the matter it interacts with, i.e., typical distances between atoms, lattice constants, fiber thicknesses, grain sizes, field curvature.[3] Unlike common 'light', γ-rays do not reflect off surfaces; they have high penetration power.

The spectrum of observed electromagnetic radiation spans more than 20 orders of magnitude on the frequency scale,[4] ranging from the low-frequency/long-wavelength regime of radio waves ($\lambda \simeq 10^2$ cm) through the infrared, optical and ultraviolet regimes up to the high-frequency/short-wavelength regime of X- and γ-rays ($\lambda \leq 10^{-13}$ cm). In the regime of γ-rays, the

---

[1] $\hbar = 1.05 \times 10^{-34}$ J s^{-1} = 6.58×10^{-22} MeV s is Planck's (reduced) constant, $h/2\pi$; $\omega = \nu/2\pi$, with ν the frequency of the radiation (Hz); per photon γ-rays are thus far more energetic than, e.g., UV radiation with typical energies of 0.5 eV. Energy conversion: 1 eV = 1.602×10^{-19} J.
[2] c = speed of light = 2.998×10^{10} cm s^{-1}.
[3] Typical structural dimensions are dust grain sizes, 10^{-4} cm; atoms, 10^{-8} cm; atomic nuclei, 10^{-13} cm.
[4] Cosmic ray energies range from $\geq$MeV up to 10^{20}eV.

particle description of electromagnetic radiation becomes more appropriate than the wave description, which works well for the less energetic radiation at radio or optical wavelengths. (The general field of high-energy astrophysics is addressed in detail by Longair 1994, in a two-volume textbook.) This helps us understand the penetration power of γ-rays: the wavelength of the radiation is short compared to the spacing of the atoms in the material, hence the radiation mainly 'sees' the atom's components, the compact nucleus and the electrons at comparatively large distances (which are seen by γ-rays as almost independent from the nucleus). The atom is mostly 'empty space' for a γ-ray photon; interactions occur upon relatively low-probability encounters with the nucleus or electrons. It is more appropriate to view γ-rays as energetic particles, their 'size' being their wavelength. In such a corpuscular description, the energy of the photon is a better characteristic quantity than the wavelength or frequency of the electromagnetic field. Energy is measured in voltage units 'eV' for electromagnetic particles (photons), 1 MeV being the energy an electron would have if accelerated by an electrostatic potential of a million volts. The γ-ray domain is the energy regime above X-rays,[5] above some 100 keV. Gamma-rays with energies up to TeV = 10^{12} eV have been observed, so that the domain of γ-rays spans about seven orders of magnitude (in frequency/wavelength/energy).[6]

Materials which should serve as 'mirrors' for X- and γ-rays would need characteristics which make a wave description (i.e., coherent scattering geometries) applicable for the γ-ray–material interactions. This becomes virtually impossible below a certain wavelength regime corresponding to X-rays of a few 100 keV. The specific instrumental techniques of γ-ray telescopes are described in Chap. 3.

What makes γ-rays so special that we need space observatories? The above description of the interaction between electromagnetic radiation and matter tells us that different effects will characterize the absorbing power of the Earth's atmosphere as we go up in frequency from radio waves to γ-rays (Fig. 2.1). Interaction cross sections are maximized when the wavelength of the radiation becomes similar to the size scale of the scattering centers of matter ('resonances'). Radio waves ($\lambda \geq$ cm) thus easily penetrate the atmosphere; ionized-air atmospheric turbulence has much larger typical scales, and atom dimensions are much smaller. At optical frequencies, the electronic shells of atoms and molecules resonate with electromagnetic radiation to efficiently absorb photon energy – the complex structure of the atmospheric absorption reflects the species' characteristics, molecules and water vapor

---

[5] Useful criteria could be, e.g., the transition from atomic to nuclear physics at tens of keV, or the rest mass of the electron of 511 keV, or the experimental techniques being beyond focussing optics for γ-rays above a few 100 keV. Any such lower limit for γ-rays corresponds to a wavelength of 5×10^{-5} µm or smaller.

[6] For comparison, our eyes see the optical regime of about 0.4–0.7 µm wavelength, from colors blue to red – a fairly modest dynamical range.

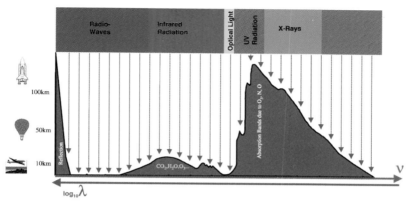

Fig. 2.1. The Earth's atmospheric transparency to cosmic electromagnetic radiation varies considerably over the range of frequencies. Transparency of the atmosphere to radio waves makes radio astronomy a ground-based discipline. Atmospheric dust, water vapor, and molecules block radiation and/or provide a bright atmospheric source in the infrared regime, which means that astronomical observations have to be made from airplanes or spacecraft. There is a narrow window again in the optical regime which allows classical optical astronomy with ground-based telescopes. This window is closed by atomic absorption in the UV through X-rays, making space-based telescopes a necessity. Also beyond, in γ-rays, continuum scattering and absorption processes with atmospheric electrons and nuclei prevent ground-based astronomy up to very high energies. Only above ∼TeV energies ($>10^{12}$ eV) can electromagnetic showers induced by cosmic γ-rays in the atmosphere be observed from the ground; indirect information about the primary γ-ray is retained, and thus ground-based astronomy becomes feasible again

absorbing in the infrared regime, atoms like oxygen and nitrogen absorbing strongly in the ultraviolet regime. A 'window' remains in what we call 'visible' light. For more energetic X-ray photons, the *inner* electrons of atoms have characteristic binding energies similar to the photon energies, and photon interactions with these may absorb X-ray radiation with characteristic spectral absorption edges. Into the γ-ray regime, scattering off individual electrons is the prime attenuation effect for a γ-ray beam, a process with intrinsically continuous spectral characteristics, and reduced in strength as the γ-rays become more energetic. This leads to an "optical depth" in γ-rays determined by the scattering cross section of γ-rays, described by the Thompson scattering cross section σ_T in the limit of low energies,[7]

$$\sigma_T = \frac{8\pi}{3} \cdot r_e^2, \tag{2.1}$$

---

[7] $E_\gamma/m_e c^2 \ll 1$, hence at the upper end of the X-ray regime ($m_e c^2 = 0.511$ MeV). The Thompson cross section corresponds to the geometrical cross section associated with the classical electron radius $r_0 = e^2/m_e c^2$, $\sigma_T = 8\pi r_0^2/3 = 0.665 \times 10^{-24}$ cm^2.

with $r_e = 2.8 \times 10^{-13}$ cm, the 'classical electron radius'. At MeV γ-ray energies, nuclear interactions become important, and spectral absorption characteristics derive from nuclear structure. Scattering off electrons (and then significantly also off protons) becomes inelastic above the X-ray regime with transfer of energy to the particle ('Compton scattering'), and is more forward-beamed than Thompson scattering; the decreasing electron scattering cross section in the γ-ray regime is described by the Klein–Nishina formula:[8]

$$\sigma_{KN} = r_e^2 \cdot \frac{\pi m_e c^2}{E_\gamma} \cdot \left[\ln\left(\frac{2E_\gamma}{m_e c^2} + \frac{1}{2} \right) \right]. \tag{2.2}$$

The different 'windows' from the surface of the Earth into space determine the techniques used for astronomy: optical, radio, and very-high γ-ray astronomy can be ground-based, while spaceborne instrumentation is required for ultraviolet, X-ray, γ-ray and infrared telescopes.

The relative penetration power of γ-rays derives from the absence of atomic interactions. In a 'thick' layer of material a γ-ray will however eventually collide with an individual atomic nucleus or electron in spite of their relatively large spacing. The γ-ray's arrival direction gets lost in repetitive interactions, the energy being redistributed among secondary particles and photons. The penetration depth for γ-rays corresponds to a few grams of material per cm^2. For a characteristic thickness of the Earth's atmosphere of 10 km, and a typical density of air of 1 mg cm^{-3} this amounts to 1000 g cm^{-2}. Thus, the Earth's atmosphere is a thick shield! Gamma-ray telescopes must therefore be operated at high altitudes, above 40 km, to be able to measure cosmic γ-rays directly.[9]

2.2 Production Processes

What are the processes that make γ-rays? (See also the general textbooks on celestial γ-rays by Fichtel and Trombka 1997; Stecker 1971; Paul and Laurent 1998, for more details.) Electromagnetic radiation may be 'thermal' or 'nonthermal'. Thermal radiation emerges from a large population of electromagnetically interacting particles in equilibrium, with their mean energy characterized by particle temperature. The spectrum of radiation intensity follows the 'blackbody' distribution, which reflects the dynamical equilibrium of state population in the radiation field and the multi-particle system:

$$I(\nu) = \frac{8\pi h \nu^3}{c^3} \cdot \frac{1}{e^{h\nu/kT} - 1} \tag{2.3}$$

---

[8] Here in the high-energy limit $E_\gamma/m_e c^2 \gg 1$.

[9] Additionally, the cosmic-ray nuclear interactions with the gas of the Earth's upper atmosphere result in its bright glow with γ-rays, much brighter than any cosmic sources. Instruments for cosmic γ-rays have to avoid viewing this luminous layer at 25–30 km altitude and be adequately constructed to not confuse γ-rays from the Earth's atmosphere below with γ-rays from celestial sources.

is the energy density[10] of radiation at frequency ν, which is a function of particle temperature T and Boltzmann's constant k. Collisions and coupling interactions between radiating material and radiation are so intense that the energy densities of both are identical. When the 'fire' becomes hotter, the blackbody distribution shifts its median towards more energetic radiation, so that each of the photons carries more energy. From Wien's law, the product of temperature and the wavelength of the peak of the radiation spectrum is a constant: $0.2898\,(\text{cm K}) = \lambda_{\max} \cdot T$. From this relation, $\lambda_{\max}$ falls into the visible regime for $\sim$6000 K, the approximate temperature of the sun's surface, the big bang residual radiation at 2.7 K temperature peaks at a wavelength below a millimeter. For thermal γ-rays of 1 MeV, the corresponding temperature of a fireball would be above 2×10^9 K.[11]

Nonthermal processes are thus more typical sources of γ-rays, with specific interactions of matter fields and radiation. A physical process may generate a γ-ray, but we now do *not* require that the entire source environment is thermalized and thus emits blackbody radiation where this γ-ray emission dominates all radiation processes. In general, all elementary particles which take part in an electromagnetic interaction may be sources of γ-rays, if accelerated in some way through external fields of any kind. Likewise, if a system of particles (including electromagnetically interacting particles) changes its state to another one with different energy, the energy difference (or parts thereof) may be radiated in the form of electromagnetic field quanta, hence also γ-rays.

Several processes can be distinguished as shown in Fig. 2.2. (For a more thorough treatment of electromagnetic-field interactions see textbooks by Jackson 1999; Rybicki and Lightman 1979; Rauch and Rohrlich 1980; Heitler 1960).

2.2.1 Charged Particles in Strong Electric or Magnetic Fields

The motion of a charged particle (e.g. electron) can be viewed as a charge current along its trajectory. The particle's charge produces a Coulomb field; its motion thus corresponds to an electromagnetic field, varying as the charged particle moves. Any acceleration of the charged particle now can be viewed as dynamic modification of this electromagnetic field, at the expense of the charged particle's energy of motion. Thus kinetic energy is transformed into energy of the electromagnetic field. Again, for the high energies of such electromagnetic energy loss into γ-rays, the particle picture becomes more appropriate, and quanta of electromagnetic energy, the γ-ray photons, are emitted by accelerated charged particles, 'particle' momenta being conserved.

The motion of a charged particle in a magnetic field B is described by its 'pitch angle' θ, the angle between the particle trajectory and the direction of

---
[10] Per surface area of the blackbody (W s cm^{-2}).
[11] Nuclear fusion inside the sun occurs at 15×10^6 K, corresponding to keV's in thermal energy. In comparison, γ-ray fireballs would be even hotter.

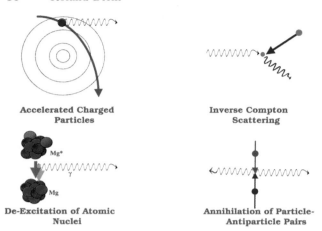

Fig. 2.2. Charged-particle acceleration results in photon emission. As an example for charged-particle acceleration the case of bremsstrahlung is illustrated (*top left*). Weak decays inside nuclei result in excited nuclear states, which often decay through γ-ray emission (*bottom left*). Likewise, the decay of unstable particles such as pions, and the annihilation of particle–antiparticle pairs constitute γ-ray source processes (*bottom right*). Soft photons of energies lower than γ-ray energies may gain energy from electromagnetic-field interactions. The inverse-Compton process on energetic electrons or protons is illustrated (*top right*) as the most important example

the magnetic field, and can be described as gyration of the particle around the field direction, with characteristic gyration frequency $\nu_g = eB/2\pi m_e$. The radiated energy originates from the velocity component perpendicular to the magnetic field, the acceleration of the charged particle being the vector product $v \otimes B$. The resulting synchrotron photon spectrum peaks at a frequency

$$\nu_c = \frac{3}{2}\gamma^2 \nu_g \sin\theta = \frac{3}{2}\gamma^2 \cdot \frac{eB}{2\pi m_e \cdot \sin\theta}. \quad (2.4)$$

Energetic electrons ($\simeq 1000\,\text{MeV}$) moving in the interstellar magnetic field[12] thus radiate 'synchrotron' photons, which can be observed in the radio regime.[13] As a scaling estimate, a high-energy (cosmic ray) electron in a typical interstellar magnetic field would radiate synchrotron photons at

$$E_{h\nu} \simeq 0.05 \cdot \left(\frac{E_e}{\text{TeV}}\right)^2 \cdot \frac{B}{(3\,\mu\text{G})}\,(\text{eV}). \quad (2.5)$$

Both higher magnetic field strength or more energetic particles shift this radiation up in energy, into optical regimes for particle accelerators in high-

---

[12] Measured in Gauss in most astrophysical literature. $1\,\text{G} = 10^{-4}\,\text{T}$; the typical magnetic field strength in the interstellar medium is a few µG.

[13] In extreme cases such as young supernova remnants, synchrotron emission may occur in the UV/X-ray regime.

energy physics laboratories, and into the γ-ray regime near the surface of neutron stars,[14] where magnetic fields are many orders of magnitude higher.

Even the curvature of magnetic field lines in the vicinity of neutron stars can provide sufficient 'bending' acceleration to charged particles that move along these field lines, so that 'curvature γ-rays' are emitted. In pulsars, where magnetic fields are 10^{10} G and more, we attribute part of the observable γ-rays to this process.

Charged-particle acceleration through electric fields is more important for production of γ-rays and results in 'bremsstrahlung' emission. For example, an electron passing very close by an atomic nucleus will experience the strong positive charge of the nucleus, so that the electron's trajectory is substantially changed by such acceleration. From this change in electron energy due to the electron–ion collision, we obtain the total intensity per unit frequency in bremsstrahlung radiation[15]

$$I_\nu(E_e) = \frac{Z^2 e^6 n}{12\pi^3 \epsilon_0^3 c^3 m_e^2 v_e} \cdot \ln\left(\frac{192 v_e}{Z^{1/3} c}\right). \tag{2.6}$$

The spectrum of bremsstrahlung radiation remains flat up to roughly the electron kinetic energy, i.e. up to[16]

$$E_\gamma = (\gamma - 1) m_e c^2. \tag{2.7}$$

It drops sharply towards zero above, as effectively all the kinetic energy of the electron has been transferred to the bremsstrahlung photon.

Depending upon the energy distribution of electrons and ions, one distinguishes 'thermal' and 'nonthermal' bremsstrahlung again; this specifies whether the charged-particle energy distribution has been thermalized through collisions, or if 'nonthermal' and being the result of, e.g., a particle-acceleration process such as Fermi acceleration. For a hot plasma, the specific radiation power or emissivity[17] in thermal bremsstrahlung may be expressed as a function of temperature

$$k_\nu = 5.443 \times 10^{-39} \cdot Z^2 \cdot g(\nu, T) \cdot \exp(-E_\gamma/kT) \cdot T^{-1/2} \cdot n_e n_p. \tag{2.8}$$

---

[14] In magnetic fields above a critical strength of $\sim 10^{13}$G, relativistic effects and the quantum phenomena of pair creation and multi-photon states result in a more complex description of the radiation field, called 'magneto-bremsstrahlung' or 'synchro-curvature radiation' (see Chap. 6).

[15] $I_\nu(E_e)$ is integrated over the collision process, in units of erg cm^{-2}. We assume the much heavier ion is at rest. $e = 1.602 \times 10^{-19}$C is the electron's charge, v_e the electron velocity and m_e its mass, $Z \cdot e$ the ion charge, n the number density of matter, and $\epsilon_0 = 8.8542 \times 10^{-3}$ C^2 erg^{-1}cm^{-1} the dielectric constant or permittivity of the vacuum.

[16] With the electron's Lorentz factor $\gamma = \frac{1}{\sqrt{1-\frac{v^2}{c^2}}}$.

[17] Per unit frequency, in W cm^{-3} Hz^{-1}, using T for the temperature of the plasma in K; $k = 8.6174 \times 10^{-5}$eV K^{-1} is Boltzmann's constant, and n the electron and ion number density per cm^3.

The 'Gaunt factor' $g(\nu, T)$ encodes the characteristic collision impact parameter distribution for a plasma of specific composition and temperature, and may be approximated in the γ-ray regime by $(E_\gamma/kT)^{1/2}$.

For nonthermal particle spectra, electron energy distributions of a power-law type

$$dN_e(E)/dE \sim E^{-\alpha} \tag{2.9}$$

are frequently found, with $\alpha \simeq 1 - 3$. Depending upon energy, ion–electron bremsstrahlung dominates electron–electron bremsstrahlung at low energies, while in the high-energy γ-ray regime the ratio of both rises to order unity. Inverse bremsstrahlung, which is the photon emission from energetic protons and ions colliding with ambient electrons or ions, is negligible in most circumstances.

Summarizing, the acceleration of charged particles of high energy by a suitable force field may result in γ-rays. Hence, observation of γ-rays may be used to study energetic particles moving in strong fields. 'Energetic' here means travelling almost at the speed of light with $\gg$MeV energies; fields must be "strong" enough to change the particle's motion by a significant amount.

2.2.2 Inverse Compton Scattering

Up-scattering of photons of lower energy through collisions with energetic particles is termed the 'inverse-Compton process'. As we employ the 'particle' description for γ-rays, we may imagine them to collide with other particles such as electrons. This is known as the Compton scattering process. Equivalently, we may also view the electromagnetic field which corresponds to a photon to penetrate into the regime of the strong electromagnetic field caused by the charged particle ('wave' or 'field' description). In 'normal' Compton scattering a γ-ray photon will collide with one of the many atomic electrons within some material and be scattered in the collision, transferring some of its energy to the electron. The inverse energetics may also apply, however, and provide a process for γ-ray production: if photons of lower energy collide with energetic electrons, these photons may gain energy in the collisions, thus being promoted in energy, e.g., from X-rays to γ-rays. This 'inverse-Compton scattering' is important in regions of high photon densities. Other examples are compact stars where an accretion disk is sufficiently hot to emit X-rays, and the compact object generates beams of charged particles in its vicinity (see below). The spectral intensity distribution of inverse-Comptonized isotropic photons of frequency ν_0 and photon density $N(\nu_0)$ from scattering off high-energy electrons of Lorentz factor γ is

$$j_\nu(\nu_0) = \frac{3\sigma_T c}{16\gamma^4} \cdot \frac{N(\nu_0)}{\nu_0^2} \cdot \nu \cdot \left[2\nu \cdot \ln\left(\frac{\nu}{4\gamma^2\nu_0}\right) + \nu + 4\gamma^2\nu_0 - \frac{\nu^2}{2\gamma^2\nu_0}\right], \tag{2.10}$$

with σ_T as the Thompson scattering cross section.[18] The typical energy of inverse-Compton-scattered photons rises rapidly with energy:

$$E_\gamma \simeq 1.3 \left(\frac{E_e}{(\text{TeV})}\right)^2 \cdot \left(\frac{E_{\text{ph}}}{2 \times 10^{-4}\,(\text{eV})}\right)\,(\text{GeV}), \qquad (2.11)$$

with an ambient photon-field[19] typical energy E_{ph}. Therefore the inverse-Compton process plays an important role for cosmic γ-ray sources. The blackbody thermal radiation of the ambient field may be expressed as the energy distribution of photons, as number density (photons cm^{-3}) of ambient photons with energy $E = h\nu$,

$$n_{\text{BB}}(E_{h\nu}) = \frac{(E_{h\nu}/m_e c^2)^2}{\pi^2 \lambda_C^3} \cdot \frac{1}{\exp(E_{h\nu}/kT) - 1}, \qquad (2.12)$$

using the definition of the Compton wavelength $\lambda_C = h/2\pi m_e c$. These photons may be up-scattered to γ-ray energies by collision with high-energy electrons of energy E_e.

2.2.3 Nuclear Transitions

The nuclear (or strong) force is responsible for the binding of protons and neutrons in the atomic nucleus. It outweighs the Coulomb repulsion between the protons closely packed in the nucleus. As a quantum system, the nucleus has specific, quantized states of energy for this compact assembly of nucleons, which can be thought of to be similar than states of electrons in atomic shells, by analogy.[20] These nuclear states have typical energy spacings of $\sim$MeV; hence any transition in states of atomic nuclei may involve absorption or emission of MeV γ-rays,

$$^n X \longrightarrow {}^{n-1}Y^* + e^+ \longrightarrow {}^{n-1}Y + \gamma. \qquad (2.13)$$

Therefore whenever some energetic interaction brings disorder into the state of a nucleus, we expect γ-ray line emission from nuclear de-excitation. En-

---

[18] At high energies, energy transfer to the electron may become important. This will imply treatment through the Compton scattering formalism, replacing the Thompson cross section with the Klein–Nishina expression.

[19] Photons from the cosmic 3 K background radiation are characterized by a typical energy 2×10^{-4} eV and energy density of 0.26 eV cm^{-3}.

[20] This corresponds to the shell model of the nucleus; for heavier nuclei a drop model better describes the nucleus (see nuclear physics literature for details). In atomic shell transitions between electron states, familiar line radiation is emitted. Inelastic collisions of unbound electrons in the vicinity of nuclei may also result in photon emission; this process is called 'free–free emission', from scattering of the electrons between 'free' or unbound atomic states; as opposed to 'bound–free' transitions, which emit UV/X-ray line radiation.

ergetic collisions from cosmic rays with interstellar gas nuclei are one example, *radioactive decay* of a freshly synthesized nucleus another. In radioactive decay, one of the particles of the nucleus transforms into one of another kind as a consequence of the 'weak force'; neutrons decay into protons ('β^--decay'), or protons can transform into neutrons ('β^+-decay'). The so-caused disorder in the tight arrangement of nucleons is unstable under the influence of the strong and electric forces, and new, stable arrangements are obtained, through emission of the energy difference, often as a γ-ray in the MeV regime. Some characteristic energy levels of first-excited states which are important for γ-ray astronomy are 4.438 MeV(^{12}C*), 6.129 MeV(^{16}O*), and 1.809 MeV(^{26}Mg*). The cross section for excitation of nuclei into these levels is maximized in resonances, i.e. when the collision energy is in the vicinity of the energy of the excited level. This implies that low-energy cosmic-ray interactions become visible in the domain of γ-ray lines. Generally, observation of characteristic γ-ray lines thus tell us about nuclear transitions in a region of either nucleosynthesis or cosmic-ray interactions.

2.2.4 Decays and Annihilation

Pion decay is somewhat similar to above nuclear processes. The pion is an elementary particle involved in the strong nuclear interaction. Pions are created during strong-interaction events such as collisions of cosmic-ray protons with ambient-gas nuclei. Neutral pions (π^0) decay rapidly into two γ-ray photons, with an energy distribution peaking at ~70 MeV, half the rest mass of the pion. Observation of a pion decay peak in a γ-ray spectrum thus provides insight into collisions of energetic (>135 MeV) protons with nuclei. The pion decay γ-ray bump is broadened as the momentum distribution of the high-energy collision adds a Doppler shift and broadening. For beamed annihilation (e.g. in jets), the peak will be Doppler-shifted towards higher energies.

Annihilation of pairs of particle and anti-particle also produces γ-rays. From equivalence of mass and energy, field energy may be converted into pairs of particles and anti-particles in sufficiently strong electromagnetic fields. Even vacuum field fluctuations would produce virtual pairs of particles and antiparticles; however in a strong field, the trajectories of both particles diverge sufficiently such that the inverse 'annihilation process' does not occur immediately. The lightest particle–antiparticle pair that can be generated is the electron and its antiparticle, the positron, each with 511 keV rest mass; creation requires a minimum energy of 1.022 MeV. The inverse process occurs when a particle encounters its antiparticle, and is called 'annihilation'; the mass of both particles is radiated as electromagnetic energy. Conservation laws demand that electron–positron annihilation will produce two (or more) photons. Distribution of the rest mass energy among these photons results in 0.511 MeV photons for two-photon annihilation, in the rest frame of the annihilation process. Positrons and electrons also may form a bound, atom-like system, which exists in two different states distinguished by the relative orientations of particle spins. The parallel-spin state forms a triplet in external fields and decays into three photons because of spin and momentum conservation, so that three photons share the 1.022 MeV total annihilation energy; this produces a γ-ray spectrum with a maximum energy of 511 keV and a continuum distribution towards lower energies.

Radioactive decay processes involving 'β-decays' also produce positrons. Through the weak interaction, a proton can decay into a positron and a neutron (plus an electron-neutrino ν_e, but we ignore this highly penetrating particle here).

Annihilation photons originate in the vicinity of radioactive decay regions, or energetic environments capable of positron production by other processes. The surface and vicinity of compact stars (neutron stars and black holes) may have such high energy density both from gravity and from strong magnetic fields compressed from the original star. Similarly, hadronic antiparticles (antiprotons and antimatter nuclei) in the universe may annihilate upon collision with normal matter and be responsible for spectral features at correspondingly higher energies in the γ-ray spectrum.

2.2.5 Charged particles bound in strong magnetic fields

Just as the electric (or Coulomb) force holds electrons close to a nucleus in the quantized system of an atom, strong magnetic fields can force electrons into orbits around their field lines, leading to quantized energy levels for allowed electron energies. Magnetic fields close to neutron stars can be sufficiently strong for the steps between such allowed electron orbit levels to be in the regime of tens of keV, the low γ-ray regime (or high X-ray regime). Electron transitions from one allowed state to another will eject or absorb photons of this characteristic energy difference, producing so-called cyclotron line radiation. Note that only the motion perpendicular to the magnetic field is quantized, the parallel velocity component is unaffected by the magnetic field. The 'Landau' energy levels $E = h\nu_c$ may be obtained from the cyclotron frequency

$$\nu_c = ZeB/2\pi\gamma m_0, \tag{2.14}$$

for a particle with charge Ze and velocity v [hence with a Lorentz factor $\gamma = 1/\sqrt{1 - (v^2/c^2)}$] in a magnetic field B. For field strength values around 10^{12} G as observed in strongly magnetized neutron stars, cyclotron lines fall in the X-ray regime,[21] from $\hbar\omega \simeq 12$ keV $B/(10^{12}$G$)$.

We have seen that the processes which create γ-rays are different in general from the more familiar thermal blackbody radiation. The relevant physical processes involve rather special conditions. Generally speaking, violent processes are at play. Observation of γ-rays enables us to study such exceptional

---

[21] The cyclotron frequency of a non-relativistic electron ($\gamma = 1$) is 2.8 MHz G^{-1}.

places in nature and thus explores a different aspect of the universe than optical observation.

2.3 Gamma-Ray Interactions, Absorption Processes

How does a γ-ray interact with matter? From the above description, we learn that, unlike optical light which is deflected from mirror surfaces, γ-rays do not interact with the surface of materials. The γ-rays mainly scatter off electrons within materials, and they randomly penetrate to some variable depth of material before their interaction. Scattering of γ-rays is itself a random process with a wide spectrum of possible results, averaging to the distributions presented above when many interactions are summed; the individual fate of a γ-ray photon in a material is not predictable. Moreover, scattering of incident γ-ray photons will generally be incoherent, i.e. the scattered photons are unrelated, and do not form, e.g., patterns of diffraction. The wave description of electromagnetic interactions with matter is inappropriate; we apply a particle interaction model instead.

The interaction processes vary with the energy of the γ-ray, and are as follows (from low to high energy; see Fig. 2.2):

- Photoelectric absorption: Atomic electrons are removed from their nuclei, thus reducing the energy of the γ-ray photon by the electron's binding energy. This is the dominant process at energies below 100 keV.
- Compton scattering: Electrons are hit by the γ-ray photons, and gain a fraction of the photon's kinetic energy in this collision. This is the dominant interaction process in the 0.1 MeV to a few MeV regime.
- Pair creation: In the presence of an electric field (usually of the atomic nucleus), the γ-ray energy may be converted into a particle–antiparticle pair, electron and positron. From momentum conservation, these two particles move in opposite directions so that the momentum component perpendicular to the γ-ray cancels; transformation into the laboratory system makes them appear to fork out of the γ-ray's incidence direction with a narrow opening angle, which decreases with energy. This interaction process cannot occur below a threshold of 1.022 MeV, and the cross section increases with energy so that it dominates over Compton scattering above several MeV. The same process can occur in dense photon fields through collisions of energetic photons (Nikischow effect; photon–photon pair production, see also Chap. 2.3).
- Quantized absorption: Systems such as atomic nuclei or charged particles captured in strong magnetic fields are characterized by states with level differences in the regime of γ-ray energies. Hence, state transitions of such systems may absorb γ-rays resonantly, e.g. forming an excited nucleus. When the reverse transition to the system ground state is inhibited (e.g. from transition selection rules, or high collision frequencies), other interactions may dissipate the excitation energy, e.g., into lattice

phonons, and no fluorescent γ-ray emission occurs; the γ-ray is absorbed. This process is analogous to formation of optical absorption lines such as Fraunhofer lines.

2.4 Cosmic Gamma-Ray Sources

According to the characteristic production processes, we may distinguish the following different types of cosmic γ-ray sources, as described in subsequent chapters in detail:

- Fireballs, such as found in γ-ray bursts, and possibly in the vicinity of black holes
- Explosive events of extreme energy density, such as supernovae and novae
- Energetic collisions, such as in the vicinity of accreting compact objects, particle jet sources (microquasars, active galactic nuclei), or cosmic ray collisions with ambient matter, or in solar flares
- Charged-particle beams in the vicinity of compact objects – hence strong gravitational or magnetic fields (quasars, active galactic nuclei)

The thermal radiation from an optically thick 'γ-ray fireball' constitutes probably the most extreme and violent source from the above. The Big Bang, but also explosions of stars in supernovae (or γ-ray bursts) in principle approximate this extreme. In explosive events, gigantic amounts of energy may be released within short times (fractions of seconds) into dense environments, and thus provide extreme heat. However, we rarely can see the γ-ray photosphere of such a fireball due to the scattering and absorption of overlying matter surrounding the fireball site. Fireball 'temperatures' may exceed several 10^9 K and cause atomic nuclei to dissolve and rearrange upon cooling down, with radioactive nuclei as by-products, whose decay may produce γ-rays. Direct γ-ray observations of sufficiently hot fireballs may be possible for events where the energy release is not covered by stellar envelopes, such as neutron star collisions and similarly extreme and rare events, which have been discussed as possible explanations of γ-ray bursts. Moreover, γ-rays may provide unique insights to such processes, even though originating from secondary, nonthermal processes, as in the case of radioactive decay.

Neutron stars are highly compact objects, with $\simeq 1$ solar mass compressed into a 10 km-sized sphere. They are known to be common sources of X-rays, mostly caused by release of gravitational energy when matter falls onto their surface. The complex path of matter accreting onto the star in strong gravitational and magnetic fields is the subject of broad astrophysical studies, involving radiation from radio frequencies to γ-rays. The extreme plasma motions near neutron stars cause complicated accretion disk configurations and beams of particles, and these in turn produce the fascinating diversity of pulsing phenomena of these objects in the X-ray regime. Nuclear excitation of infalling matter from close to the neutron star's surface can be expected to result in characteristic γ-ray line emission. Further out in the magnetosphere, γ-rays within a broad frequency range are known to be produced in isolated neutron stars whose magnetosphere is relatively undisturbed by accreting matter. Their γ-ray emission is attributed to curvature radiation of particles accelerated by large electric fields. The observed pulsing behavior varies strongly with frequency of the radiation and can be explored to diagnose the plasma acceleration and magnetic field configurations in great detail.

Charged-particle accelerators and relativistic-plasma interactions on gigantic dimensions are observed in the extremely luminous nuclei of a subtype of galaxies, the 'active galactic nuclei'. Here particle energies apparently achieve the highest values allowed from the principles of physics, γ-rays up to 10 TeV have been seen from such objects. Spectacular jets of plasma are ejected from these active galaxies, extending many thousand light years into space. Intense γ-ray emission has been observed from such galaxies (called therefore γ-ray blazars) when we view such jets directly (under viewing angles of the jet axis of a few degrees at most). We still do not know what powers the inner cores of these galaxies to make them far more luminous than the entire population of the 10^{10} stars of normal galaxies. From detailed studies of nearby accreting compact stars we may learn how gravitational energy can be converted into such jet-like plasma beams. Extreme magnetospheres can be studied in γ-ray pulsars. Combining these lessons, we may have a better understanding of the extreme phenomena in active galactic nuclei and near black holes.

We speculate that the combined γ radiation from active galactic nuclei may comprise the bulk of the diffuse sky brightness in γ-rays, called the cosmic diffuse γ-ray background. Gamma-ray lines from the radioactivity of cosmological supernovae superimpose (with their characteristic redshift) to add to the cosmic diffuse γ-ray background in the $\simeq 1$ MeV regime; potentially this contribution can be identified from its distinct spectral edges.

The bright large-scale γ-ray emission observed from our Galaxy is caused by the interaction of cosmic rays with interstellar gas. Bremsstrahlung and pion decay processes cause a diffuse glow of the Galaxy in γ-rays from MeV energies up to 10^4 MeV, supplemented by inverse-Compton γ-rays from starlight boosted by cosmic rays. This γ-ray glow provides a unique tracer to study cosmic rays indirectly throughout the Galaxy. The sources of cosmic rays, in particular the origin of charged particles with energies up to 10^{20} eV,[22] are still unknown. Acceleration in the expanding remnants of supernova explosions has been proposed as the most likely source. The composition of accelerated material is investigated through direct cosmic-ray isotope measurements and nucleosynthesis arguments, to eventually reveal the injection

---

[22] Ultra-high-energy cosmic-rays (UHECR).

process into such particle acceleration regions.[23] The Sun itself and the γ-ray emission from solar-flare particle interactions in the upper solar atmosphere provides a nearby laboratory to study details of such interactions with superior obervational resolution.

2.5 The Fate of Gamma-Rays Between Source and Detector

On their journey from the source region to our detectors, γ-rays may traverse environments with substantial interaction probabilities, both in the vicinity of the source region as well as close to our telescope platform. Absorption and scattering will be parts of our measurement, and we must account for them in our studies.

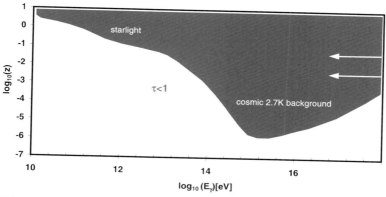

Fig. 2.3. The horizon of high-energy γ-rays. The shaded region marks the regime of large optical depth, i.e., γ-rays at these energies from sources at these redshifts will not reach us. The basis for this curve is an ambient photon distribution and intensity composed of the cosmic microwave background and average starlight. The contribution of each component to the absorption from $\gamma - \gamma$-pair production is indicated

Low-energy γ-rays traverse long paths of interstellar space without scattering or absorption. The interstellar gas and dust absorbs optical radiation readily at equivalent column densities of 10^{23} hydrogen atoms cm^{-2} and prevents useful measurements; this much material amounts to less than $0.1\,\mathrm{g\,cm^{-2}}$, which is practically transparent to γ-rays. Such an effective material thickness corresponds roughly to occultation by a sheet of paper: optically, we cannot see through easily, but γ-rays hardly notice this material.

---

[23] This explanation appears inadequate for energies above 10^{15} eV; thus plasma-jet-related acceleration sites similar to active galactic nuclei may be needed for higher energies.

We note however that even in 'empty' intergalactic space, γ-ray source photons may suffer energy losses over the large distances[24] from cosmic sources to γ-ray telescopes. Over such large distances, even a normally unimportant interaction process may become crucial. On a large scale, the universe is dominated by radiation, with major components from starlight and from the cosmic-diffuse 2.7 K microwave glow remaining from the Big Bang. Gamma-ray photons may interact with these ambient photons, producing e^+e^- pairs. Therefore the horizon for γ-rays is limited by this pair production process, despite their penetrating nature. Photon–photon pair production is possible only above a threshold energy given by the rest mass of the pair,

$$E_{\text{th-pp}} = \frac{2m_e^2 c^4}{[1-\cos(\phi)](1+z)^2 E_\gamma} \sim 1 \left(\frac{1+z}{4}\right)^{-2} \cdot \frac{30\,\text{GeV}}{E_\gamma} (\text{eV}) \qquad (2.15)$$

(for head-on collisions, ϕ is the photon collision angle). We see that above ∼30 GeV the energy loss of γ-rays in interstellar space from scattering with starlight becomes significant and limits the horizon to ∼500 Mpc at 1 TeV, while at higher energies scattering on cosmic background photons effectively cuts the visibility distance to the few nearest galaxies only (Fig. 2.3). Conversely, sources at distances above $z \sim 2$ cannot be seen directly above this energy. The γ-ray horizon is shown in Fig. 2.3 for the upper regime of the γ-ray domain in terms of redshift of the source. Note that photon–photon pair production results in high-energy charged particles; these will inverse-Compton scatter on the same photons, and redistribute the high-energy γ-ray energy to lower γ-ray energies, according to

$$E_{\gamma,\text{IC}} \sim 10 \cdot \left(\frac{1+z}{4}\right) \left(\frac{E_e}{30\,\text{GeV}}\right) (\text{MeV}), \qquad (2.16)$$

thus producing a diffuse-cosmic continuum spectrum in the form of a power-law $I \propto E^{-\alpha}$ with index $\alpha \sim 2$.

In dense environments, interactions with charged particles are more important than photon–photon processes, and γ-rays will mostly scatter off electrons, losing part of their energy to the plasma in this way. This Compton scattering will therefore modify the shape of the spectrum of γ-rays, and we must account for the plasma conditions in order to reveal the original source γ-ray spectrum through proper calculation. Note that photons may also gain energy in such plasma collisions, if the plasma is more energetic than the photons; this results in observation of 'Comptonized' γ radiation (see Chap. 2.2 and Fig. 2.2 above). Line features in spectra can be significantly distorted by these effects, and scattering too. We can use such distortions to diagnose

---

[24] The nearest galaxies are at distances of 100 kpc, while quasar distances are more appropriately expressed in terms of Doppler redshift of spectral lines, expressed as fractional redshift from $z = 0$ (= no shift, nearby, ∼Mpc) out to $z \simeq 4$, the most distant quasars observed so far, corresponding to 10^5 Mpc. The details of the conversion from redshift to distance depend on the cosmological model.

the plasma conditions, from the measured profile of γ-ray lines. Even if unaffected by additional scattering, the line profile will tell us about the relative motion of the γ-ray source and our observing telescope: Doppler shift of the original frequency tells us about the kinetic energies within the source, such as the expanding motion of radioactive matter after a supernova explosion, translating velocities of $\simeq$1000's of $km\,s^{-1}$ into a line width of $\sim$0.1 MeV for a 1 MeV line energy (normally, this line would be narrower than the best instrumental resolution of $\simeq$1 keV). Note that the gravitational field of compact sources such as neutron star surfaces can also result in substantial changes in line energies: photons have a hard time leaving the star against this strong gravitational force, and can lose $\simeq$20% of their energy, with a corresponding downward frequency shift.

Getting close to the detector, γ-rays again encounter material in the upper atmosphere of the Earth. Interactions in the upper atmosphere result in electromagnetic showers with large numbers of secondaries. At very high γ-ray energies, these are directly used for ground-level detection of the incident γ-ray. Telescopes for cosmic γ-ray measurements at lower energies have to operate at least at altitudes above $\sim$40 km, i.e. with residual-atmosphere thicknesses below $\sim$3 g cm^{-2}. For these measurements, secondaries produced in the upper atmosphere and within the generally massive spacecraft platform of the telescope provide undesired background. This, and even more so the enormous bombardment of the spacecraft by charged particles from the radiation belts and from cosmic radiation, results in a glow of γ-rays from such spacecraft and instruments. The next chapter illustrates that γ-ray telescopes have to be built with complex detector configurations and triggering techniques; additionally, powerful analysis algorithms must be employed to discriminate the primary cosmic γ-rays from the instrumental background.

References

Fichtel, C.E., Trombka, J.I., 1997, *Gamma-Ray Astrophysics*, NASA-1386
Heitler, W., 1960, *The Quantum Theory of Radiation*, Clarendon Press, Oxford
Jackson, J.D., 1999, *Classical Electrodynamics*, John Wiley, New York
Longair, M.S., 1994, *High-Energy Astronomy*, Cambridge University Press, Cambridge
Paul, J., Laurent, P., 1998, *Astronomie Gamma Spatiale*, Gordon & Breach, Paris
Ramana Murthy, P.V., Wolfendale, A.W., 1981 *Gamma-Ray Astronomy*, Cambridge University Press, Cambridge
Rauch, M.M., Rohrlich, F., 1980 *The Theory of Photons and Electrons*, Springer, Berlin
Rybicki, G.B., Lightman, A.P., 1979 *Radiative Processes in Astrophysics*, Wiley, New York
Stecker, F.W., 1971 *Cosmic Gamma-Rays*, NASA SP-249

3 Instruments

Giselher Lichti and Robert Georgii

3.1 Detection of Gamma-Rays from Space

The observer who wants to perform astronomical observations in the γ-ray regime below about 100 GeV is confronted with three fundamental problems:

- the opaqueness of the atmosphere for γ-rays
- the low cosmic γ-ray fluxes
- the high intrinsic background of γ-ray telescopes

These problems and the consequences for experimental γ-ray astronomy will be discussed in more detail in the next three sections.

3.1.1 The Opaqueness of the Atmosphere

Although the density of air is rather low (1.293 kg m^{-3} at ambient pressure) and although γ-rays are highly penetrating particles, the atmosphere is opaque for γ-rays because its integrated matter density amounts to $\sim$1000 g cm^{-2}. Since the mass-attenuation coefficient for air at 1 MeV is $\sim$0.00642 cm^2 g^{-1} the absorption probability for a 1 MeV γ-ray is $>$ 99.8%. So cosmic γ-rays cannot be observed directly at the surface of the Earth. Therefore direct-measurement γ-ray telescopes have to be brought above the atmosphere in order to allow the successful observation of celestial γ-rays. Consequently, traditional γ-ray astronomy can only be performed with balloon-borne γ-ray telescopes or with γ-ray telescopes mounted on a satellite platform. Only above $\sim$100 GeV can indirect terrestrial observations be performed with ground-based γ-ray telescopes (see Chap. 3.3.5).

3.1.2 The Low Cosmic Gamma-Ray Fluxes

Observations of celestial γ-rays are additionally aggravated by the fact that the fluxes of cosmic γ-rays are rather low. This is related to the energy of the γ-rays; they carry more than 1 million times the energy of ordinary optical photons. Therefore less γ-rays than optical photons are needed to transport the energy being produced in the processes which are responsible for γ-ray production. Consequently, although these processes belong to the most-energetic ones known in the universe and although most of the energy production takes place at these energies, the number of γ-rays generated is small compared to other wavelength regions. This leads to low γ-ray fluxes, which for the closest known γ-ray sources are of the order of $\sim 10^{-2}$–10^{-3} γ-rays cm^{-2} s^{-1} MeV^{-1}. In order to be able to measure such low fluxes γ-ray telescopes must have a large effective area, and long observation times are needed to collect enough γ-rays. Because of their small cross sections γ-rays are highly penetrating particles. Therefore thick detectors (with an area matter density of the order of 10 g cm^{-2}) with a high stopping power are needed to achieve a sufficiently high detection probability. So normally γ-ray telescopes consist of large heavy detectors. This has two consequences: on the one side it leads to a large weight of the instruments, which renders their launch with a balloon or a rocket difficult, on the other side it leads to a severe background problem.

3.1.3 The Background of Gamma-Ray Telescopes

The material of a γ-ray telescope is permanently irradiated by the particles of the cosmic-rays (protons and heavier atomic nuclei) and, depending on the orbit of the space-borne telescope, of the trapped particles (mainly protons and electrons) of the radiation belts of the Earth. These particles can interact with the material of the γ-ray telescope. A result of these interactions are excited nuclei and spallation products, which in turn decay, emitting prompt γ-rays and neutrons and delayed γ-rays. Especially the secondary neutrons are problematic because they are often indistinguishable from γ-rays and have the disadvantage that they can produce long-lived radioactive nuclei via neutron activation.

Besides this instrument activation the natural radioactivity of the γ-ray telescope material may make an important contribution to the intrinsic γ-ray background. In the past this background contribution was negligible compared to the activation background. But for the γ-ray telescopes planned in the future this background may become dominant.

The Different Background Types

The Charged-Particle Background. In general two background types can be distinguished. These are the charged-particle background and the background induced by neutral particles (i.e. neutrons and γ-rays). Fortunately charged particles ionize the atoms along their path when they penetrate matter. This behavior is used to reject these particles by surrounding the γ-ray detector with a thin ($\sim$1 cm) plastic scintillator. Each time a charged particle penetrates the scintillator a light pulse is emitted which is measured by photomultipliers and used as a veto signal for the detector (anticoincidence method). Since charged particles and γ-rays from prompt nuclear reactions can be efficiently suppressed by such an anticoincidence method this type of background is not considered further here. Instead we concentrate on the delayed γ-ray and neutron background.

However we should mention here that such an anticoincidence method does not suppress all charged particles. A certain generally tiny fraction is not vetoed due to electronical time jitter, leading to a violation of the coincidence condition. These particles can therefore leak through the anticoincidence, increasing the background. This is called the electronic leakage-rate background.

The Background from Neutral Particles. The background from neutral particles can be divided into the "external" and the "internal" background, depending on where the background photons are created. We understand the external background to be the background from neutral particles which fall from outside onto the γ-ray telescope, whereas the internal background is created within the matter of the instrument itself. In addition one has to distinguish between γ-rays and neutrons. Primarily the γ-ray astronomer is interested in γ-rays, not in neutrons. However, neutrons can only be distinguished from γ-rays with difficulty. We shall explain later how this can be achieved.

1. The external background consists of atmospheric and extraterrestrial parts. By the bombardment of the atmosphere with the primary particles of cosmic radiation, secondary particles, mainly neutrons and γ-rays, are created. So the atmosphere itself is a source for these particles. This type of background is therefore especially important for balloon telescopes, but it can also make a significant contribution to the background for satellites at low altitudes. For the early omni-directional γ-ray detectors this background was a big problem, whereas for a γ-ray telescope with small angular-resolution elements [see (3.8)] this background contribution can be small (at least for a satellite telescope).

Sometimes these types of background radiation are themselves a topic of interest. Knowledge of these background types is not only needed for a proper understanding of the astrophysically interesting data, but it is also interesting from a physical point of view in itself. The processes within the atmosphere have attracted the interests of physicists since the early seventies [Beuermann (1971); Stecker (1973)] and several groups have performed measurements on atmospheric γ-rays and neutrons [Fichtel et al. (1969); Scheel and Röhrs (1972); White (1973); Thompson (1974); Kanbach et al. (1974); Schönfelder and Lichti (1975)]. Nowadays a good understanding of the processes which generate the γ-rays and neutrons in the atmosphere has been achieved (for more details see Chap. 2).

2. The internal γ-ray background is true background radiation (in the sense that it has not the intrinsic interest of the external background), and its suppression is a challenge for the instrument builder. It is created within the material of the instrument via two main processes:

- activation of atomic nuclei by bombardment of the instrument material with particles from the cosmic radiation and from the radiation belts and with atmospheric neutrons, and
- decay of natural radioactive elements.

When a particle (proton, neutron, α-particle, etc.) with a high-enough energy hits the passive or active material of an instrument, it may undergo a nuclear interaction resulting in prompt spallation products, whereby neutrons can be emitted, and/or excited nuclei which decay, emitting, for example, electrons, positrons, α-particles and γ-rays. The reaction rates depend on the amount of matter and on the reaction cross sections. So the aim of the instrument builder has to be to use as little passive material as possible and to select those passive materials which have small reaction cross sections.

In summary, in each γ-ray telescope background radiation is generated, and therefore sophisticated background-suppression methods have to be applied in order to reduce the background-count rate to an acceptable level. In the next section the most common methods used will be introduced.

Background-Suppression Methods

In general two fundamental background-suppression methods can be distinguished: passive and active methods.

Passive Background Suppression. Passive shielding around the active detector elements, which are made from heavy materials such as lead or tungsten, absorb γ-rays quite effectively: a 5 cm thick lead shield absorbs more than $\sim$92% of the γ-rays (the minimal absorption of 92% is reached at $\sim$3.5 MeV). So one could easily build a γ-ray telescope with a passive collimator made from one of these two materials. But with such a telescope successful measurements could be performed only in an ideal environment which is free of moving energetic particles. In realistic environments, however, more γ-rays are produced within the material of the collimator via nuclear reactions than are suppressed. So the intrinsic background radiation counteracts the shielding effect of the collimator and makes passive collimators useless for the γ-ray astronomer.

Active Background Suppression. For the above reason the active background-suppression methods were developed. The use of thin plastic scintillators to suppress the charged-particle background has already been mentioned. However, with plastic scintillators it is very difficult to build a collimator because of the high penetration capability of γ-rays (if one wants to absorb a 1 MeV γ-ray in plastic with a probability of 90%, one needs a thickness of at least 33 cm). Instead of a plastic scintillator one can of course use a scintillator material with a higher stopping power. Such scintillator materials are CsI, NaI and bismuth-germanate (BGO). BGO is especially well suited for this purpose because of its high density (7.13 g cm^{-3}). Gamma-ray telescopes have been built with all three types of scintillators [the large γ-ray telescope of the University of New Hampshire (Forrest et al. 1975); the γ-ray instrument on HEAO-A (Metzger 1973; Matteson 1974); the γ-ray instrument of the Solar Maximum Mission (Forrest et al. 1980); the balloon-borne telescope HEXAGONE (Matteson et al. 1990)].

Another approach which has been used in the past is to combine the passive and the active shielding methods by constructing a sandwiched collimator. Such a collimator consists of several layers of a passive shielding material interleaved with thin shields of an active scintillator. However, again the γ-rays produced within the shield itself have a low probability to interact in the plastic scintillator of the shield and, therefore, can not be rejected, leading to a higher γ-ray background with the collimator than without.

We mentioned already that γ-rays and neutrons, being neutral particles, cannot be distinguished by the methods mentioned above. Since the neutron-induced γ-ray background can be a significant fraction [up to 50%, see White and Schönfelder (1975)] of the measured signal, its suppression is important. Fortunately the interaction mechanisms of γ-rays and neutrons with matter are different. Whereas γ-rays interact via electromagnetic processes with the material of the scintillator, neutrons can react via nuclear reactions. In both processes different particles are responsible for the scintillation process: electrons in the case of electromagnetic processes and protons in the case of nuclear reactions. Because of the difference in mass of these two particles, the decay time of the photon-creation processes within specially-treated organic scintillators is different for the two interactions. The light decay-time difference of the created light pulses can be measured electronically and thus allows a suppression of the neutron-induced γ-ray background. This so-called pulse-shape discrimination (PSD) has been widely used and is of special importance for Compton telescopes. It has successfully been applied by Schönfelder et al. (1981) among others.

3.2 Gamma-Ray Detection Techniques

3.2.1 Scintillation Techniques

Gamma-rays are highly penetrating particles. Since they do not ionize matter – as charged particles do – along their path, they can only be observed indirectly. This means that γ-rays can only be detected after they have interacted with matter through one of the following three processes:

1. Photoelectric absorption (for $E \lesssim 300$ keV);
2. Compton scattering (for ~ 300 keV $< E \lesssim 8$ MeV); and
3. Electron–positron pair creation (for $E \gtrsim 8$ MeV).

In all three processes charged particles (electrons, positrons) are generated which can be measured in particle detectors.

So a γ-ray detector has to possess the following two distinct properties: it needs a sufficiently high density in order to have a high interaction probability for a γ-ray, and it has to transform a significant fraction of the energy of the electrons/positrons into a measurable quantity. It turns out that the scintillation process in which optical photons are created is a good method to detect γ-rays. So scintillators are well suited for this purpose. They are normally dense enough to lead to a γ-ray interaction and to stop the secondary electron within a reasonable small volume, thus allowing the measurement of the total deposited energy, and they are transparent for the created optical photons. Their main disadvantage, however, is their relatively low energy resolution, which results from the fact that about 100 eV or more are needed to create a photoelectron. Therefore the energy resolution of a scintillation detector is limited by statistical fluctuations, and with these detectors energy resolution in the few-percent range can only be reached in the best cases. In the following the main properties of scintillators will be discussed. An excellent review of the theory and applications of scintillators has been written by Birks (1964).

Organic Scintillators

In organic scintillators the emission process is based on transitions within the energy-level structure of a single molecule. The emission process is therefore independent of the physical phase of the scintillator (solid, liquid or gaseous). The scintillation process itself is the prompt emission of optical photons via fluorescence (fast component) and phosphorescence (slow component) after an external excitation by γ-rays or ionizing radiation. In general organic scintillators have a low density and a low light yield, since their main constituents are atoms with a low Z-value. Because such atoms have small cross sections for the three relevant interactions of γ-rays with matter, these scintillators are relatively poor γ-ray absorbers. But they possess two properties which make them interesting for the γ-ray astronomer: they have fast response times (several ns) and some of them allow the application of the PSD technique (see Sect. 3.1.3). So organic scintillators in plastic form are often used for anticoincidence shields for charged particles in γ-ray telescopes, whereas special liquid organic scintillators are used in Compton telescopes for PSD.

Inorganic Scintillators

The scintillation process in an inorganic scintillator is based on the lattice structure of the crystal, which is characterized by the valence band and the conduction band. Under normal conditions the electrons are bound to the atoms of the lattice, and the conduction band, where electrons can move freely, is empty. A charged particle moving through a crystal can excite an electron from the valence to the conduction band creating an electron–hole pair. After some time the electron will recombine with a hole, emitting its energy as a photon. Since, however, in pure crystals this recombination is an inefficient process and since in addition, because of the large gap width, the wavelength of the emitted photon lies outside the visible wavelength region, pure crystals are poor scintillators (a pure CsI crystal has only ~5% of the light output of a doped CsI(Tl) crystal).

It was in 1948 that Robert Hofstadter discovered that special impurities added to the melt can enhance the scintillation process significantly (Hofstadter 1948). The atoms of these impurities, called activators, modify the lattice structure of the crystal and create energy states which lie in the band gap between the valence and the conduction band. The excitation of these activator states occurs more efficiently, and furthermore, if one has chosen the right material, the photon emitted from the de-excitation process lies in the visible wavelength region. Apart from the fact that the doping of the crystal makes the scintillation process more effective, it has another consequence: it suppresses self-absorption. If the emission were the result of the recombination of an electron of the conduction band with the hole of the valence band, alone the energy of the emitted photon would be sufficient to elevate another electron from the valence band to the conduction band, thus leading to self-absorption. Since the photons of the activator states have less energy, they are not absorbed by the optical absorption band of the crystal and the crystal is transparent to these photons.

In this way many inorganic scintillator crystals have been developed and those which are most important for the γ-ray astronomer are listed in Table 3.1 together with their main characteristic properties. The most important ones are the density, the index of refraction, the decay-time constants and the total light yield. The density determines the absorption probability for γ-rays, and denser crystals are the better γ-ray absorbers. However, it is not only the absorption probability which needs to be high, but also the light output. Statistically the higher the light yield, the better the energy resolution of a detector. Although BGO has a very high density, resulting in a high stopping power, its light yield is rather small compared to the most commonly used crystals, such as NaI(Tl), CsI(Tl) and CsI(Na). So BGO is very well suited for an anticoincidence shield, but less so for a γ-ray spectrometer. For the latter one prefers the less-dense scintillators with their higher light yield. The index of refraction should be close to that of glass (~1.5) for a good optical coupling between the crystal and the glass of the photomultiplier (in this sense BGO is a bad scintillator because its high refractive index hampers the photons from leaving the crystal).

For many applications the timing behavior of the scintillation-light creation in a crystal is crucial. It is determined by the characteristic time of the emission process. Since in many cases the scintillation light originates from several emission processes, the time scales of the light creation are different. Fortunately most of the scintillation light originates from one process only, and the other components can often be neglected. Sometimes these weak components, called afterglow, can last for several milliseconds. This undesired effect can lead to very long detector dead times at high counting rates. In Table 3.1 the decay constants and the pulse-rise times are shown for the main component only.

Table 3.1. Properties of the most common inorganic scintillators. [from Knoll (1989)]

Parameter	NaI(Tl)	CsI(Tl)	CsI(Na)	BGO	BaF$_2$	CaF$_2$
Density (g cm^{-3})	3.67	4.51	4.51	7.13	4.89	3.19
λ_{max} (nm)	415	540	420	~500	310	435
Index of refraction	1.85	1.80	1.84	2.15	1.49	1.44
Decay constant (ns)	230	1000	630	300	620	900
Pulse-rise time (10–90%) (µs)	3.67	4.51	4.51	7.13	4.89	3.19
Afterglow (after 3 ms)	5%	5%	5%	0.1%	–	<0.3
Light yield (in photons MeV^{-1})	38000	52000	39000	8200	10000	24000
Energy resolution (FWHM at 662 keV)	7%	9%	8.5%	9.5%	~10%	–
Chemical formula	NaI	CsI	CsI	Bi$_4$(GeO$_4$)$_3$	BaF$_2$	CaF$_2$

In Fig. 3.1 the emission spectra of several inorganic crystals are shown. The main emission is in the range from 250 to 650 nm. In order to measure this light with a high efficiency, one has to use a sensor which is sensitive to these photons. Normally photomultiplier tubes (PMTs) are used for this purpose, but in recent years large-area photodiodes (PIN-diodes) have been developed which can be used for the same purpose.

PMTs are devices which are able to convert the extremely weak optical signal of a scintillator into a measurable electronic signal. The photons have firstly to be converted into charge carriers (normally electrons) which then have to be amplified in order to yield an electronically measurable charge. Therefore a PMT consists of two main components: a photosensitive layer (the photocathode) and a charge-amplification structure. Photons impinging onto the photocathode knock out low-energy electrons (a few eV) via the photoelectric effect (the transformation efficiency, called quantum efficiency, ranges from ~10% to ~30%). These electrons are accelerated and focused via electric fields towards the first dynode. Dynodes are metal electrodes with a high secondary-electron emission capability. Each electron which strikes the surface of a dynode knocks out further electrons (of the order of 5), which are themselves accelerated and guided by electric fields to the next dynode, and so on. At the end of the dynode structure the amplified charges are collected at the anode. Since conventional PMTs consist of about 10

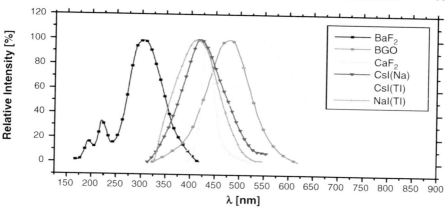

Fig. 3.1. The emission spectra of several inorganic scintillators (compiled from Bicron Corporation)

dynodes the amplification factors achieved are of the order of $5^{10} \approx 10^7$ (actual amplification factors – depending on PMT type and applied high voltage – range from $10^5 - 10^8$). Therefore one photoelectron leads finally to a charge of 1.6×10^{-12} C, a charge which can be measured by conventional electronics. For more details we refer the reader to the excellent book of Knoll (1989).

The light of a scintillator can also, instead of using a PMT, be transformed into an electrical signal by the use of photodiodes. Photodiodes are semiconducting solid-state devices with a band-gap width of 1–2 eV. Therefore the photons of a scintillator, which have an energy of 3–4 eV, are able to generate electron-hole pairs. The charges of these pairs can be measured electronically. In comparison with PMTs photodiodes have the advantages of small size, low power consumption (no high voltage is needed) and a higher quantum efficiency. In addition they are insensitive to magnetic fields. These advantages and the fact that large-area photodiodes have been developed in the last decade (Markakis 1988) make photodiodes, and newer versions, such as avalanche or drift photodiodes, a realistic alternative with respect to PMTs. Their disadvantages (small light-collection area and large electronic noise) have however prevented their application in γ-ray astronomy so far.

3.2.2 Solid-State Detectors

Solid-state detectors are superior to scintillation detectors because of their better energy resolution. Whereas in scintillators about 100 eV or more are needed to create a photoelectron, only a few eV are necessary for the generation of an electron-hole pair in a semiconductor (because of the smaller gap energy). Therefore in solid-state detectors the number of charge carriers is about two orders of magnitude larger than in scintillation detectors, and one can achieve energy resolutions which are at least a factor of ten better.

Apart from this very desirable property solid-state detectors offer other important advantages: they are small in size, thus allowing the construction of compact detectors; they have a fast response time allowing measurements with good time resolution; and they have a high absorption probability. On the other hand they are sensitive to radiation damage and usually have to be cooled to low temperatures, which makes their application in space complicated.

Historically the development of solid-state detectors depended on the availability of high-purity semiconductors, which have become available since the early 1960s. Since then a lot of semiconducting materials have been developed. The most important ones are silicon and germanium, but more exotic ones have also been developed. Among these are lithium-drifted silicon detectors, cadmium telluride, mercuric iodide and gallium arsenide. We shall not describe all these detectors, but only those which are currently used in γ-ray astronomy. These are high-purity Ge (HPGe) detectors.

In Table 3.2 the most important properties of Si and Ge are given. The band-gap of Ge is ~ 0.7 eV. This rather small band-gap energy has the negative effect that at room temperature quite a large number of electron-hole pairs is generated by thermal excitation, leading to a high leakage current. For Ge this current is higher than that produced by the γ-rays one wants to measure. Therefore one has to cool HPGe detectors down to temperatures below ~ 120 K in order to reduce the leakage current, thus requiring an expensive cooling mechanism. In addition a high voltage ranging from 2 to 9 kV has to be applied across the HPGe detector, thus making these type of γ-ray detectors complicated.

Table 3.2. Properties of silicon and germanium. [from Knoll (1989)]

Parameter	Si	Ge
Atomic number	14	32
Atomic weight	28.09	72.60
Density at 300 K (g cm^{-3})	2.33	5.33
Atoms cm^{-3}	4.96×10^{22}	4.41×10^{22}
Energy gap at 300 K (eV)	1.115	0.665
Energy gap at 0 K (eV)	1.165	0.746
Intrinsic carrier density at 300 K (cm^{-3})	1.5×10^{10}	2.4×10^{13}
Intrinsic resistivity at 300 K (Ω cm)	2.3×10^5	47
Energy per hole-electron pair at 300 K (eV)	3.62	-
Energy per hole-electron pair at 77 K (eV)	3.76	2.96
Gamma-ray attenuation length (1/e) at 100 keV(mm)	2.3	3.5
Gamma-ray attenuation length (1/e) at 1 MeV(mm)	65	29

In general HPGe detectors as used in γ-ray astronomy are cylindrical in shape with the core cut out along the central axis. Typical dimensions of a large detector are $\sim$6 cm in diameter and 5–10 cm in length. The diameter of the central hole is normally <1 cm. The high voltage is applied between the outer surface and the inner hole via conductive contacts on the outer and inner surfaces. The conductive surface layers are produced by implanting Li atoms (n^+ layers) and B atoms (p^+ layer). Detectors where the n^+ layer is on the outer surface are called p-type detectors, and those where the p^+ layer is on the outer surface are called n-type detectors. Since n-type detectors are less susceptible to radiation damage than p-type detectors (Pehl et al. 1979), n-type detectors are normally used in balloon-borne and satellite γ-ray telescopes. More details about solid-state detectors can be found in the article by Gehrels et al. (1988) and in the book by Knoll (1989).

3.2.3 Spark Chambers

For high-energy γ-rays (E $\gtrsim$ 30 MeV) the dominant interaction process of γ-radiation with matter is the pair-production process. In this process a γ-ray converts its complete energy in the electric field of an atomic nucleus into an electron–positron pair. Since the conversion probability scales with Z^2, a spark chamber must consist of a conversion medium with a high Z-value in which the conversion process can take place and a pair-tracking device in which the tracks of the electron–positron pair are visualized.

The conversion layer is composed of a stack of thin metal plates made from heavy metals like tungsten or tantalum. In one of these plates the conversion takes place and the electron–positron pair is created. These two particles then move through the pair-tracking device, which is a gas-filled spark chamber composed of several layers of thin metal plates or metal wires. Below the spark chamber the trigger telescope is mounted; this consists of two thin plastic-scintillator plates separated by several tens of centimeters and viewed by photomultipliers.

The working principle of a spark chamber is as follows: After the conversion of the γ-ray the electrons and positrons move through the spark chamber, ionizing the gas along their flight path. Then they penetrate the two plastic scintillators of the trigger telescope, where they produce photons which are registered by the phototubes. Applying a coincidence measurement one can produce a trigger pulse. This pulse is used to fire the spark chamber by applying a high voltage to its wires or metal plates. Because of the inherent inertia of the ionization of the spark-chamber gas a spark will break through along the ionization path. The positions of these sparks can now be recorded in two ways, either via an optical readout with a camera or via an electronic readout. In order to reconstruct the tracks of the electrons and positrons in three dimensions, the readout has to be made from two orthogonal directions (in the case of the optical readout one or two cameras are needed, whereas for the electronic readout two wire planes which are rotated by 90 degrees are sufficient). From the V-shaped tracks of the electron–positron pair the direction of the infalling γ-ray can be reconstructed from the bisector of the angle.

One disadvantage of spark chambers, however, should be mentioned: they have a limited lifetime when used as particle detectors on satellites. Ingredients of the spark-chamber gas, needed for quenching (i.e. suppressing) the UV-light emitted by the sparks, are cracked by the sparks. The cracking products and other not completely understood processes degrade the performance of spark chambers. Consequently the tracks become less well traced and reconstruction becomes difficult and less accurate, thus reducing the sensitivity of the instrument. This 'aging' depends on the number of recorded events and can be 'cured' to a large extent by exchanging the gas at intervals. The lifetime of spark-chamber instruments is therefore limited by the refill gas available on board.

3.2.4 Cherenkov Detectors

Cherenkov detectors are charged-particle detectors which are occasionally used in γ-ray telescopes. Therefore we discuss shortly their main physical properties. Whenever a charged particle moves through a transparent medium faster than the local speed of light within this medium ($\beta n > 1$, with β being the particle velocity relative to the speed of light and n the refractive index of the medium) then Cherenkov photons are emitted. These photons can be measured with normal phototubes. Therefore Cherenkov detectors are similar to common scintillation detectors. However, they show characteristic features which are different from those of scintillation detectors and make them unique. They possess, for example, an inherent threshold, which depends only on the refractive index:

$$E_{\text{thr}} = m_e c^2 \cdot \left(\frac{n}{\sqrt{n^2 - 1}} - 1 \right). \tag{3.1}$$

Only charged particles exceeding this threshold can create Cherenkov photons. In addition these photons are created within a very short time (i.e. the time required to slow down to a velocity below the threshold velocity), which is for solids or liquids of the order of picoseconds, making Cherenkov detectors very fast. Furthermore the photons are emitted in a cone along the direction of the particle velocity. The corresponding cone angle is given by

$$\cos \Theta = \frac{1}{\beta n}. \tag{3.2}$$

Especially this latter property makes these types of detectors interesting for γ-ray telescopes, because they allow the distinction between downward- and upward-moving charged particles by appropriate positioning of the photomultiplier tubes. Because of this feature Cherenkov detectors were normally used in spark-chamber telescopes (see Sect. 3.3.4). The Cherenkov process is also used in γ-ray telescopes which utilize the atmosphere as the detecting medium for the measurement of very high-energy γ-rays (see Sect. 3.3.5).

3.2.5 Drift Chambers

Wire-drift chambers are charged-particle devices similar to spark chambers. They are gas- or liquid-filled volumes with thin wires to which a high voltage is applied [we should mention that drift chambers using silicon instead of a gas can also be built (Gatti et al. 1984)]. A charged particle traversing the volume ionizes the gas along the particle track. Because of the applied electrical potential the electrons drift to the anode wires, whereas the ions drift in the opposite direction with a much slower drift speed than the electrons (with the consequence that the ions can be neglected). The charge pulse collected by the anode wire and its arrival time can be measured electronically. From the time difference Δt between the charge-pulse arrival time and the passage time of the particle, which has to be measured with another method (for example, a trigger telescope) the position of the particle trajectory can be calculated. This method, however, works only when the electric field across the active volume is constant, because the drift velocity v_D at constant pressure is essentially proportional to the electric field. By arranging the wires carefully, the requirement of a constant electric field can be approximately fulfilled. The drift distance d can then simply be calculated as follows:

$$d = v_D \cdot \Delta t. \qquad (3.3)$$

From the achievable accuracy of timing measurements (~ 1 ns) and drift velocities (~ 4 cm µs^{-1}) one can obtain spatial resolutions of the order of ~ 0.04 mm. Drift chambers are mainly used in nuclear-physics applications [see, for example, Salabura et al. (1996)]. However, with this technique a balloon-borne γ-ray telescope has also been developed, with which successful flights have been performed (Aprile et al. 1998).

3.3 Gamma-Ray Telescopes

A γ-ray detector becomes a γ-ray telescope when it is able to measure the direction, energy and arrival time of celestial γ-rays. This distinguishes our field from laboratory physics, where the direction of the incident radiation is known in most cases. Normally a γ-ray detecting device is surrounded by an active anticoincidence shield, which has to effectively suppress the charged-particle background (see Sect. 3.1.3). Sometimes the shield also defines the field of view, which is normally quite large.

The sensitivity of a γ-ray telescope (i.e. the minimal flux which can be measured with it) can be derived from the statistical variation of the measured counts which are given by the sum of possible unknown additional source counts $S(E)$ and the background counts $B(E)$. For small count numbers Poisson statistics holds and the minimal detectable flux $F_{\min}$ is:

$$F_{\min}(E,\Theta,\Phi) = \frac{n \cdot \sqrt{S(E)+B(E)}}{A_{\text{eff}}(E,\Theta,\Phi) \cdot T_{\text{obs}}}. \quad (3.4)$$

n defines the number of standard deviations of the background fluctuations and is usually to be taken as 3 (i.e. the probability that the observed signal is due to a statistical fluctuation is <0.3%). $A_{\text{eff}}(E,\Theta,\Phi)$ is the effective detector area, which depends on the energy and the direction, and T_{obs} is the effective observation time.

The source counts $S(E)$ for a source at the sensitivity limit depend themselves on $F_{\min}$: $S(E) = F_{\min} \cdot A_{\text{eff}}(E,\Theta,\Phi) \cdot T_{\text{obs}}$. Therefore one has to solve (3.4) for $F_{\min}$:

$$F_{\min}(E,\Theta,\Phi) = \frac{n}{2 \cdot A_{\text{eff}}(E,\Theta,\Phi) \cdot T_{\text{obs}}} \cdot (n + \sqrt{n^2 + 4B(E)}). \quad (3.5)$$

For most instruments the relation $n^2 \ll 4B(E)$ or $n \ll 2 \cdot \sqrt{B(E)}$ holds, leading to

$$F_{\min}(E,\Theta,\Phi) = \frac{n \cdot \sqrt{B(E)}}{A_{\text{eff}}(E,\Theta,\Phi) \cdot T_{\text{obs}}}. \quad (3.6)$$

This is the formula which is usually used in γ-ray astronomy to calculate the sensitivity of γ-ray telescopes. More comprehensive considerations especially about the statistical meaning of (3.6) can be found in Zhang and Ramsden (1990).

The number of background counts can be calculated from the background spectrum:

$$B(E) = \frac{\mathrm{d}F_B(E)}{\mathrm{d}E} \cdot \Delta E \cdot \Delta\Omega \cdot A_{\text{eff}}(E) \cdot T_{\text{obs}}. \quad (3.7)$$

Here $\mathrm{d}F_B/\mathrm{d}E$ is the energy spectrum of the background radiation (in units of $\mathrm{cm}^{-2}\,\mathrm{s}^{-1}\,\mathrm{sr}^{-1}\,\mathrm{MeV}^{-1}$), ΔE can be the energy resolution (full width at half-maximum, FWHM) of the detector or any other energy interval over which the flux is to be determined and $\Delta\Omega$ is the angular-resolution element. Inserting (3.7) into (3.6) one obtains:

$$F_{\min}(E,\Theta,\Phi) = n \cdot \sqrt{\frac{\frac{\mathrm{d}F_B(E)}{\mathrm{d}E} \cdot \Delta E \cdot \Delta\Omega}{A_{\text{eff}}(E,\Theta,\Phi) \cdot T_{\text{obs}}}}. \quad (3.8)$$

Using (3.8) we shall now discuss what properties a γ-ray telescope must have in order to make $F_{\min}$ as low as possible. It is obvious that the background intensity $\mathrm{d}F_B(E)/\mathrm{d}E$ has to be as small as possible. The background consists of two parts, the external and the internal background. The external background is given and cannot be influenced by the telescope builder. However the internal background can be minimized, and the relevant background-suppression methods have already been discussed in Sect. 3.1.3. Therefore, if one wants to build a sensitive γ-ray telescope, one has to obey the rules which have been outlined there. It is also clear from (3.8) that a better energy resolution leads to a higher sensitivity for narrow energy bins (for broad energy bins the resolution has no effect!). One can minimize the sensitivity further by making the effective area A_{eff} as large as possible and the effective observation time T_{obs} as long as possible. However, for very long T_{obs} systematic effects eventually dominate and provide a limitation for T_{obs}.

One important consequence of (3.8) is that the background spectrum has to be well known. Information about this spectrum can be obtained by two principal methods: the chopper technique and on/off measurements.

3.3.1 The Chopper Technique

In this technique the source signal is suppressed by blocking the field of view with a so-called occulter. The occulter is a block of heavy metal (often Pb or W) which covers the field of view for a certain time (often in a periodic manner). The thickness of the occulter has to be so large that most of the γ-rays coming from the source are absorbed. In the blocked mode background counts are therefore predominantly measured. When the occulter is removed both source and background counts are recorded. Information about the source can now be gained by subtracting the background spectrum from the spectrum measured in the unblocked mode. In practice the occulter in the chopper technique is moved back and forth at regular time intervals, thus modulating the signal with the corresponding time pattern. With this method time variations of the background (especially important for instruments which are moving in a changing background environment as in balloons and satellites) can be filtered out. However we should point to the following inherent disadvantage of this method: since the geometrical arrangement of the instrument is not the same for the two measurements (the occulter is located at two different places) the background in the unblocked mode is different from the blocked one. This difference has to be known and corrected for. Information about it is normally obtained via Monte-Carlo simulations. Especially at γ-ray energies this difference may be large, since the occulter itself can be a source of γ-rays!

3.3.2 The On/Off Technique

In this procedure information about the background spectrum is obtained by observing alternately two regions of the sky, the source and the background region. Also this procedure has – as is obvious – inherent drawbacks: the background region has to be absolutely source free and the background should be stable (i.e. no variations with time and direction)!

Imaging via Active and Passive Collimators

One important feature of a γ-ray telescope is its imaging capability. For hard X-rays, for instance, imaging is achieved by using passive collimators which define a field of view (normally several square degrees wide). These collimators are arrays composed of rectangular, hexagonal or round tubes whose walls are made from passive-absorbing materials. They are mounted in front of the detector so that only X-rays with paths more or less parallel to the tube walls can reach the detector. Inclined X-rays will hit the walls and will therefore be absorbed. Within the field of view defined by the collimator, no information about the direction of the infalling X-rays is available. However, one can use the point-spread function (i.e. the spread of the measured data from a γ-ray point source in the data space) of the instrument to locate an X-ray source with a better accuracy than is defined by the field of view when the source is moving across it.

Because of the high penetration ability of γ-rays and of the physics of absorption processes (mainly Compton scattering at low energies and pair-production at high energies) this type of collimator cannot be used at γ-ray energies. Too much matter would be needed to stop the γ-rays and would be accompanied by the creation of a high background radiation within the material of the collimator itself, diminishing the shielding effect. Therefore, very sophisticated measurement techniques have to be applied in order to obtain a γ-ray telescope with a good angular resolution [and since the sensitivity depends according to (3.8) on the angular resolution, the importance of a good angular resolution becomes obvious].

In summary, at γ-ray energies passive collimators do not work satisfactorily for the following two reasons: first thick collimator walls are needed to absorb the γ-rays, and second the collimator itself is a background source (see Sect. 3.1.3). Therefore in the early days of γ-ray astronomy the angular resolution of γ-ray telescopes was very poor. The field of view was defined only by the geometrical arrangement of the detectors, thus leading to very wide fields of view. Therefore passively collimated instruments are normally not used in γ-ray astronomy. However actively collimated γ-ray telescopes have been built and successfully used. Yet the point-source location accuracy achieved was poor (several degrees only). Within this field of view no further directional information was available. OSO-3, the γ-ray experiment of the Solar-Maximum Mission (SMM), the Gamma-Ray Imaging Spectrometer (GRIS) and the Oriented Scintillation-Spectrometer Experiment (OSSE) were γ-ray telescopes of this kind.

OSO-3

The γ-ray detector OSO-3 was a counter telescope consisting of a multilayer conversion scintillation detector, a directional Cherenkov counter and a sandwiched energy calorimeter of three tungsten plates and two NaI(Tl) crystals. This stack of detectors was surrounded by a ~2.5 cm thick plastic-scintillator shield (Kraushaar et al. 1972). The identification of a γ-ray event was achieved by requiring coincident signals of the three detectors (conversion detector, Cherenkov counter and energy calorimeter) and an anticoincidence signal of the plastic-scintillator shield. This γ-ray telescope was sensitive to γ-rays above ~50 MeV, its FWHM opening angle was ~25° and its effective area for an incidence angle of 0° and an energy of 100 MeV was ~2.5 cm^2. This instrument measured a celestial γ-ray event on average every 2.2 hours. This low counting rate illustrates the problem the γ-ray astronomy was confronted with in its early days. Nevertheless the existence of a galactic component of cosmic γ-rays was confirmed with this telescope.

The Gamma-Ray Experiment of the Solar-Maximum Mission (SMM).
The γ-ray experiment of SMM was an actively shielded multi-crystal scintillation γ-ray spectrometer with a wide field of view which was sensitive to γ-rays in the energy range from ~300 keV to ~100 MeV. It consisted of the following subsystems:

- the γ-ray spectrometer;
- a high-energy calorimeter;
- shielding elements;
- an automatic electronic gain-stabilization system; and
- an auxiliary X-ray detector.

The γ-ray spectrometer was composed of seven cylindrical NaI(Tl) crystals with a diameter of 7.6 cm and a height of 7.6 cm. Each of these crystals was viewed by a PMT. These crystals were surrounded on the sides by a 2.5 cm thick CsI annulus. On the rear side a circular CsI crystal with a thickness of 7.6 cm and a diameter of 25 cm was placed. The seven NaI (Tl) crystals and this CsI crystal formed together the high-energy detector. In order to suppress charged particles, the front and rear sides were covered with sheets of plastic scintillators, thus shielding the main detector from all sides. An auxiliary detector for the measurement of X-rays in the energy range 10–140 keV was attached on the outside of the main-detector housing. A schematic of this instrument is shown in Fig. 3.2.

Close to the seven main crystals three time-tagged ^{60}Co calibration sources were mounted, which were used to actively stabilize the gain of the PMTs via an electronic gain-control loop. Such gain-control systems utilize the fact that during a ^{60}Co decay two γ-rays and an electron are emitted. The electron can be measured with a small scintillation detector, thus indicating that two γ-rays were emitted by the calibration source. Performing a coincidence measurement between the signal from the calibration source and the detectors, one is able to filter out only the γ-rays from the calibration source. One can measure electronically the count rate for these calibration events within a certain energy range and control this rate via a change of the high voltage

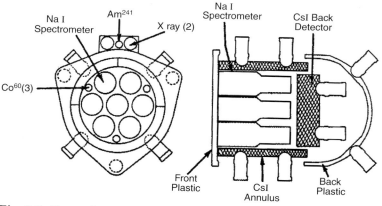

Fig. 3.2. Top and cross-sectional views of the SMM γ-ray spectrometer

of the PMTs attached to the detectors (Forrest et al. 1972). The γ-ray spectrometer had a very wide field of view of ~120°, an energy resolution of 7% at 662 keV and an effective area ranging from 20–200 cm².

The Gamma-Ray Imaging Spectrometer (GRIS)

is a balloon-borne high-resolution γ-ray spectrometer consisting of an array of seven high-purity Ge detectors surrounded by an active NaI(Tl) anticoincidence shield. It measures γ-rays in the energy range 20 keV to 8 MeV, with an energy resolution of 1.8 keV FWHM at 511 keV. In its final configuration it is equipped with a wide-field active collimator (100° × 75° FWHM field of view) and with a 15 cm thick NaI occulter for the background measurement, especially important for the observation of diffuse γ-ray sources.

In order to increase the sensitivity, a technique to suppress the β-decay background component of Ge is used in GRIS. For this each Ge-detector is segmented into several 1 cm thick slices, each of which yields an individual signal. Since an electron from a β-decay event deposits its energy very locally within the Ge crystal, whereas a γ-ray normally undergoes several Compton scatterings, depositing its energy at multiple locations, a distinction between these two event types is possible. With this background-suppression technique, the sensitivity of GRIS is improved by a factor of two.

The Oriented Scintillation-Spectrometer Experiment (OSSE)

OSSE, one of the four instruments on board of the Compton Gamma-Ray Observatory (CGRO) – one of NASA's "Great Observatories"– (see Fig. 3.3) uses active and passive collimators for the definition of the field of view. A schematic of one OSSE detector is shown in Fig. 3.4, and a detailed description of OSSE can be found in Johnson et al. (1993). OSSE consists

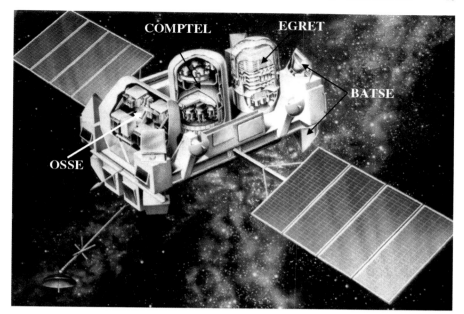

Fig. 3.3. Schematic view of CGRO, which was launched on April 5, 1991

of four identical detector systems, which are normally coaligned with the other CGRO telescopes. Each of these detectors can be independently rotated by 192° around an axis perpendicular to the nominal pointing direction of CGRO. This allows the observation of secondary targets and the application of the on/off background-determination technique as described above.

The main element of each of the four detectors is a phosphor-sandwich (phoswich) detector consisting of a NaI(Tl) crystal with a diameter of 33 cm and a thickness of 10.2 cm to which a 7.6 cm thick CsI(Na) crystal is attached at the rear. In front of the NaI(Tl) crystal a passive tungsten collimator is mounted, which defines the final FWHM field of view of 3.8° × 11.4°. The two phoswich crystals and the tungsten slat collimator are surrounded by an annular shield of 8.5 cm thick NaI(Tl) crystals which are operated in anticoincidence to suppress background radiation. The opening of the collimator is covered with a thin plastic-scintillator sheet (0.6 cm thick) for rejection of charged particles. All scintillators are viewed by PMTs.

On the lower side of the phoswich detector seven 3.5 inch PMTs are attached, measuring the light from both phoswich crystals. In the event-processing electronics a pulse-shape analysis is performed. This uses the different scintillation decay-time constants of NaI and CsI to distinguish events which interact in the NaI crystal from those which occur in the CsI crystal. Since the probability is higher that a γ-ray entering the phoswich from below interacts in the CsI crystal than in the NaI crystal (and vice versa for

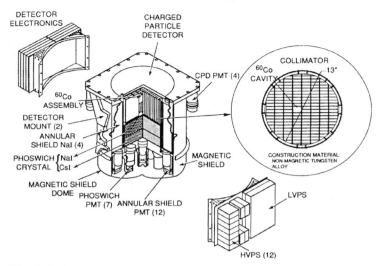

Fig. 3.4. A cross-sectional view of one of the four OSSE γ-ray detectors

a γ-ray entering the phoswich from above), γ-rays coming from below can be rejected.

In order to achieve an optimum spectral resolution, the gain of the seven PMTs is precisely controlled. As with SMM this is done via control of the high voltage of the PMTs. As a reference the light of a stabilized light-emitting diode, which is optically coupled to the CsI crystal, is measured and compared with a reference voltage. Whenever this voltage differs from the PMT signal, the corresponding high voltage is adjusted. The absolute gain of the phoswich is monitored by using an internal weak tagged radioactive ^{60}Co source. OSSE is sensitive to γ-rays in the 0.05–10 MeV range. It has an effective area of $\sim 400\,\mathrm{cm}^2$ around 1 MeV and the typical energy resolution of a NaI(Tl) crystal (5–10% depending on energy). Its 3σ sensitivity for narrow γ-ray lines at

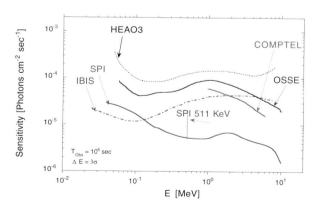

Fig. 3.5. A comparison of the 3σ narrow-line sensitivity for different γ-ray instruments

1 MeV for an observation with $5 \cdot 10^5$ s live time with all the four detectors is $9 \cdot 10^{-5}$ $\gamma\mathrm{cm}^{-2}$ s^{-1} (see Fig. 3.5).

Imaging via Modulation Techniques

One can improve the angular resolution of a γ-ray instrument with a wide field of view significantly by modulating the signal from the source to the detector in a characteristic way. This modulation is obtained by intercepting the paths of the radiation from a source with a mass (an absorber or occulter) which blocks the γ-rays nearly completely. From the measured modulation pattern one can derive the position of a γ-ray source in the sky. This modulation can be achieved either temporally or spatially.

Imaging via Temporal Modulation

In this technique information about the position of a γ-ray source is obtained by intercepting the path from the source to the detector by moving the occulter across the field of view, thus eclipsing the source completely. From the positions of the detector and the occulter and from the time when the source signal vanishes one can gain information about the arrival direction of the incident radiation.

The Occultation Technique

Even with an omnidirectional detector one can achieve high positional resolution when the distance between the detector and the occulter is large. This is normally the case for the earth and the moon. From the known positions of the detector and the occulter (in our case earth or moon) and from the time when the source signal vanishes, one can derive an arc on the sky on which the source must lie. If one can derive in a similar way another arc which is rotated by a certain angle with respect to the first one, the cross point of the two arcs determines the source position. Especially the earth is a good occulter for earth-orbiting satellites because it obscures a large region of the sky (and therefore many sources) at regular time intervals, while the moon, because of its small size, is less suitable. (However moon occultations yield much better source positions than earth occultations.) The earth-occultation technique was successfully applied by SMM and especially BATSE [see Sect. 3.3.6, Purcell et al. (1989) and Harmon et al. (1992)], resulting in the first BATSE bright-source catalog.

Fourier-Transform Telescopes

Since imaging via the occultation technique is complicated and poses severe constraints on the observability of γ-ray sources, another imaging method was developed in which the incident radiation from the sky is temporally modulated. In this method the signal is modulated by means of collimators or grids rotating across the field of view, thus modulating the signal. Using a Fourier algorithm one can reconstruct an image of the sky from the measured time sequence. Telescopes using this imaging technique are called Fourier-transform imagers. Although this method has the advantage that it does not require a position-sensitive detector, it has, however, not established itself in γ-ray astronomy.

Imaging via Spatial Modulation: Coded-Mask Telescopes

The spatial modulation is obtained by the use of an occulter which consists of opaque and transparent elements and which covers the field of view. A γ-ray point source throws therefore a shadow on the detector. With a position-sensitive detector the shape of the shadow can be measured. From the measured shape of this shadow one can derive the position of a source which is inside the field of view.

When the occulter consists of an array of opaque and transparent elements which are arranged in a regular pattern, one calls this occulter a coded mask. The imaging through such a mask is called the coded-aperture imaging technique. This imaging technique works as follows: A point source casts a shadow of the coded mask onto the detector (the so-called shadowgram). From the measured pattern of the shadow one can reconstruct the position of the point source using deconvolution algorithms [see the overview by Caroli et al. (1987)]. However, in order to be able to do this one needs a position-sensitive detector. If a pixelized detector is used for the position determination, each pixel of the detector must not exceed in size the pixel size of the mask elements, otherwise the shadowgram cannot be reconstructed uniquely. However, the smaller the detector-pixel size, the better the reconstruction.

For a point source another advantage lies in the simultaneous determination of the background and source flux, thus facilitating the data evaluation significantly. This is due to the fact that half of the field of view is occulted by the opaque mask elements (allowing for the measurement of the momentary background in the detector) and half of the field of view is open for source counts to be detected.

Several instruments using the coded-mask principle have been built or designed [for example, the balloon telescope GRIP of Althouse et al. (1985)].

SIGMA

The first successful satellite instrument using a coded mask in the γ-ray domain was the French telescope SIGMA aboard the French–Russian satellite GRANAT (Paul et al. 1991). The mask was made from tungsten elements, an Anger-camera (see Sect. 3.3.3) was used as the position-sensitive detector and a shield of CsI scintillator crystals read by PMTs was used as the aperture-defining element. This instrument was launched in 1989 and observed mainly the Galactic-Center region in the energy range 35 keV–1 MeV.

INTEGRAL

The INTErnational Gamma-Ray Astrophysics Laboratory (INTEGRAL) mission of the ESA [Carli et al. (1998); Winkler (1998)], which will be launched in 2002, will harbor two similar γ-ray telescopes, IBIS and SPI. Both instruments use coded-aperture masks for imaging purposes based on the SIGMA design. They will be operated in the energy range between 15 keV and 10 MeV. Two additional monitor instruments, JEM-X, an X-ray instrument for measurements between 3 keV and 35 keV, and OMC, an optical telescope observing at 500–850 nm, will also be flown on this satellite. They will perform simultaneous measurements of the objects being observed in the γ-ray regime, allowing for multi-wavelength observations. A picture of the INTEGRAL satellite with the two main telescopes is shown in Fig. 3.6. The two main telescopes are explained below and their key features are listed in Table 3.3. Their 3σ sensitivities are shown in Fig. 3.5 in comparison with other γ-ray telescopes.

1. IBIS (Imager on Board of the Integral Satellite): This instrument will be able to detect γ-ray sources with a very good angular resolution of 12 arc seconds within a fully coded field of $9°$ (achieved by a modified uniformly redundant array with 53×53 opaque and transparent elements). It consists of two planes of detectors. The lower layer consists of 4096 CsI scintillator bars of dimensions $9 \times 9 \times 30$ mm^3 (giving a total area of 3318 cm^2) for the detection of γ-rays from 200 keV to 10 MeV with a typical energy resolution of 6%. The upper plane is made from 16384 CdTe pixels of $4 \times 4 \times 2$ mm^3 (giving a total area of 2621 cm^2) used to measure between

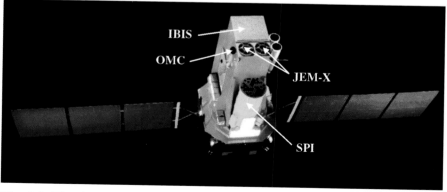

Fig. 3.6. The INTEGRAL satellite with the two main instruments (SPI and IBIS) and the two monitor instruments (JEM-X and OMC)

Table 3.3. The properties of the two main instruments aboard INTEGRAL

Parameter	SPI	IBIS
Energy range	20 keV–8 MeV	15 keV–10 MeV
Energy resolution at 1 MeV [keV]	~2	~60
Angular resolution	~2.5°	~12 arc min
Narrow-line sensitivity at 1 MeV for 10^6 s [cm^{-2} s^{-1}]	~5 × 10^{-6}	~4 × 10^{-4}
Continuum sensitivity at 1 MeV for 10^6 s [cm^{-2} s^{-1} MeV^{-1}]	~10^{-4}	
Source-location accuracy	~1°	<1 arc min
Timing accuracy	100 µs	<1 ms

15 keV and 400 keV with an energy resolution of about 7%. The mask with tungsten elements is mounted 3.2 m above the upper detector plane. The aperture is defined through an active BGO shield around and below the detector and a passive tungsten collimator between the mask and the BGO shield. A detailed description of IBIS can be found in Ubertini et al. (1999).

2. SPI (SPectrometer on Integral): SPI uses, as IBIS, a coded-aperture mask with 63 opaque and 64 transparent elements. Its main detector consists of an array of 19 high-energy-resolution Ge crystals. This will allow very precise energy measurements of γ-rays ($\Delta E/E \approx 0.2\%$) between 20 keV and 8 MeV. Due to this extraordinary energy resolution the sensitivity for narrow γ-ray lines [see (3.8)] will be at least a factor of 10 better than for other γ-ray telescopes (see Fig. 3.5). Since these Ge detectors have to be cooled to 85 K by a Stirling cooler, they are housed in a cryostat made from beryllium. This material was chosen in order to reduce the background due to the activation of the passive material. The whole detector and most of the aperture are surrounded by an active BGO shield of mass 511 kg viewed by 181 PMTs. The mask is mounted at a distance of 1.71 m above the Ge-detector plane in order to match its pixel size to the Ge detectors. Due to the limited number of Ge detectors the position resolution within the fully coded field of view of ~15° will only be of the order of 2°, while the position accuracy – depending on the source-detection significance – can be much better than 1°. More details about SPI can be found in Lichti et al. (1996) and Vedrenne et al. (1999).

3.3.3 Compton Telescopes

Since the attenuation coefficient which describes the interaction probability of γ-rays in matter reaches a minimum in the energy range from 1 to 10 MeV (see Fig. 3.7), the imaging methods described in Sect. 3.1 do not work satisfactorily in this energy range. The dominant interaction mechanism of γ-rays with matter at these energies is the Compton process. In this process a γ-ray

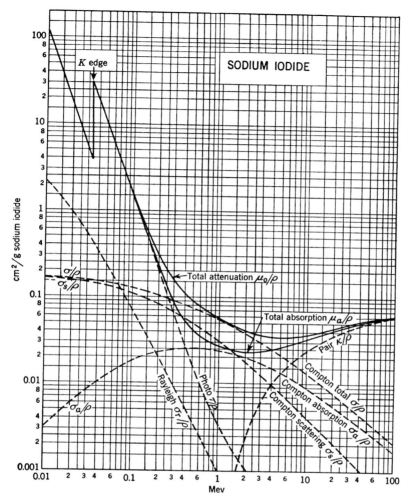

Fig. 3.7. Attenuation coefficient for γ-rays in matter (here sodium iodide) versus energy

is scattered off from an electron, transferring only a part of its energy to the electron (the recoil energy); the γ-ray continues with reduced energy and changed direction. The recoil energy E_1 depends on the γ-ray energy E and the scatter angle $\bar{\varphi}$ according to the Compton scattering formula:

$$E_1 = \frac{E^2 \cdot (1 - \cos \bar{\varphi})}{E \cdot (1 - \cos \bar{\varphi}) + m_e c^2}. \tag{3.9}$$

m_e is the rest mass of the electron and c is the speed of light. It is obvious from (3.9) that from a measurement of E_1 alone one is not able to deduce the value of $\bar{\varphi}$. Only if the energy E is known in addition can (3.9) be solved for $\bar{\varphi}$. The goal must therefore be to measure the total energy E as well. This is done in the conventional Compton telescopes which were developed in the early seventies. An even better way would be to determine the direction of the recoil electron in addition to the energy.

A Compton telescope consists of two detector planes, the scatter-detector plane (D1) and the absorption-detector plane (D2), both of which are separated by a certain distance (~ 2 m). In the scatter detector Compton scattering is favoured. The Klein–Nishina formula, which describes the cross section for the Compton process, scales linearly with the atomic number Z, whereas the corresponding cross sections for photoelectric absorption and pair production scale as $\sim Z^n$ with $n \geq 2$. Therefore the Compton process is favoured in low-Z materials. For this reason the scatter detectors have so far been made from organic scintillators. This has another advantage: one can use the PSD technique to suppress the neutron-induced background (see Sect. 3.1.3). The absorption detector has to absorb the energy of the scattered γ-ray with a high probability. Therefore this detector has to consist of materials with high Z-values. So far CsI or NaI crystals have been used. In the future germanium may also be used to improve the energy resolution.

In the ideal case the measurement of the energy losses in the scatter detector (E_1) and the absorption detector (E_2) allows the calculation of the total energy E and the scattering angle $\bar{\varphi}$ of the infalling γ-ray according to (3.9):

$$E = E_1 + E_2, \tag{3.10}$$

$$\bar{\varphi} = \arccos\left[1 - m_e c^2 \cdot \left(\frac{1}{E_2} - \frac{1}{E_1 + E_2}\right)\right]. \tag{3.11}$$

However – as is obvious from Fig. 3.8 – without knowing the scatter direction (χ, ψ) of the scattered γ-ray, one cannot determine the incoming direction of the γ-ray. The interaction locations, from which the scatter direction can be derived, can be determined in two ways: either one constructs the two detector planes from small scintillator modules and records the module in which the interaction took place, or one assembles them from larger scintillator modules and determines the interaction location within such a module by applying the Anger-camera technique (see below).

We should mention – as is obvious from Fig. 3.8 – an inherent disadvantage of present Compton telescopes: the infalling direction of a single γ-ray is not uniquely defined! The measurement of E_1 alone does not yield any information about the azimuth of the scattering process. Thus the arrival direction of a γ-ray is only known to lie on a circle (called an event circle) whose radius is given by the scatter angle $\bar{\varphi}$. Only when the direction of the recoil electron is measured can the arrival direction be uniquely defined. Instruments with this capability are described in Sect. 4.2.1.

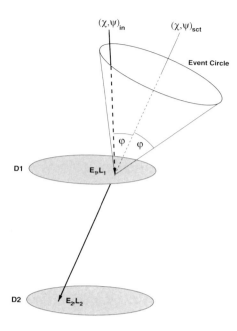

Fig. 3.8. Sketch of the measurement principle of a Compton telescope. (Courtesy of G. Weidenspointner)

For the determination of the correct scatter direction the interaction in the scatter detector must occur before the interaction in the absorption detector. This sequence is guaranteed by performing a time-of-flight measurement. In practice one performs a delayed-coincidence measurement between the two detectors. (One delays the pulse from the scatter detector by the time the scattered γ-ray needs to travel from the upper to the lower detector and then performs a coincidence measurement. With this method one can suppress all γ-rays coming from below and can thus reduce the background significantly.)

The three measured quantities χ, ψ and $\bar{\varphi}$ span a three-dimensional data space in which each event is represented by a dot. For a point source these data points are distributed around the mantle of a cone with the vertex located at the source position χ_0, ψ_0 and with a cone half-opening angle of $45°$ (the measured scatter angle $\bar{\varphi}$ is statistically distributed around the true scatter angle). Due to the fact that the energy measurements are in many cases incomplete (due to multiple scatterings in D1 and due to incomplete absorption in D2), the scatter angle $\bar{\varphi}$ according to (3.11) is not exactly known. Its error can be calculated from the energy resolutions of both detectors in the following way, by differentiating (3.9):

$$\Delta\bar{\varphi} = \frac{m_e c^2}{\sin\bar{\varphi}} \cdot \sqrt{\left(\frac{\Delta E_1}{E^2}\right)^2 + \left(\frac{E_1(E_1 + 2E_2)\Delta E_2}{E^2 \cdot E_2^2}\right)^2}. \tag{3.12}$$

In addition the scatter direction is also not known exactly because of the imperfect scatter-direction determination. The error σ_s in the scatter direction is for the case when the detectors are composed from small modules given by the spatial dimensions which are in this case independent of energy. However, if the Anger-camera principle is used for the determination of the interaction location within a detector, then σ_s may depend slightly on energy. The total uncertainty of the scatter angle $\bar{\varphi}$ is given by (with $\sigma_E = \Delta\bar{\varphi}$):

$$\sigma_{\bar{\varphi}} = \sqrt{\sigma_E^2 + \sigma_s^2}. \qquad (3.13)$$

$\sigma_{\bar{\varphi}}$ describes the statistical broadening of the data distribution around the abovementioned cone mantle, for which it is a coarse measure. This distribution is unsymmetrical and called the point-spread function for a given energy or energy interval. It has its maximum at or close to the cone mantle. The general behavior of the angular resolution of a Compton telescope is shown in Fig. 3.9.

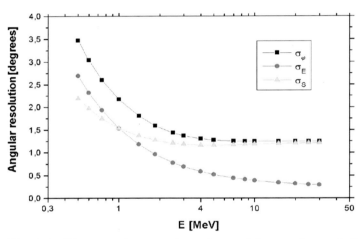

Fig. 3.9. The general energy dependence of the angular resolution of a Compton telescope

An important property of Compton telescopes should be mentioned: they normally have a large field of view of ~1 sr, which makes them very powerful for performing all-sky surveys!

Telescopes based on the above-described measurement principle were developed in the early 1970s in Germany and the U.S. The first instruments were balloon telescopes with poor to moderate energy and angular resolutions. In these the modular concept as described above was used to measure the scattering direction. Since the detector modules of these instruments were fairly large (several thousand cm^3) they had rather poor angular resolutions. The first Compton telescope ever built (Schönfelder et al. 1973) had an angular resolution of only 30° FWHM. The Compton telescope of Herzo et al.

(1975) had ~16° FWHM, whereas that of Schönfelder et al. (1982) had ~10° FWHM. The energy resolutions were ~50% FWHM and ~25% FWHM, respectively. A breakthrough in low-energy γ-ray astronomy was only achieved with the application of the Anger-camera principle, which was first used in the Compton telescope COMPTEL. Since with COMPTEL it was for the first time possible to perform astrophysically interesting measurements with a Compton telescope, it is described here in more detail.

COMPTEL

The Compton telescope COMPTEL is an advancement of the MPE balloon-borne Compton telescope and was the first Compton telescope flown on a satellite, the CGRO (see Fig. 3.3). It measures γ-rays in the energy range from ~700 keV to ~30 MeV. COMPTEL consists of two detector arrays, a scatter-detector array (D1) of low-Z material (liquid organic scintillator NE 213A allowing the application of the PSD technique) and an absorption-detector array (D2) of high-Z material [NaI(Tl) crystals] separated by 1.5 m. Each detector array is composed of several (7 in D1 and 14 in D2) circular detector modules. Each of these modules is viewed by PMTs (8 in the case of D1 and 7 in the case of D2). This allows the application of the Anger-camera principle, according to which the interaction location within a scintillation detector can be derived from the relative pulse heights of all the PMTs viewing a detector module (Anger 1958). With this method the interactions can be located within a circle with a diameter of ~2 cm, leading to an uncertainty in the scatter direction of only ~0.76°. Because of the rather good energy resolutions of the two detectors (the 1σ resolutions vary from 3%–15% for D1 and from 2.3%–3.7% for D2) the width of the point-spread function according to (3.13) is approximately given by the fit function

$$\sigma_{\bar{\varphi}}(\text{degree}) = \frac{1.247}{1 - \exp\left(-0.854 \cdot E_{\text{MeV}}^{0.9396}\right)} \cdot \quad (3.14)$$

It varies from ~3.5° at 500 keV to ~1.25° at 10 MeV. The 1σ energy resolution of the whole telescope can be described by

$$\sigma_{\text{MeV}}(E) = 0.01 \cdot \sqrt{14.61 \cdot E_{\text{MeV}} + 2.53 \cdot E_{\text{MeV}}^2}. \quad (3.15)$$

To suppress the charged-particle background each detector array is entirely surrounded by thin, dome-shaped plastic scintillators. The background is further reduced by performing a time-of-flight measurement and a PSD in the D1 detector. Halfway between the two detectors at opposite sides two tagged ^{60}Co sources are placed. These are used as electronically gated γ-ray calibration sources.

Although the total geometrical area of the D1 detector is 4188 cm^2 and that of the D2 detector 8744 cm^2, the effective area of the whole telescope ranges only from $\sim 10\,\mathrm{cm}^2$ to $\sim 50\,\mathrm{cm}^2$ (depending on event-selection criteria and energy). Thus the point-source narrow-line sensitivity of COMPTEL for a 2-week observation is $\sim 5 \times 10^{-5}\,\mathrm{cm}^{-2}\,\mathrm{s}^{-1}$ at 1 MeV and $\sim 10^{-5}\,\mathrm{cm}^{-2}\,\mathrm{s}^{-1}$ at around 10 MeV (see Fig. 3.5). A detailed description of COMPTEL can be found in Schönfelder et al. (1993).

3.3.4 Pair-Tracking Telescopes

For the measurement of high-energetic γ-rays (E $\gtrsim$ 30 MeV) pair-tracking telescopes are used. The heart of such a telescope is a pair-tracking device. In most cases this is a spark chamber (see Sect. 3.2.3). However, in order to make such a pair-tracking device or such a spark chamber a γ-ray telescope it has to possess in addition the following elements:

- a Cherenkov detector or a time-of-flight measurement system;
- a calorimeter; and
- an anticoincidence shield.

The working principle of a spark chamber has already been described in Sect. 3.2.3. A Cherenkov detector is normally integrated with the trigger telescope in order to be able to distinguish between upward- and downward-moving electrons and positrons. For the measurement of the energy of the γ-rays a calorimeter is mounted beneath the trigger telescope which is composed of thick inorganic crystals [e.g., NaI(Tl)] in which the electrons and the positrons deposit their final energy. (Another method would be to derive the total energy from the measured energy loss $\frac{dE}{dx}$, but this leads to good results only at lower energies.)

The whole array of the pair-tracking telescope (spark chamber, trigger telescope and calorimeter) is surrounded on the top and the sides by a thin dome-shaped plastic scintillator for the rejection of charged particles. This anticoincidence shield is open at the bottom to avoid the self-veto effect (i.e., the rejection of a cosmic γ-ray because its electron–positron pair triggers the anticoincidence).

Several pair-tracking telescopes have been built and flown in the past. They used optical read-out systems as well as electronic ones to record the electron–positron tracks within the spark chamber. In the γ-ray telescope S-133 of the ESRO-satellite TD-1 (Voges et al. 1974) an optical read-out system was used, whereas the satellite γ-ray telescopes SAS-II (Derdeyn et al. 1972), COS-B (Bignami et al. 1975) and EGRET (Kanbach et al. 1988) used an electronic read-out system. Schematic cross-sectional views of the latter three telescopes are shown in Fig. 3.10.

The development of pair-tracking telescopes was characterized by great difficulties and setbacks. One of the first of these telescopes (S-133 on TD-1) suffered from a low signal-to-noise ratio. One had to learn first how a spark chamber could be shielded efficiently against the background radiation. The

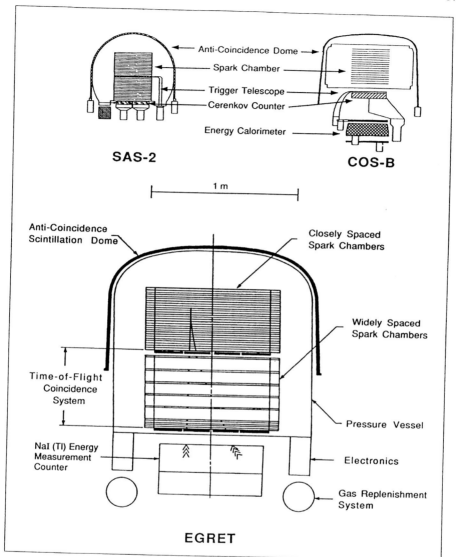

Fig. 3.10. Schematic cross-sectional views of the SAS-II, COS-B and EGRET pair-tracking telescopes (displayed with equal scale)

pair-tracking telescope of the TD-1 satellite, for example, was saturated by background because the anticoincidence hood was not extended far enough (Voges et al. 1974). A first breakthrough came with the satellite γ-ray telescope SAS-II, which was launched in 1972 (Derdeyn et al. 1972). Unfortunately this experiment survived only for about half a year. However, it demonstrated for the first time that one can successfully measure high-energy cosmic γ-rays with this technique. The final breakthrough in high-energy γ-ray astronomy was then reached with the successful operation of the European COS-B satellite, which was launched in August 1975. Since this mission was very effective and since COS-B showed all the characteristic features of a pair-tracking telescope, this experiment is described in more detail here.

COS-B

The central part of COS-B was a gas-filled wire-grid spark chamber with 16 grid planes and an electronic read-out system. Below its 12 top grid gaps thin tungsten sheets were mounted as a converter. Beneath the spark chamber the trigger telescope with the Cherenkov detector in between was mounted. The energy calorimeter was composed of CsI(Tl) crystals with a thickness which corresponded to 4.7 radiation lengths. Since this was not sufficient to stop γ-rays with energies >300 MeV, a 1 cm thick plastic scintillator was mounted beneath the calorimeter, yielding information about the leakage rate of these γ-rays. This counter led to a refined energy measurement and helped in recognizing proton-induced γ-ray background. The whole spark chamber and a part of the trigger telescope were shielded against the charged-particle background with a dome-shaped 1 cm thick plastic scintillator.

COS-B was sensitive to γ-rays in the energy range from ∼30 MeV to several GeV over a wide field of view of ∼2 sr. The energies could be measured with a moderate energy resolution only (∼10% FWHM at 100 MeV and ∼100% at 1 GeV). The 1σ angular resolution improved from ∼10° at 30 MeV to ∼2.5° at 2 GeV. The 2σ sensitivity limit of COS-B for a 4-week observation for E > 50 MeV was about 10^{-6} cm^{-2} s^{-1}. In this time about 2200 counts were detected from a Crab-like point source on the axis.

EGRET

The Energetic Gamma-Ray-Experiment Telescope (EGRET) is an improved version of the SAS-II γ-ray telescope. It is one of the four γ-ray telescopes of CGRO (see Fig. 3.3) and is sensitive to γ-rays in the energy range 20 MeV to 30 GeV.

EGRET features as its central unit a multilevel wire-grid spark chamber with interspersed tantalum conversion layers which is triggered by a trigger telescope consisting of plastic-scintillator sheets integrated into the lower part of the spark chamber. A time-of-flight measurement discriminates between upward- and downward-moving charged particles. This and the fact that the spark chamber is split into two parts (an upper part with 28 closely-spaced wire grids and a lower part with 8 widely spaced wire grids) distinguishes it from the otherwise very similar SAS-II and COS-B telescopes. As in these telescopes an energy calorimeter, consisting of 20 cm thick NaI(Tl) crystals which are mounted below the spark chamber, allows the measurement of the

Table 3.4. Main characteristic properties of the four CGRO instruments

Parameter	OSSE	COMPTEL	EGRET
Energy range (MeV)	0.1–10.0	1–30	20–30 000
Energy resolution (FWHM)	12.5% at 0.2 MeV 6.8% at 1 MeV 4% at 5 MeV	8.8% at 1.27 MeV 6.5% at 2.75 MeV 6.3% at 4.43 MeV	~20%
Effective area (cm^2)	2013 at 0.2 MeV 1480 at 1 MeV 569 at 5 MeV	25.8 at 1.27 MeV 29.3 at 2.75 MeV 29.4 at 4.43 MeV	1200 at 100 MeV 1600 at 500 MeV 1400 at 3000 MeV
Position location	10 arc min	8.5 arc min	5–10 arc min
Field of view	$3.8° \times 11.4°$	~3 sr	~0.6 sr
Max. effective geometric factor (cm^2 sr)	13	30	1050 (~500 MeV)
Narrow-line sensitivity (for 10^6 s) ($cm^{-2} s^{-1}$)	$(2-5) \times 10^{-5}$	$(0.3-3) \times 10^{-5}$	n. a.
Continuum sensitivity (for 10^6 s) ($cm^{-2} s^{-1}$)	2×10^{-7} keV^{-1} at 1 MeV	1.7×10^{-4} (1–3 MeV)	5×10^{-8} (>100 MeV)

Parameter	BATSE Large Area	BATSE Spectroscopy
Energy range (MeV)	0.03–1.9	0.015–110
Energy resolution (FWHM)	32% at 0.06 MeV 27% at 0.09 MeV 20% at 0.66 MeV	8.2% at 0.09 MeV 7.2% at 0.68 MeV 5.8% at 1.17 MeV
Effective area (cm^2)	1000 at 0.03 MeV 1800 at 0.1 MeV 550 at 0.66 MeV	100 at 0.3 MeV 127 at 0.2 MeV 52 at 3 MeV
Position location	1°	n. a.
Field of view	4π sr	4π sr
Max. effective geometric factor (cm^2 sr)	15 000	5000
Continuum-source sensitivity (for 10^6 s) [erg/cm^2]	6×10^{-8} (10 s burst)	n. a.

energy of the electron–positron pair and thus the determination of the energy of the infalling γ-ray. For background suppression the top and two-third of the sides of the spark chamber are surrounded by an anticoincidence hood made from a 2 cm thick plastic scintillator. From the geometrical arrangement of the trigger telescope, γ-rays up to incidence angles of ∼40° with respect to the instrument axis can trigger the telescope, leading to a field of view of ∼1.5 sr. A full description of EGRET is given by Kanbach et al. (1988).

EGRET has an energy resolution of 20–25% FWHM. Its angular resolution improves from ∼10° at 60 MeV to ∼0.5° at 10 GeV. With an effective area of ∼1000 cm^2 at several hundred MeV on the axis, EGRET reached a 3σ sensitivity limit of ∼10^{-7} cm^{-2} s^{-1}, an order of magnitude better than obtained with COS-B. The main characteristic properties of EGRET are compared with those of the other three CGRO instruments in Table 3.4.

3.3.5 Air-Shower Detection Techniques

General Considerations

The higher the energies to be measured, the lower the γ-ray fluxes will be (see Sect. 3.1.2), therefore requiring larger effective detector areas. Due to the opaque character of the atmosphere at γ-ray energies below 100 GeV, all instruments discussed so far had to be flown on balloons or were carried by satellites. Due to space and mass limitations in space they are limited in size.

For energies above 100 GeV these limitations change drastically, since now the atmosphere itself can be used as the sensitive medium of the detector. Gamma-rays above 100 GeV are so highly penetrating that they will only be absorbed in the lower parts of the atmosphere, mainly by creating large particle showers (so-called secondary air showers) of electrons and positrons. Whereas these particles reach the ground only for primary γ-rays above 10 TeV (and can then be used for detection themselves), their Cherenkov light can be used for their detection. Thus the registration of primary γ-rays can be achieved even for energies lower than those required to reach the ground physically. This detection technique is called the atmospheric-Cherenkov technique (ACT) (Cawley and Weekes 1995).

In Table 3.5 an overview of the existing and of some proposed instruments is given. Some of them will be explained in greater detail in the following.

Cherenkov Telescopes

Gamma-rays of high energy (above 100 GeV) interact with the atmosphere via creation of electron–positron pairs, which themselves have energies above the Cherenkov threshold (21 MeV at sea level). Above this threshold they propagate faster than the local velocity of light in air ($v_{e^-,+} > c/n$, with c being the speed of light and n the refractive index) and will therefore emit a flash of Cherenkov light in a cone with opening angle Θ (see Sect. 3.2.4),

Table 3.5. An overview of very high-energy γ-ray experiments

Experiment	Technique	Energy threshold (TeV)	Location
Existing experiments			
Whipple[a]	IACT	> 0.25	Arizona, USA
HEGRA-IACT[b]	IACT-array	> 0.50	Canary Islands, Spain
CAT[c]	ICAT	> 0.25	Themis, France
CANGAROO[d]	IACT	> 1	Woomera, Australia
CANGAROO II[e]	IACT	> 0.1	Woomera, Australia
CELESTE	ACT	> 0.02	Themis, France
Mark VI[f]	IACT	> 0.25	Narrabri, Australia
HEGRA-AIROBICC[g]	Wavefront sampling	> 15	Canary Islands, Spain
Themistocle[h]	Waveforrnt sampling	> 3	Themis, France
ASGAT[i]	Wavefront sampling	> 0.6	Themis, France
STACEE[j]	Solar Tower ACT	> 0.05	Albuquerque, USA
CELESTE[k]	Solar Tower ACT		Themis, France
CASA-MIA	Air shower	> 90	Dungway, USA
EAS-TOP[l]	Air shower	> 120	Gran Sasso, Italy
CYGNUS[m]	Air shower		Los Alamos, USA
MILAGRO, MILAGRITO[n]	Air shower		Los Alamos, USA
Under construction			
HEGRA-CLUE[o]	UV-IACT	> 1	Canary Islands, Spain
PACT[p]	IACT array		Pachmarhi, India
HESS[q]	IACT array	> 0.04	Gamsberg, Namibia
VERITAS[r]	IACT array	> 0.05	Arizona, USA
MAGIC[s]	IACT	> 0.01	Canary Islands, Spain

[a] Weekes et al. (1989)
[b] Daum et al. (1997)
[c] Barrau et al. (1998)
[d] Tanimori et al. (1998)
[e] http://icrhp9.icrr.u-tokyp.ac.jp/c-ii.html
[f] Chadwick et al. (1996)
[g] Karle et al. (1995)
[h] Baillon et al. (1993)
[i] Goret et al. (1993)
[j] Chantell et al. (1998)
[k] http://borsu8.in2p3.fr/Astroparticule.html
[l] Aglietta et al. (1992)
[m] Alexandreas et al. (1992)
[n] http://www.lanl.gov/milagro/home.htm
[o] http://www.pi.infn.it/clue
[p] http://www.tifr.res.in/ hecr/vhe.html
[q] http://www-hfm.mpi-hd.mpg.de/HESS/HESS.html
[r] http://earth.physics.purdue.edu/veritas/
[s] http://www.hegra1.mppmu.mpg.de

typically of the order of 1°. This flash lasts for about 5 ns and yields about 50 photons m^{-2} within 100 m around the axis of a primary γ-ray of about 1 TeV. The spectrum of this flash peaks in the blue to near-UV. The very diffuse nature of the flash together with the distance to the event (about 20 km) requires large optical mirrors for the collection of all the light from the flash. In order to detect this flash it has first to be distinguished from the background of normal light in the atmosphere. Typically the night-sky photon flux is about 10^{12} photons m^{-2} s^{-1} sr^{-1} between 330 nm and 450 nm. Since the opening angle of the Cherenkov flash is rather small, the field of view of the telescope has to be matched to this angle. Therefore during the time of the flash (5 ns) and within the field of view of the telescope, the night sky will yield about 1–2 photons m^{-2}, well below the yield of the Cherenkov flash (about 50 photons m^{-2}). Furthermore the maximum emission of the light from the sky lies towards the red end of the spectrum enabling one to choose the PMT response of the ACT instrument accordingly.

The discrimination against showers initiated by other particles like hadrons, which are created as secondary particles from cosmic-rays, is the crucial technique in this domain of γ-ray astronomy. It hampered the detection of sources with the first generation of telescopes considerably (Chudakov et al. 1965). In the late 1980s the use of selection criteria for the observables of the air showers, such as intensity, angular spread and other shower parameters, led to the detection of sources in the TeV range (Weekes et al. 1989). Two approaches were used: the wave-front sampling technique, where one tries to measure the propagation of the Cherenkov wave front for discrimination between hadronic and electronic showers, and the more successful imaging technique. In the following, two instruments using the latter technique are described in more detail.

Whipple

The Whipple telescope in Arizona, USA, used this so-called imaging atmospheric Cherenkov technique (IACT) to detect the first TeV γ-rays from the Crab Nebula in 1989 (Weekes et al. 1989). This method allows one to differentiate between hadron and electron showers via the different shape of their images in the telescope. Since the electron–positron showers from γ-rays are smaller and better confined than showers initiated by hadrons (they are subjected to larger scattering in the atmosphere), their images in the focus of the telescope are narrower and easier to reconstruct. This leads to a suppression of 99.7% of the unwanted hadronic background, and this is sufficient to detect the strongest TeV sources.

HEGRA-IACT

Another telescope using the IACT technique is HEGRA-IACT (high-energy gamma-ray astronomy imaging air Cherenkov telescopes), being built and run by a collaboration of Max Planck Institutes and European universities. It is located on the Canary Island of La Palma, Spain. A picture of one detector is shown in Fig. 3.11. This instrument consists currently of 5 mirrors with arrays of photomultipliers at their focus. This allows stereoscopic imaging of the Cherenkov showers, giving lower trigger thresholds (0.5 TeV), better energy resolution and better hadron separation than single-mirror Cherenkov telescopes (Daum et al. 1997).

Existing Cherenkov telescopes only operate above 300 GeV, due to the difficulty in discriminating the Cherenkov light from the background light at lower energies (the energy threshold scales only with the square root of the inverse effective light-collection area). On the other hand, due to small effective areas, satellite observations above 100 GeV are very time consuming and their results are statistically insignificant. Therefore an observational gap exists between these two energies. To overcome this problem a new concept for Cherenkov telescopes has been proposed, and prototypes are already under construction. The main idea is to enlarge the collection area of the ground-based telescopes in order to collect more light from the Cherenkov flashes.

Fig. 3.11. A photo of one of the HEGRA-IACT on La Palma, Spain

This can be achieved by using the mirrors, each one observed by a PMT, of former solar energy plants. Such a concept will allow the energy threshold of Cherenkov telescopes to be reduced to about 50 GeV, thus giving an overlap with satellite measurements. In Table 3.5 three of those experiments are listed.

For energies of primary γ-rays above 10 TeV, where the secondary particles reach the ground, a differentiation between the showers can be performed via particle detectors on the ground working as anticoincidence counters. A Cherenkov flash from γ-rays will not be accompanied by muons, in contrast to showers produced by cosmic-rays. This can be used as an additional parameter for the reduction of the hadronic background.

The IACT technique, powerful as it is for γ-ray astronomy above 100 GeV, has some inherent limitations. By using the Cherenkov light being produced by the secondary electrons and positrons the field of view has to be matched to the angular size of the flash, thus hampering all-sky surveys. Furthermore only dark clear nights (without the moon) can be used for effective observations, restricting the observation time of the instruments.

Air-Shower Arrays

Some of the IACT limitations can be overcome with air-shower arrays operating above 10 TeV. They rely on the direct detection of the particles from the secondary air showers using scintillation detectors. The direction of the primary particle can be inferred from an exact timing of the arrival times of the particles in the different scintillation detectors of the array. The energy of the corresponding event can be deduced from the total number of detected particles. Since these arrays do not use the Cherenkov light from the showers, they can be operated during daylight as well. Furthermore their field of view is not strongly limited so all-sky surveys are feasible. On the other hand, due to the large extent of the shower (about $0.5\,\text{km}^2$) these detectors need to be huge in order to detect all the particles of a shower and are therefore rather expensive. This is partially compensated by the fact that those arrays can also be used for the detection of very high-energetic cosmic-rays, thus doubling the scientific output. A description of such an air-shower array is given in the following.

CYGNUS

The CYGNUS array is located near the Los Alamos Meson Physics Facility (LAMPF) in Los Alamos, New Mexico, USA (Alexandreas et al. 1992). It consists of 204 plastic-scintillator counters each with an active area of $1\,\text{m}^2$ viewed by 2 and 3 inch photomultipliers. The detectors are spread over a total area of $86\,000\,\text{m}^2$. In addition, 30 large scintillator detectors, ($0.75\times3\times3\,\text{m}^3$), made from the former central detector of the LAMPF experiment 225, were used for the detection of muons from the showers. These muon detectors are viewed by two 5 inch photomultipliers. The total area of all 30 detectors is 70 m². They are used for differentiation between electron–positron showers from γ-rays and hadron showers from cosmic-rays.

3.3.6 Gamma-Ray Burst Instruments

Gamma-ray burst instruments need to have a large field of view, since the duration of the bursts is of the order of seconds and their rate of occurrence is only of the order of one per day over the whole sky. Due to the short duration of a burst a fast orientation of the satellite in the direction of the burst would be required for a successful observation. Therefore only an omnidirectional detector can detect nearly all bursts (since some can still be hidden by the earth!). Furthermore, the determination of the direction of the bursts with high accuracy during the short duration is very difficult.

In the early days γ-ray bursts were measured with omnidirectional γ-ray detectors only. Therefore the direction to the γ-ray burst source could be derived only via triangulation, for which the γ-ray arrival time at several differently located γ-ray burst detectors was needed. Measurements performed with the Vela satellites, with Helios-2, the Konus instrument, the Pioneer-Venus Orbiter, the Venera and Ulysses spacecrafts and Phebus on board of GRANAT successfully contributed among others to these measurements. However, with the launch of CGRO and its BATSE instrument a new era of burst measurements began (see Chap. 15).

BATSE

The Burst And Transient Source Experiment (BATSE) is one of the four instruments aboard CGRO. It is a full-sky monitor for γ-ray bursts and is also used for the observation of transient sources (Fishman et al. 1989). It consists of eight thin scintillation modules, one on each corner of CGRO (see Fig. 3.3). Each detector plane is oriented in a different direction. From the relative intensity measured in these eight modules, the direction of a γ-ray burst can be deduced from the relative counting rates, with an accuracy of 1° to 10°, depending on the intensity of the burst. Each of the modules contains two detectors, the Large-Area Detector (LAD), optimized for sensitivity and directional response, and the Spectroscopy Detector (SD), optimized for broad energy coverage and energy resolution. The LAD is a NaI scintillator-crystal disk, 50.4 cm in diameter and 1.27 cm thick, having a sensitive area of 2025 cm². The light is collected in the housing of the crystal and transferred to a 5 inch (= 12.7 cm) PMT. Additionally, the crystal is surrounded by a thin lead and tin layer at its back and side to reduce the instrumental background. A 6.35 mm thick plastic scintillator in the front is used as an active anticoincidence shield for suppressing the charged-particle background. The energy range of the LAD is 20 keV–1.9 MeV. The SD is a NaI(Tl) crystal with a diameter of 12.7 cm and a thickness of 7.62 cm, thus having a sensitive area of 127 cm². It is directly coupled to a 5 inch PMT. The housing is again a passive lead/tin shield at the side and back of the crystal and a 0.68 mm thick beryllium window in the front, allowing good transmission down to ~7 keV (to reduce the rates during solar flares, the window extends only to 3 inches instead of 5 inches). Due to the relatively thick crystal the SD can measure up to 100 MeV using the pair-production process. The sensitivity for a burst duration of 1 s is 10^{-8} erg cm^{-2}.

3.4 Specific Calibration and Data-Analysis Problems

Measurements with γ-ray telescopes are difficult due to low numbers of detected source photons and high instrumental background. Therefore data analysis for γ-ray telescopes is characterized by some unusual complexities. The signal-to-background ratio is often as low as a few percent, so that measurement or modelling of the background is crucial. Additionally, the response function of γ-ray telescopes spreads data from a single point source over wide ranges of measured parameters, i.e. the point-source response is highly non diagonal.

In general, measurements performed with an event-type recording telescope can mathematically be expressed as

$$D = R \cdot I + B + N, \tag{3.16}$$

with R being the instrument-response function, I the celestial source intensity, B the instrumental background, N the noise of the measurement, and D the measured data. We call (3.16) the imaging equation. Since many parameters may be measured per detected photon, the response matrix R is multi-dimensional and the data-analysis problem resembles finding complex structures in a multi-dimensional data space.

The response function of the instrument may be obtained through ground calibrations with γ-rays produced either from radioactive decays or from nuclear reactions in the target of an accelerator facility. It is very difficult to obtain 'clean' calibration sources with a monochromatic and well-defined photon spectrum; secondary γ-rays, low-source intensities, and large room background γ-rays all provide severe contaminations of the calibration source. Additionally, the number of γ-ray energies over the γ-ray regime is limited by the feasibility of nuclear reactions or the availability of excited nuclear states. Only at higher γ-ray energies do continuous and tunable sources become available again, e.g. from laser backscattering on an electron beam.

Therefore an essential tool for generation of the instrument-response function is the use of simulation codes, where all physical interactions within the instrument are traced. Monte Carlo techniques then allow a sampling of expected measurements for a given γ-ray incidence. These software tools include tabulations of reaction cross sections, and often approximations are used for different energy regimes of primary and secondary particles. The instrument mass model must be accurate in geometry and chemical composition in order to properly trace all nuclear interactions. One example of such a software system is CERN's GEANT system, widely used for high-energy detector construction and calibration.

Instrumental background arises from a variety of nuclear reactions of cosmic-ray particles or atmospheric neutrons, which bombard the instrument and its platform and initiate electromagnetic cascades with abundant γ radiation. The study of such background again heavily relies on Monte-Carlo simulations of cosmic-ray interactions in a mass model of the instrument. In addition a calibration to the highly variable cosmic-ray flux must be made through measurements, which is additionally complicated by build-up of radioactivity within the instrument with decay time constants from seconds to years. Most difficult are those times which match observational periodicities, such as orbit durations or re-pointing cycles. Charged-particle detectors therefore are part of most telescopes; the analysis of their data in terms of cosmic-ray activation presents a formidable data-analysis problem in itself. Empirical determinations of background from regions in data space which are 'off-source' therefore comprise an important analysis technique that is feasible and mostly used for point sources.

The inversion of (3.16) is a fundamental problem for experimental astronomy [see e.g. Skinner and Ponman (1995) or Lucy (1994)], since this equation can rarely be solved in algebraic form: the inverse of the response matrix is ill-defined for typical γ-ray telescopes, and the statistical-noise component N adds in addition sufficient distortion so that the derived signal above the large background is subject to large oscillations, making the solution of (3.16) highly unstable. Therefore the inversion can only be performed using the methods of statistical estimation theory like the maximum-likelihood method. Also iterative solutions of (3.16) are common, including additional regularization terms to achieve stability. One example of such iterative approaches is the maximum-entropy method. Other approaches iteratively remove the forward-folded contribution of the strongest estimated source; a test of the residuals may suggest other source contributions or be in satisfactory agreement with background and statistical noise only, at which point the iterations are terminated. One characteristic of these analyses is that a unique solution of (3.16) cannot be obtained; rather, the solution reflects the prior knowledge built into the convergence path and criteria, which have been selected for the astronomical question that is posed as the data-analysis problem. Although in principle these considerations hold for any data analysis, γ-ray data and their results are more obviously affected by their consequences due to the probabilistic interaction of γ-ray photons with the telescope components.

3.5 Current Developments and Future Instrumentation

3.5.1 Advanced Compton Telescopes

Conventional Compton telescopes, as explained in Sect. 3.4, suffer from three disadvantages:

- the ambiguity of the γ-ray arrival direction (event circle);
- the coarse determination of the scatter angle $\bar{\varphi}$ according to (3.12); and
- the coarse determination of the scatter direction.

If one wants to build a more sensitive Compton telescope one has to improve, according to (3.8), the angular-resolution element which depends on the aforementioned quantities. Thus one has to improve the determination of the scatter angle $\bar{\varphi}$ and the scatter direction, and one has to gain, if possible, some information on the azimuth of the Compton-scatter process. In the following, techniques which aim for these improvements, together with the corresponding instrument developments, will be described.

Time-Projection Drift-Chamber Compton Telescopes

For the measurement of a Compton interaction the drift-chamber principle (see Sect. 3.2.5) can be used. In order to obtain a sufficiently high γ-ray interaction probability, one needs a volume filled with a dense-enough liquid. An ideal material for this purpose is liquid xenon. It has a high density ($3\,\mathrm{g\,cm^{-3}}$) and a high atomic number ($Z = 54$). In addition the xenon atoms, when ionized and excited by an interaction, produce a large number ($\sim 64\,000\,\mathrm{e^-\,MeV^{-1}}$) of electron–ion pairs. Furthermore xenon acts as a scintillator, and the scintillation photons can be recorded with PMTs. By applying a uniform electric field across the volume the electrons drift towards the anode. From the measured drift time the vertical coordinate of the interaction can be derived by means of (3.3). The other two spatial coordinates can be obtained from the charge signals produced in a pair of mutually orthogonal sets of parallel sense-wire electrodes placed above the anode (time-projection chamber). The overall-deposited energy can be inferred from the total charge collected at the anode. For a multiple Compton interaction the charges at the anode arrive at different times, leading to characteristic charge steps. Each of these steps is proportional to the energy deposited at the respective interaction site. The right sequence of the interactions (first the Compton scatter and then the absorption) has to be derived from the relative scattering probability.

A Compton telescope based on this technique has been built by a group at Columbia University (Aprile et al. 1998). This Liquid Xenon Gamma-Ray Imaging Telescope (LXeGRIT) features a total volume of liquid Xe of 10 l and a drift gap of 7 cm, yielding a sensitive area of $400\,\mathrm{cm^2}$. This time-projection chamber is optimized for γ-rays in the 0.3–10 MeV range, where Compton scattering is the dominant interaction process. It has an energy resolution of better than 6% FWHM at 1 MeV. Its effective area at 1 MeV is ~80 cm^2 giving a continuum sensitivity of ~2×10^{-7} cm^{-2} s^{-1} keV^{-1} for a 3×10^4 s observation in the 1–3 MeV energy range.

Due to the fact that the distance between the first and second interactions is short (of the order of 10 cm), the angular resolution is dominated by the uncertainty of the scattering direction, although the interaction-position resolution in the horizontal plane is better than 1 mm, and in the vertical direction could be even better than 0.34 mm. So only a moderate angular resolution of ~$2°$ is obtained with this telescope. The feasibility of this technique has been proven during two balloon flights constituting an important milestone for the application of this technology in γ-ray astronomy.

Silicon-Tracker Compton Telescopes

The angular resolution and therefore the sensitivity of a Compton telescope could be significantly improved if information on the azimuth of the Compton-scattering process were available [see (3.8)]. The application of the technique of silicon-strip detectors allows the measurement of the direction of the recoil electron.

Silicon-strip detectors are semiconductor devices which allow positional measurements of charged particles with a high precision. These detectors were developed in the 1980s for nuclear-physics experiments. In recent years they have also become interesting for γ-ray astronomy because these detectors allow the accurate measurement of the recoil electron after a Compton collision in a Compton telescope. Therefore the functional principle of a silicon-strip detector will be shortly discussed.

A silicon-strip detector consists of a stack of thin plates of n-doped Si-crystals onto which thin (~10 μm) strips of p-type Si are implanted, doping the Si with boron and ions thus forming strips of p–n junctions (Si is used because of its larger band-gap energy compared to Ge [see Table 3.2], which leads to low leakage currents at room temperature). To minimize the multiple Coulomb scattering the crystals are made as thin as possible. However a limit on the thickness is set by the signal-to-noise ratio, which is determined by the number of electron–hole pairs created and by the capacitance C of the p–n junctions. Since C is inversely proportional to the thickness of the depletion layer and since the noise increases with C, one cannot make the crystal thinner than ~100 μm. A typical detector plate has a thickness of ~300 μm. Since the average energy loss for electrons in Si is ~390 eV μm^{-1} and since about 3.6 eV are needed for the creation of an electron–hole pair, a charged particle penetrating such a crystal plate creates ~3.2×10^4 electron–hole pairs, enough to produce a detectable electronic signal. This signal will normally appear on those strips which are nearest to the interaction location (in general more than one strip collects the produced charge, forming pulse-height clusters in adjacent strips of the detector plate), thus yielding spatial information in one coordinate. If one has another detector plate which is mounted close to the first but rotated by 90°, one has a detector module which allows a measurement of the interaction location in two dimensions. Several such detector modules stacked together then form a position-sensitive charged-particle detector. Spatial resolutions better than 0.01 mm have been achieved with such detectors (Hyams et al. 1983). The actual manufacturing of a silicon-strip detector is a complicated issue and requires a deep knowledge of semiconductors. For more details see Kemmer (1984).

Instruments using this concept are currently under development at the University of California in Riverside (TIGRE) and at the Max-Planck-Institut für extraterrestrische Physik in Garching (MEGA). As an example the instrument which is developed in the latter institute is described in more detail here.

The Medium-Energy Gamma-ray Astronomy (MEGA) telescope will consist of a stack of silicon-strip detectors which act as a scatter detector, D1, and an array of pixelized CsI-scintillators, read out via PIN diodes, which are used as the absorption detector, D2. This concept allows the complete determination of the event kinematics and restricts the event position to a small arc on the sky. This will especially improve the sensitivity and angular resolution of the telescope. Furthermore it gives the possibility to use the silicon-strip detectors as pair detectors at higher energies, thus enhancing the energy range of the telescope. The energy range of MEGA will be designed to be 300 keV–50 MeV.

Since the angular resolution of a Compton telescope depends, according to (3.12), on the energy resolution of the two detectors, D_1 and D_2, the CsI crystals were optimized to reach good energy resolutions (the energy resolution of the silicon-strip detectors is good enough in any case) and for the ability to determine the direction of the recoil electron. The Klein–Nishina equation (see Chap. 2), which gives the probability of the photon being scattered in a specific direction, shows that low-energy events are mainly scattered through angles larger than 45°, whereas for high-energy events the forward direction is preferred. This led to the use of short blocks ($\sim$3 cm) with a large pixel size of $\sim 1 \times 1$ cm^2 and a good energy resolution at the four sides of the silicon-strip stack, i.e. in the scatter direction of low-energy events. Accordingly, long blocks of CsI $\sim$7 cm with a small pixel size ($\sim 5 \times 5$ mm^2), but with a high efficiency and a moderate energy resolution were chosen below the silicon-strip stack.

Compton Telescopes with Germanium Planar-Strip Detectors

Another possibility to improve Compton telescopes is to use detectors with a good spectral and spatial resolution. Germanium planar-strip detectors have been proposed for this purpose by Johnson et al. (1995). They have an energy resolution of 2–3 keV and a spatial resolution of $\sim$2 mm. According to (3.12) this gives an uncertainty in the scatter angle of $\sim$0.5°. When the two detector planes are ~1 m apart the uncertainty in the scatter direction is only ~0.3° (compared to ~1.5° in the case of COMPTEL). This will provide an imaging capability in the arc-minute range. If each detector layer is composed of ~400 detector elements, as described by Kroeger et al. (1995), a detector area of ~1 m^2 is achieved. With a detection efficiency of 1–3% this leads to an effective area of ~100–300 cm^2. With such an instrument the 3σ narrow-line sensitivity at 1 MeV for a 10^6 s observation is ~10^{-7} photons cm^{-2} s^{-1}, much better than that of OSSE and still better by a factor of 20 or so than that of INTEGRAL.

If one places above the top detector a coded mask which is opaque to hard X-rays (20–300 keV) but transparent for higher-energy photons (>300 keV), a good imaging capability for X-rays can be obtained.

3.5.2 Advanced Pair-Production Telescopes

The proposed Gamma-ray Large-Area Space Telescope (GLAST) mission (Bloom et al. 1998) is designed to continue the successful observations of the EGRET telescope, albeit with much improved sensitivity. It will measure from 20 MeV–300 GeV, reaching a point-source sensitivity of better than 4×10^{-9} photons cm^{-2} s^{-1} for an observation time of one year above 100 MeV. GLAST will function by the same physical process as EGRET, but using modern technical concepts. It will be a pair-tracking telescope and will consist of an array of towers of pair-conversion chamber stacks arranged on a grid of 5×5 stacks. Each stack has a dimension of 33.38×33.28 cm^2, giving an effective detector area of 7000 cm^2 above 1 GeV. In contrast to EGRET the stacks will be made of silicon-strip detectors, having a much higher efficiency over a wide angle of incidence, better precision and stability and no need of consumables. Similar to EGRET there will be thin tungsten plates between the layers of silicon-strip or glass-fibre detectors for conversion of the γ-rays to e^+e^- pairs. The calorimeter will consist of 80 CsI(Tl) crystals per stack, being viewed by PIN photodiodes. The stack size will be approximately 2.3×3×31cm^3. The crystals will be arranged on a grid in each stack, allowing for a position-sensitive detection of the total shower produced by the incident γ-rays. The anticoincidence charged-particle veto shield will be a segmented plastic scintillator. The segmentation will reduce the self-veto effect at higher energies which is due to the calorimeter albedo (leakage of particles from the calorimeter). Furthermore, since no time-of-flight measurement is performed, the field of view of GLAST will be greatly increased over that of EGRET. Thus a better all-sky monitoring ability will be achieved.

3.5.3 The Gamma-Ray Lens

In the energy regime below a few MeV the present concepts (Compton telescopes and coded-mask instruments) allow for higher sensitivity and better angular resolution only by enlarging the detector area and therefore the size of the instruments. Due to the increasingly important internal background contribution (see Sect. 3.1.3), these concepts have natural limitations. In order to reach sensitivities below 10^{-6} photons cm^{-2} s^{-1} and resolutions better than arc minutes different concepts have to be used.

All designs for satellites discussed so far use the same volume for the collection and the detection of the γ-ray photons. The sensitivity of the instrument can then be found from (3.8), being inversely proportional to the square root of the effective detection (= collection) area. Thus an enlargement of this area leads not only to a small increase in sensitivity, but also to a higher background due to the activation through cosmic-rays and due to the internal background of the instrument. If one uses instead a concept where the collection and detection of photons is separated, one could in principle overcome this problem, since the enlargement of the collection area does not then increase the background in the detection volume, which can be kept small and be placed further away from the collection area.

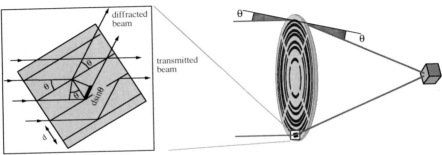

Fig. 3.12. The principle of a γ-ray lens using von-Laue diffraction

This concept is realized by the so-called γ-ray lens (Smither 1982), which utilizes the crystal diffraction to collect the γ-rays at a focal point. Either one can use the Bragg reflection of γ-rays on the surface of a single crystal or the von-Laue diffraction in the crystal volume. The latter method has the advantage of higher refractive power, thus allowing for larger deflection angles and therefore for measurement with higher energies and efficiencies. In a γ-ray lens (see Fig. 3.12) a large number of small crystals is arranged in several concentric rings and each of these rings uses a different diffraction plane. This leads to a concentration of the γ-rays in a focal point behind the lens. Here a small Ge detector is mounted for the detection of the γ-rays. Thus, the background in the detector can be kept small, whereas the collection area can be large. The angular resolution is defined through the field of view of the lens and therefore, for point-like sources, through the angular width of the rocking curve (i.e. the number of counts as a function of the angle Θ) of the crystals. Using single crystals would lead to a far too small angular resolution, on the order of the Darwin width (i.e. the natural linewidth) of the crystals. Therefore so-called mosaic crystals are being used. These crystals are composed of a large number of small single crystals with their crystallographic axes oriented in slightly different directions, thus giving a rocking width of some arc seconds. A prototype of such a system has been built at the Argonne National Laboratory by Smither et al. (1995). Such a system can only be designed for one single small energy band, such as the 511 keV annihilation line.

For astrophysical applications a much broader energy band would be desirable. Therefore the concept of a tunable γ-ray lens for a space-borne experiment (see Fig. 3.13) has been proposed by von Ballmoos et al. (1995). Here the diffracting crystals on the ring are positioned via small piezo electric elements such that they fulfil Braggs law for different energies between 200 keV–1300 keV. Since for each energy the focal length of the lens has to be adjusted accordingly, a deployable boom is used to position the detector at the focal point. This detector is a segmented Ge detector allowing for simultaneous measurement of the signal and the background of the system. Such a lens could have an energy bandwidth of 10 keV at 847 keV, an energy resolution of about $\Delta E/E \approx 0.2\%$, an angular resolution of 15" FWHM and a sensitivity of a few 10^{-7} photons cm^{-2} s^{-1}.

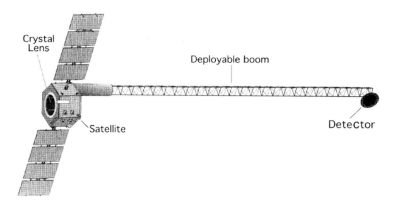

Fig. 3.13. A possible design for a space-borne γ-ray lens

References

Aglietta, M., Alessandro, B. et al., 1993, Nuov. Cimento **15**, 723
Alexandreas, D.E. et al., 1992, Nucl. Instr. Meth. in Phys. Res. **311**, 350
Althouse, W.E., Cook, W.R. et al., 1985, Proc. 19th Int. Cosmic-Ray Conf. **3**, 299
Anger, H.O., 1958, Rev. Sci. Instr. **29**, 27
Aprile, E., Egorov, V. et al., 1998, Nucl. Instr. Meth. **A 412**, 425
Baillon, P. et al., 1993, Astropart. Phys. **1**, 341

Barrau, A., Bazer-Bachi, R., Beyer, E. et al., 1998, Nucl. Instr. Meth. A, **416**, 278
Beuermann, K.P., 1971, J. Geophys. Res. **76**, 4291
Bignami, G.F., Boella, G., Burger, J.J. et al., 1975, Space Sci. Instr. **1**, 245
Birks, J.B., *The Theory and Practice of Scintillation Counting*, Pergamon Press, Oxford, 1964
Bloom, E. et al., *A proposal for the Gamma-ray Large Area Space Telescope*, Stanford, 1998
Carli, R. et al., 1998, Astrophys. Lett. Commun. **39**, 317 (785)
Caroli, E. et al., 1987, Space Science Rev. **45**, 349
Cawley, M.F., Weekes, T.C., 1995, Experimental Astronomy **6**, 7
Chadwick, P.M., Dickinson, J.E., Dickinson, M.R. et al., 1996, A&A Supl. **120** C657
Chantell, M.C., Bhattacharya, D., Covault, C.E. et al., 1998, Nucl. Instr. Meth. A **408**, 648
Chudakov, A.E. et al., 1965, Translation Consultants Bureau, P.N. Lebedev Phys. Institute **26**, 99
Daum, A., Hermann, G., Hess, M., Hofmann, W. et al., 1997, Astropart. Phys. **8**, 1
Derdeyn, S.M., Ehrmann, C.H., Fichtel, C.E., Kniffen, D.A., Ross, R.W., 1972, Nucl. Instr. Meth. **98**, 557
Fichtel, C.E., Kniffen, D.A., Ögelmann, H.B., 1969, Ap. J. **158**, 193
Fishman, G.J. et al., 1992, NASA Conf. Publ. **3137**, 26
Forrest, D.J., Higbie, P.R., Orwig, L.E., Chupp, E.L., 1972, Nucl. Instr. Meth. **101**, 567
Forrest, D.J., Chupp, E.L., Gleske, I.U., Internal report of the Department of Physics, University of New Hampshire, Durham, 1975
Forrest, D.J., Chupp, E.L., Ryan, J.M. et al., 1980, Solar Physics **65**, 15
Gatti, E., Rehak, P., Walton, J.T., 1984, Nucl. Instr. Meth. **226**, 129
Gehrels, N., Crannell, C.J., Forrest, D.J., Lin, R.P., Orwig, L.E., Starr, R., 1988, Solar Physics **118**, 233
Goret, P., Palfre, T., Tabary, A., Vacanti, G., Bazer-Bachi, R., 1993, A&A **270**, 401
Harmon, B.A. et al., in *The Compton Observatory Science Workshop*, ed. by Shrader, C.R., Gehrels, N., Dennis, B., NASA **CP-3137**, 1992, 69
Herzo, D., Koga, R., Millard, W.A. et al., 1975, Nucl. Instr. Meth. **123**, 583
Hofstadter, R., 1948, Phys. Rev. **74**, 100
Hyams, B., Koetz, U., Belau, E. et al., 1983, Nucl. Instr. Meth. **205**, 99
Johnson, W.N., Kinzer, R.L., Kurfess, J.D. et al., 1993, Ap. J. Suppl. **86**, 693
Johnson, W.N., Kroeger, R.A. et al., 1995, Hard X-ray and γ-ray imaging systems utilizing Germanium strip detectors. Proc. Int. Workshop on Imaging in high-energy astronomy in Anacapri, 329
Kanbach, G., Reppin, C., Schönfelder, V., 1974, J. Geophys. Res. (Space Physics) **79**, 5159
Kanbach, G., Bertsch, D.L., Favale, A. et al., 1988, Sp. Sci. Rev. **49**, 69
Karle, A., Merck, M., Plaga, R. et al., 1995, Astropart. Phys. **3**, 321
Kemmer, J., 1984, Nucl. Instr. Meth. **226**, 89
Knoll, G.F., *Radiation Detection and Measurement*, John Wiley & Sons, New York, 1989
Kraushaar, W.L., Clark, G.W., Garmire, G.P. et al., 1972, Ap. J. **177**, 341

Kroeger, R.A., Johnson, W.N. et al., *Spatial and spectral resolution of a Germanium strip detector*, Proc. Int. Workshop on Imaging in high-energy astronomy in Anacapri, 325, 1995

Lichti, G., Schönfelder, V., Diehl, R. et al., 1996, SPIE Proc. **2806**, 217

Lucy, L.B., 1994, Astron. Astrophys. **289**, 983

Markakis, J., 1988, IEEE Trans. Nucl. Sci. **NS-35**(1), 356

Matteson, J.L., Proc. of Conf. on Transient Cosmic Gamma- and X-Ray Sources (ed. Strong, I.B.), LA-5505-C, Los Alamos Scientific Laboratory, Los Alamos NM, 237, 1974

Matteson, J.L. et al., 21st Int. Cosmic-Ray Conf. (Adelaide), **2**, 174, 1990

Metzger, A.E., *Gamma-Ray Astrophysics* (ed. by Stecker, F.W., Trombka, J.I.), NASA **SP-339**89, 97, 1973

Paul, J., Mandrou, P., Ballet, J. et al., 1991, Adv. Sp. Res. **11/8**, 289

Pehl, R.H., Madden, N.W., Elliot, J.H. et al., 1979, IEEE Trans. Nucl. Sci. **NS-26**(1), 321

Purcell, W.R., Ulmer, M.P., Share, G.H., Kinzer, R.L., SMM observations of interstellar ^{26}Al: a status report, in: Proc. of the Gamma-Ray Observatory Science Workshop, ed. Johnson, N., 4–327, 1989

Salabura, P. et al., 1996, Acta Physics Polonica **B27**, 421

Scheel, J., Röhrs, H., 1972, Z. Physik **256**, 226

Schönfelder, V., Hirner, A., Schneider, K., 1973, Nucl. Instr. Meth. **107**, 385

Schönfelder, V., Lichti, G., 1975, J. Geophys. Res. (Space Physics) **80**, 3681

Schönfelder, V., Graser, U., Diehl, R., 1981, A&A **110**, 138

Schönfelder, V., Graser, U., Diehl, R., 1982, A&A **110**, 138

Schönfelder, V., Aarts, H., Bennett, K. et al., 1993, Ap. J. Suppl. Ser. **86**, 657

Skinner, G.K., Ponman, T.J., 1995, Inverse Problems **10**, 655

Smither, R.K., 1982, Rev. Sci. Instruments **44**, 131

Smither, R.K. et al., in: *Imaging in high-energy astronomy*, ed. by Bassani, L., di Cocco, G., Kluver Academic Publishers, Dordrecht, 47–56, 1995

Stecker, F.W., 1973, Nature (Phys. Sci.) **242**, 59

Tanimori, T., Sakurazawa, K., Dazeley, S.A. and et al., 1998, Ap. J. **492**, L33

Thompson, D.J., 1974, J. Geophys. Res. **79**, 1309

Ubertini, P., Lebrun, F., DiCocco, G. et al., 1999, Astrophys. Lett. Comm. **39**, 331

Vedrenne, G., Schönfelder, V., Albernhe, F. et al., 1999, Astrophys. Lett. Comm. **39**, 325

Voges, W., Pinkau, K., Koechlin, Y. et al., Gamma-Ray Observations of the Sky from the TD-1 Satellite, Proc. of 9th ESLAB Symp. (Frascati, ed. Taylor, B.G.), ESRO **SP-106**, 205–209, 1974

von Ballmoos, P. et al., in: *Imaging in high energy astronomy*, ed. by Bassani, L., G. di Coco, Kluver Academic Publishers, Dordrecht, 239–245, 1995

Weekes, T.C. et al., 1989, The Astrophysical Journal **342**, 379

White, R.S., 1973, Rev. Geophys. Space Phys. **11**, 595

White, R.S., Schönfelder, V., 1975, Astrophys. Space Sci. **38**, 19

Winkler, C., 1998, Astrophys. Lett. Comm. **39**, 309 (777)

Zhang, S.N., Ramsden, D., 1990, Experimen. Astronomy **1**, 145

4 Summary of the Gamma-Ray Sky

Volker Schönfelder

4.1 Introduction

The last decade of the 20th Century provided the first complete all-sky surveys in γ-ray astronomy in the energy ranges from about 1 MeV to more than 1 GeV. The surveys, which were performed by the telescopes COMPTEL (Compton Telescope 1–30 MeV) and EGRET (Energetic Gamma-Ray Experiment Telescope >100 MeV) on board NASA's Compton Gamma-Ray Observatory (CGRO), provided a complete overview of the γ-ray sky not only in continuum, but also in line emission (e.g., at 1.809 MeV [radioactive ^{26}Al], 1.157 MeV [radioactive ^{44}Ti], and 2.23 MeV [neutron capture]) at unprecedented sensitivities.

In other spectral ranges (<1 MeV, and at TeV energies) no surveys exist yet. However, measurements of special regions of the sky have been performed with high sensitivity and/or excellent angular resolution (e.g., by SIGMA on board the Russian GRANAT mission, by OSSE [Oriented Scintillation-Spectrometer Experiment] and BATSE [Burst and Transient Source Experiment] on board CGRO, and by several ground-based TeV telescopes).

From all these measurements we know that γ-ray emission is a phenomenon of the most compact and energetic objects in the Universe. In the Milky Way these are especially spin-down pulsars, accreting X-ray binaries, and supernova remnants. Spin-down pulsars are isolated rotating neutron stars. Accreting X-ray binaries are systems in which a collapsed star – a neutron star or a black hole – accretes matter from its normal stellar companion. A special subclass of accreting X-ray binaries are the superluminal jet sources. These are supposed to be accreting, rapidly spinning stellar black holes that feature a double-sided jet structure at radio wavelengths emanating from the compact core – similar to that found in quasars but on a much larger scale. Supernovae and their remnants are probably the sources of the bulk of cosmic rays in the Galaxy, which produce intense, though spatially diffuse emission of γ-radiation in interstellar space through energetic particle interactions with interstellar matter and fields. Our knowledge about the diffuse interstellar emission from our own Galaxy is complemented by γ-ray measurements from our neighboring galaxies, the Large and Small Magellanic Clouds. Supernovae and their remnants are also the sites of nucleosynthetic line production: within the Galaxy, lines from radioactive ^{26}Al (1.809 MeV) and ^{44}Ti (1.157 MeV) have been detected, and outside the Galaxy, lines from radioactive ^{56}Co and ^{57}Co have been found.

In extragalactic space, active galactic nuclei (in the form of Seyfert galaxies, radio galaxies and γ-ray blazars [a subclass of quasars]) are the most common γ-ray-emitting objects. They are in absolute terms extremely compact and luminous objects and are supposed to contain massive black holes of 10^6 to 10^9 $M_\odot$ in their nuclei. These objects probably explain also part or all of the isotropic cosmic γ-ray background radiation.

Since 1997 we have known that at least some (probably all) classical γ-ray bursters are of extragalactic origin and possibly are the result of the merging process of two compact stars (like neutron stars, stellar black holes or white dwarfs) and/or the collapse of a massive star leading to the formation of a hyperaccreting black hole.

Finally, it should be mentioned that also the Sun – due to its proximity – can be an intense source of γ-radiation in the form of line and continuum emission during times of strong solar flares. Even the moon has been detected as a high-energy (100 MeV) γ-ray source. Here the γ-rays are produced by cosmic-ray interactions with the lunar surface matter (Thompson et al. 1997).

4.2 All-Sky Maps in the MeV–GeV Region

Three γ-ray all-sky maps are shown in Fig. 4.1 in Galactic coordinates: these are continuum maps from COMPTEL (1–30 MeV) and EGRET (>100 MeV), and the 1.809 MeV map from COMPTEL. The concentration of the emission along the Galactic Plane is the most striking aspect of all three maps. The plane stands out clearly against the rest of the sky, indicating that most of the measured γ-ray fluxes come from regions or objects inside the Galaxy.

The dominant Galactic continuum emission comes from interstellar space and is visible as diffuse Galactic radiation. Superimposed on the large-scale Galactic emission are point-like sources (a few of which are identified and indicated by their names in the COMPTEL and EGRET maps, such as Crab, Geminga, Vela, and Cygnus); however, many of the Galactic point sources remain unidentified.

Perhaps the most remarkable feature of the EGRET map is the many discrete sources at medium and high latitudes, a few of which are indicated in Fig. 4.1. Most of these have been identified as γ-ray blazars, and a few of them have been seen by COMPTEL as well (e.g., 3C 273, 3C 279). Some of the extragalactic COMPTEL and EGRET sources are not visible in the two continuum maps, because they flare up only occasionally; on average they are too weak to be visible in the all-sky maps.

When the COMPTEL and EGRET continuum maps are compared with each other, there are similarities (e.g., the overall structure), but also marked differences: there are sources which are detected only at low energies and not at high energies, and vice versa.

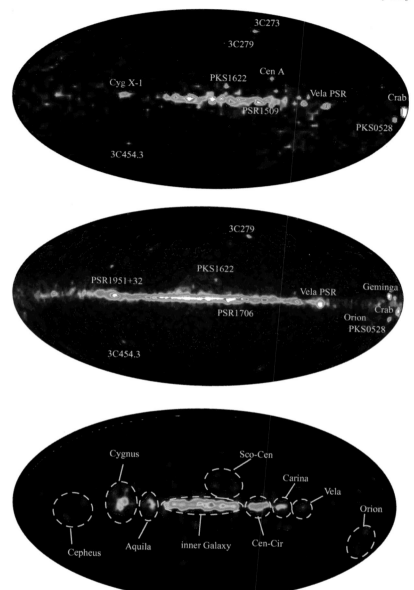

Fig. 4.1. All-sky maps in γ-ray astronomy. The top map is from COMPTEL in the 1–30 MeV range (Strong et al. 1999), the middle map is from EGRET (>100 MeV) (Reimer 1998), the bottom map is the COMPTEL 1.809 MeV map from radioactive ^{26}Al (Oberlack 1997). The coordinate system is the Galactic one with longitude and latitude as axes. The origin $(l, b) = (0, 0)$ is the center in the three maps.

The most remarkable feature of the 1.809 MeV ^{26}Al all-sky map is its irregularity along the Galactic Plane and its asymmetry around the Galactic Center: in the inner Galaxy there is more emission coming from the plane at negative longitudes than at positive longitudes. The reason may be a few nearby sources, especially in the Vela, Carina, and Cygnus regions. Also tangential projections to the spiral arms are regions of enhanced emission. These findings suggest that ^{26}Al is produced in regions of young, massive star associations.

The three γ-ray all-sky maps of Fig. 4.1 can be compared with three all-sky maps from other branches of astronomy (Fig. 4.2). The top map of Fig. 4.2 is the 408 MHz radio map of synchrotron radiation, which reflects the cosmic-ray electron component and magnetic field of the Galaxy and which, therefore, is of relevance for the electron-induced component of the diffuse Galactic γ-ray emission. In the middle of Fig. 4.2, the distribution of neutral and molecular hydrogen in the Galaxy derived from 21 cm line measurements of neutral hydrogen and 2.6 mm line measurements of CO molecules (a tracer of molecular hydrogen) is shown. The hydrogen distribution determines the matter-induced components (via π°-decay and electron bremsstrahlung) of the diffuse Galactic γ-ray emission. The bottom panel of Fig. 4.2 is the 2.2 µm infrared map, which reflects the stellar photon density in the Galaxy and which, therefore, is of relevance for the inverse Compton component of the diffuse Galactic γ-ray emission. These three maps provide the basis for an understanding of the interstellar γ-ray emission from the Galaxy.

Superimposed on the large-scale Galactic component are many discrete γ-ray sources. An all-sky map of the COMPTEL and EGRET sources is shown in Fig. 4.3. Table 4.1 lists all presently known COMPTEL and EGRET sources types. The main EGRET sources (Hartman et al. 1999) are spin-down pulsars and γ-ray blazars (nearly one third of all sources). Two thirds of the EGRET sources are still unidentified. The COMPTEL source categories (Schönfelder et al. 2000) are spin-down pulsars, accretion-driven X-ray binaries, γ-ray blazars, and special γ-ray line sources (supernova remnants and regions of massive star associations).

Two sources in the 1.157 MeV line from radioactive ^{44}Ti at the position of Cas A and the new supernova remnant RX J0852-4622 (discovered by the Röntgensatellit ROSAT) are the most remarkable features in the COMPTEL all-sky map at this energy (Iyudin et al. 1999). The 2.223 MeV COMPTEL all-sky map is practically featureless and shows only one single source candidate, which happens to coincide with an unusual nearby white dwarf (RE J0317-853) (McConnell et al. 1997).

4.3 The Sub-MeV Domain

In the sub-MeV domain no all-sky maps yet exist. But special regions of the sky have been studied with high sensitivity. The arc-minute resolution

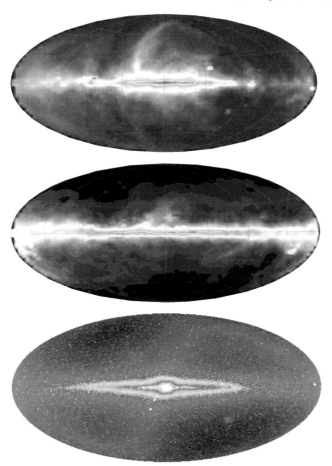

Fig. 4.2. All-sky maps from other branches of astronomy: the top map shows the radio-synchrotron emission at 408 MHz (based on data from Haslam et al. 1982), the middle map shows the distribution of hydrogen (neutral and molecular) throughout the Galaxy (see Strong et al. 1988), and the bottom map is the DIRBE-infrared map at 2.2 µm (see Arendt et al. 1994). Reproduced with permission from *Astronomy and Astrophysics* (top and middle map) and the American Astronomical Society (bottom map). I also acknowledge permission for reproduction from the authors

of SIGMA in the 35 keV to 1.3 MeV range has led to the identification of several sources. Table 4.2 summarizes the objects detected by SIGMA (from Vargas et al. 1997). Except for four active galactic nuclei and one unidentified high galactic latitude source, all sources listed are located along the Galactic Disk, most of them in the Galactic Bulge region (a sphere of about 10° radius around the Galactic Center). The sources listed in Table 4.2 have been grouped into "hard" and "soft" sources according to their hardness ratio

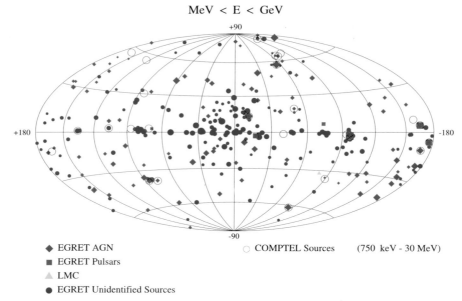

Fig. 4.3. All-sky map of gamma-ray sources detected by EGRET and COMPTEL

(HR), which is defined here as the flux ratio between the 75–150 keV and 40–75 keV energy bands.

Most of the "hard" sources are supposed to be accreting X-ray binaries with stellar black holes. Here we distinguish between persistent and transient emission (the latter sources are also called X-ray novae). The "soft" sources are mostly binaries with accreting neutron stars. Accreting pulsars have a high-mass stellar companion, X-ray bursters have a low-mass stellar companion. The fact that black-hole binaries have "hard" spectra, whereas neutron-star binaries have "soft" spectra is easily understood: we know that the accreting material forms a ring of plasma around the compact star (the accreting disk). The inner part of the disk releases intense thermal radiation in the X-ray band. Gamma-rays are expected to be produced by the Compton upscatter process, in which hot disk electrons near the compact object scatter and boost the disk X-ray photons to higher energies. In the case of a neutron-star system, low-energy X radiation from the surface of the neutron star exists in addition to the disk radiation. The effect of this surface radiation is to cool the electrons in the disk so that they can no longer upscatter the disk photons into the γ-ray range. In the case of a black-hole system this surface radiation does not exist, and therefore the cooling of the electrons does not take place; hence the disk radiation can be upscattered into the γ-ray range.

Table 4.1. Source Types detected by the COMPTEL and EGRET all-sky surveys

Source type	COMPTEL detections (1–30 MeV) no. comments		EGRET detections (>100 MeV) no. comments	
Spin-down pulsars	3	Crab, Vela, PSR 1509-58	6	Crab, Vela, Geminga, PSR 1786-44, PSR 1055-52, PSR 1951+32
Other Galactic sources $\|b\| < 10°$	7	Cyg X-1, Nova Persei 1992, GRO J1823-12, GRO J2228+61, GRO J0241+6119, Crab Nebula, Carina/Vela region	2	Cen X-3, Crab Nebula
Normal galaxies	–		1	LMC
Radio-loud quasars and BL Lac objects	9	Mostly variable	77	Mostly variable
Radio galaxies	1	Cen A	1	Cen A
Gamma-ray line sources	7	SN 1991T (^{56}Co), SNR RX J0852-4642 (^{44}Ti), Cas A (^{44}Ti), Vela (^{26}Al), Carina (^{26}Al), Cyg regions (^{26}Al), RE J0317-853 (2.223 MeV)		
Unidentified sources	5	(At high latitudes)	186	
Total number of sources	32		273	

Figure 4.4 illustrates two things: first SIGMA's power in image resolution, and second the variability of the γ-ray sky at these energies (from Gehrels and Paul 1998). Two maps in the 40 to 150 keV band taken at different times separated by about six months are presented. The persistent accreting black-hole source identified as 1E 1740.7-2942 is visible in both maps at

Table 4.2. Schematic representation of the compact sources detected by SIGMA. The hard and soft sources are distinguished by their hardness ratio (HR; see Sect. 4.3). (a) sources with HR > 0.8, (b) sources with HR < 0.8, (c) sources with variable HR, (d) sources showing radio jet emission, (e) high-energy transient emission, (f) fast transient sources, (g) uncertain classification

SIGMA Sources

Hard Sources (a)

Isolated pulsars
Crab Pulsar
PSR 1509–58

Accreting black holes with persistent emission
Cyg X–1
GX 339–4 (c)
GRS 1915+105 (d)
1E 1740.7–2942 (d, e)
GRS 1758–258 (d)

Sources detected in the field of the Galactic Bulge

Black hole X-ray Novae
Nova Ophiuchi
GRS 1730–312
GRS 1739–278
Nova Muscae (e)
Nova Persei
TrA X–1
GRS 1009–45

Soft Sources (b)

Accreting pulsars
4U 1700–37
OAO 1657–429
GRS 0834–429
Vel X–1
GX 1+4
GRO J1744–28

X-ray bursters
Terzan 2
GX 354–0
Terzan 1
A 1742–294
KS 1731–260 (f)

Unidentified sources
SLX 1735–269
GRS 1743–290
GRS 1734–292 (f)
GRS 1741.9–2953 (f)

Active Galactic Nuclei	
NGC 4388	NGC 4151
Cen A	GRS 1227+025 (f, g)
3C 273	

$l = 359, b = 0$. The upper map illustrates the discovery of the X-ray nova GRS 1730-312 (also named Nova Scorpii 1994b), and the lower one that of GRS 1739-278 (Nova Ophiuchi 1996).

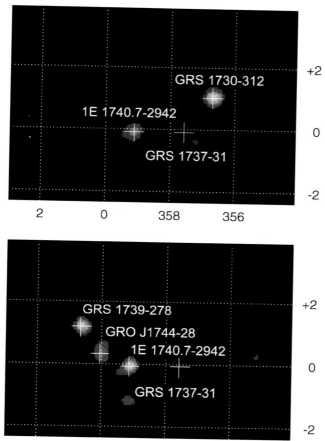

Fig. 4.4. SIGMA map in the 40–150 keV band of the Galactic Bulge field. The upper map contains data from September 22–26, 1994, the lower one from March 13–31, 1995. Colors represent the detection confidence level (*yellow*: high, *blue*: low). Coordinates are Galactic longitude (abscissa) and latitude (ordinate) in degrees. (Reproduced with permission from Gehrels and Paul 1998)

As can be judged from Table 4.2, the main achievement of SIGMA is the detection and localization of accreting stellar black holes and neutron stars. Those which exhibit transient temporal properties are particularly interesting. Nova Muscae 1991 was the first of this class of object; it showed a strong spectral feature around 500 keV for just one day (January 20–21, 1991). The most interesting of these sources is probably the famous black-hole candidate 1E 1740.7-2942 (see also Fig. 4.4). It showed a broad feature around 511 keV for a few hours, but more interestingly, correlated observations using the ground-based Very Large Array at radio wavelengths have shown a double-sided jet structure emanating from the central source. This was the first "microquasar" detected in the Galaxy. Since this first observation, other systems with similar properties have been detected.

Since the launch of CGRO in 1991, BATSE has continuously monitored the sky above 20 keV in order to search for source types dicovered by SIGMA. Several new transients have been discovered in this way (six X-ray novae between 1991 and 1998), and major outbursts of previously known objects were seen first by BATSE. This is reflected in more than 170 BATSE IAU telegrams. OSSE was very often able to improve the spectral analysis of transients discovered by BATSE. A famous example of this is the X-ray binary system A 0535+26, which showed a dramatic outburst in February 1991 and reached an intensity more than eight times that of the Crab. The subsequent OSSE observation resulted in a much-improved hard X-ray spectral determination, revealing a cyclotron resonance line at 110 keV, which implies a magnetic field strength of about 1.1×10^{13} G (Kurfess 1996).

In extragalactic space OSSE detected at least 33 active galactic nuclei in the sub-MeV range (50 keV to 1 MeV). Most of these sources (23 out of 33) are Seyfert galaxies. The others are γ-ray blazars (mostly counterparts of EGRET blazars). One of the important findings of OSSE in this field is that Seyfert galaxies are not strong γ-ray sources. They show spectral breaks at about 100 keV and become very soft at higher energies (Kurfess 1996).

Perhaps the most remarkable observation of OSSE in the sub-MeV domain is the mapping of 511 keV emission in the Galactic Center region, which was made through the accumulation of sufficient exposure by means of a uniform scanning strategy around the Galactic Center.

A full-sky map in the annihilation line does not yet exist. But the results obtained by OSSE from a number of pointings along the Galactic Plane are consistent with a two-component model of the 511 keV emission: a narrow disk component and a central bulge component (Purcell et al. 1997).

4.4 The TeV Gamma-Ray Sky

The TeV γ-ray field has become interesting during the past decade with the production of verifiable results using the atmospheric Cherenkov telescope technique. Four TeV sources have been detected with certainty (the remnants surrounding the pulsars Crab and PSR B1706-44 and the two BL Lac objects Markarian 421 and Markarian 501). Another ten TeV-radiation-emitting objects have been detected by individual groups, but need further confirmation (see Fig. 4.5, which is based on data from Weekes 2000).

The progress made in this field is impressive: angular resolutions of typically 0.1° are achieved, and detection significances of 20σ are obtained in less than an hour from extragalactic objects. The presently existing source detections have demonstrated that TeV γ-ray astronomy is not simply an extension of the MeV–GeV range, but a viable discipline of its own: a variety of different objects exists in this domain (see Weekes et al. 1997).

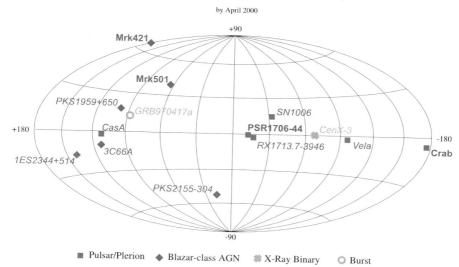

Fig. 4.5. Sky map of presently known TeV sources as observed by various Cherenkov telescope observatories around the world. Crab, PSR 1706-44, Mrk 421, and Mrk 501 are detected with certainty, the other source detections need further confirmation. (Reproduced with permission from Reimer 2000)

On the other hand, early attempts to discover objects radiating at PeV energies (from about 100 TeV to 10 PeV) have not been successful (1 PeV = 10^{15} eV). Early claims of detections could not be confirmed and no single firm source detection exists at these energies at present (Cronin et al. 1993).

4.5 Gamma-Ray Bursts

The γ-ray bursters are by far the brightest celestial γ-ray sources at the time of their outburst. They are detected by BATSE roughly once per day at an unpredictable time and position in the sky. The duration of bursts ranges from fractions of seconds to several hundreds of seconds; the energy flux usually peaks at photon energies of several hundred keV, but is still measurable at much higher energies.

Up to 1998 BATSE detected more than 2000 bursts, with a positional accuracy of typically 5°. The distribution of these sources is remarkably isotropic (see Fig. 4.6). From the intensity distribution of bursts one can conclude that the Universe is not filled uniformly with burst sources: beyond a certain distance the number density of sources decreases. Up to 1997 this distance scale was unknown, and therefore different classes of models for the

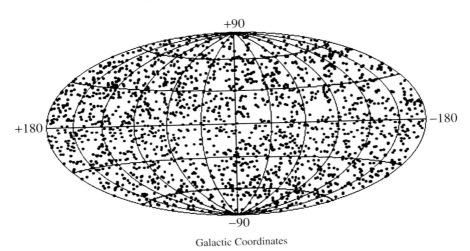

Fig. 4.6. BATSE map in Galactic coordinates of the first 2000 γ-ray bursts measured since the launch of CGRO. Each dot corresponds to a single γ-ray burst. The distribution is consistent with an isotropic burst population

origin of bursts were discussed, such as a solar system origin, a Galactic Halo origin, and a cosmological origin.

The breakthrough after more than 25 years of speculation came in 1997, when the Italian–Dutch X-ray satellite Beppo Sax was able to identify several burst sources within a few hours after the γ-ray flash. The first of these bursts was detected on February 28, 1997 (GRB 970228); it showed a transient X-ray counterpart which was fading fast (Costa et al. 1997). The location of this source could be fixed to within less than an arc minute, and subsequent optical observations with ground-based telescopes and the Hubble Space Telescope were able to identify a faint optical counterpart. For the bursts on May 8, 1997, and December 14, 1997, it was even possible to measure optical absorption lines of the counterparts which had red shifts of 0.835 and 3.42, respectively (Metzger et al. 1997; Kulkarni et al. 1998). Based on these observations, the question of an extragalactic origin of these bursts was settled, and the resulting luminosities are 7×10^{51} erg and 3×10^{53} erg, respectively, if the emission is assumed to be isotropic. The latter value is close to the entire rest-mass energy of the Sun (1.6×10^{54} erg). Another milestone was reached on January 23, 1999, when bright optical emission was detected for the first time from a burst (GRB 990123) while it was still in the progress in γ-rays (Akerlof et al. 1999). For this burst a luminosity of 3.4×10^{54} erg was derived (assuming isotropic emission). Such enormous energy releases might be produced during the coalescence of compact objects (like neutron stars or black holes) or during the collapse of a massive star leading to the formation of a hyperaccreting black hole. If the emission is beamed, the energy requirement would be relaxed by an amount which depends on the beaming factor. But even then, the burst sources would be the most luminous objects in the Universe at the time of their outbursts.

A separate class of γ-ray bursts are the so-called soft gamma-ray repeaters (SGR), which show several features that distinguish them from classical γ-ray bursts. First, they definitely repeat (unlike the classical bursts) with a recurrence time ranging from seconds to years, and with no correlation between the intensity and the time between the bursts. Second, the spectra are generally softer than those of classical bursts. In 1998 five SRG were known to exist. There are indications that all of them have correlations with Supernova remnants either in the Milky Way or in nearby galaxies. It appears that they are new manifestations of neutron stars with extremely strong magnetic fields (Smith 1997).

Finally, it is worth mentioning that the BATSE search for cosmic γ-ray bursts led to the unexpected discovery of intense γ-ray flashes of atmospheric origin (Fishman et al. (1994)), which are possibly produced as bremsstrahlung of relativistic (>1 MeV) runaway electron beams accelerated in an avalanche process by quasi electrostatic thunderstorm fields (Lehtinen et al. 1996).

References

Akerlof, C., Balsano, R., Barthelmy, S. et al., 1999, Nature, **398**, 400
Arendt, R.G., Berriman, G.B., Boggers, N. et al., 1994, Astrophys. J. Letters, **425**, L85
Costa, E., Frontera, F., Heise et al., 1997, Nature **387**, 783
Cronin, J.W., Gibbs, K.G., Weekes, T.C., 1993, Ann. Rev. Nucl. Part. Sci. **43**, 883
Fishman, G.J., Bhat, P.N., Mallozzi, R. et al., 1994, Science **264**, 1313
Gehrels, N. and Paul, J., 1998, Physics Today, Febr. 1998, p. 26
Hartman, R. et al. (1999), Astrophys. J. Suppl. **123**, 79
Haslam, C.G.T., Stoffel, H., Salter, C.J., Wilson, W.E., 1982, Astron. Astrophys. Suppl. **47**, 1
Iyudin, A. et al., 1999, Proc. of the 3rd INTEGRAL Workshop, Taormina, Italy, Sept. 14-18, 1998, eds. Palumbo, G., Bazzano, A., Winkler, C., Astrophys. Letters and Communications **38**, 383
Kulkarni, S.R., Djorgovski, S.G., Ramaprakash, A.N. et al., 1998, Nature **393**, 35
Kurfess, J., 1996, Astron. Astrophys. Suppl. **120**, 5-12
Lehtinen, N.G, Walt, M., Inan, U.S., Bell, T.F., Pasko, V.P., 1996, Geophys. Res. Letters, Vol. **23**, No 19, 2645
McConnell, M., Bennett, K., Bloemen, H. et al., 1997, Proc. of 25th Int. Cosmic Ray Conf., Durban, South Africa, eds. MS Potgieter, BC Raubenheimer and DJ van der Walt, Vol. **3**, 94-96
Metzger, M.R., Djorgovski, S.G., Kulkarni, S.R. et al., 1997, Nature **387**, 878
Oberlack, U., 1997, PhD-Thesis, Technische Universität München
Petry, D., 1997, PhD-Thesis, Max-Planck-Institut für Physik, München, MPI-Ph E/97-27

Purcell, W.R., Cheng, L.-X., Dixon, D.D. et al., 1997, Astrophys. J. **491**, 725
Reimer, O., 1998, personal communication
Reimer, O., 2000, personal communication
Schönfelder, V., Bennett, K., Blom, J.J. et al., 2000, Astron. Astrophys. Suppl. **143**, 145-179
Smith, I.A., 1997, Proc. of 4th Compton Symp. (eds. Dermer, C.D., Strickman, M.S., Kurfess, J.D.), AIP Conf. Proc. **410**, 110
Strong, A.W., Bloemen, J.B.G.M., Dame, T.M. et al., 1988, Astron. Astrophys. **207**, 1
Strong, A.W. et al., 1999, Proc. of the 3rd INTEGRAL Workshop, Taormina, Italy, Sept. 14-18, 1998, eds. Palumbo, G., Bazzano, A., Winkler, C., Astrophys. Letters and Communications **39**, 209
Thompson, D.J., Bertsch, D.L., Morris, D.J., Mukherjee, R., 1997, J. Geophys. Res. **102**, no A 7, 14 735
Vargas, M., Paul, J., Goldwurm, A. et al., 1997, Proc. of 2nd INTEGRAL Workshop, St. Malo, ESA-SP-382, 129 edit.: Winkler, C., Courvoisier, T.J.-L., Durouchoux, P.
Weekes, T.C., Aharonian, F., Fegan, D.J., Kifune, T., 1997, Proc. of 4th Compton Symposium, Williamsburg (eds. Dermer, C.D., Strickman, M.S., Kurfess, J.D.), AIP Conf. Proc. **410**, 361
Weekes, T.C., 2000, Proc. of the Workshop GeV-TeV Astrophysics: "Towards a Major Atmosphere Cherenkov Telescope VI", Snowboard, Utah, in press

5 The Sun as a Gamma-Ray Source

Erich Rieger and Gerhard Rank

5.1 Introduction

The first measurements of high-energy particles following solar flares in space and on the ground were performed during the 1940s, and by the 1960s the experiments were accurate enough to deduce spectral parameters. It was recognized that γ-ray continuum and, most importantly, γ-ray lines should be observable at high altitudes if a sufficiently large fraction of these particles interacted in the solar atmosphere. Early suggestions were made by Biermann et al. (1951) and Morrison (1958). Later, more detailed calculations (Chupp 1964; Dolan and Fazio 1965; Lingenfelter and Ramaty 1967) predicted line emission at 0.511 MeV from positron annihilation, at 2.2 MeV from neutron capture on protons, at 4.4 MeV and 6.1 MeV from deexcitation of carbon and oxygen nuclei, respectively, and a continuum at higher energies from pion decay. The discovery of high-energy bremsstrahlung photons was made in 1958 during the International Geophysical Year by a balloon experiment (Peterson and Winckler 1958), and the unambiguous discovery of nuclear lines from two large flares in August 1972 by the OSO-7 satellite (Chupp et al. 1973) followed. A more detailed description of historical issues can be found in the review by Chupp (1996).

Due to its proximity, the flaring sun can be the brightest γ-ray source in the sky. It even outshines the most intense cosmic γ-ray bursts by about a factor of ten, and it dwarfs the Crab pulsar, the brightest celestial γ-ray source at MeV energies by several orders of magnitude. This dominance even at very high photon energies (>100 MeV) is impressively demonstrated in the images in Fig. 5.1, obtained by the EGRET/CGRO instrument; when the sun began to have a high degree of activity in June 1991, it was declared a target-of-opportunity for the Compton Gamma-Ray Observatory (CGRO). For the duration of about one week, CGRO was pointed to the sun while it passed through the constellations Taurus and Gemini near the Galactic Anticenter. Three major flare events could be recorded during this observation. The full exposure of one week is displayed in the upper part of Fig. 5.1 and shows the path of the Sun and three well-known γ-ray sources: the Crab pulsar, Geminga and the blazar PKS 0528+134 . The sun was close to the Crab pulsar when flaring and can be identified as an elongated structure. The lower picture shows the same area in the sky but with an exposure of only two and

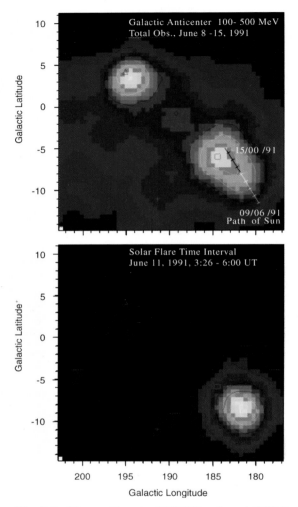

Fig. 5.1. Observations by EGRET onboard CGRO during June 1991. Three celestial γ-ray sources are indicated: the Crab pulsar (□), the Geminga pulsar (△) and the blazar PKS 0528+134 (+). The path of the sun is indicated (see Sect. 5.1 for details)

a half hours and was recorded about two CGRO orbits (about three hours) after the powerful flare on June 11, 1991. Although the flare is already in decline, the sun is still by far the brightest object, while the celestial sources are hardly identifiable. This makes it plausible that the first γ-ray lines from outer space that could be detected originated from the active sun.

After the first detection of γ-ray lines by the γ-ray spectrometer onboard the OSO-7 satellite, the sun was not routinely monitored, therefore only three more flares with γ-ray emission could be observed up to 1979 (Hudson et al. 1980; Chambon et al. 1981; Prince et al. 1982).

The number of observations increased considerably when the γ-ray spectrometers on the dedicated solar instruments Solar Maximum Mission (SMM, launched 1980) and Hinotori (launched 1981) were brought into orbit. After SMM, which ceased operations in late 1989, the sun was occasionally observed at γ-ray energies as a target of opportunity by the PHEBUS detector on board the French/Russian spacecraft GRANAT (launched 1989), by the Russian spacecraft GAMMA-1 (launched 1990, operational until 1992) and by CGRO (launched April 1991). In August 1991 the Japanese mission Yohkoh, which monitors the sun with its Wide-Band Spectrometer (WBS) was launched.

There are reasons to assume that also the quiet sun could be a source of γ-rays, though with much lower intensity. A number of different mechanisms to produce γ-rays on the quiet sun have been proposed: small flare events (called *microflares* or nanoflares) are thought a viable candidate for the heating of the corona (see also Sect. 5.3.1). Such events have been observed at hard X-ray and UV energies but it is not known what the accelerated particle spectrum looks like, how far it extends to higher energies or if non-thermal particle acceleration takes place at all. Another possibility is the decay of flare-produced long-lived radioactive isotopes, such as ^{54}Mn or ^{56}Co, which may emit detectable γradiation long after the flare has taken place. Also the interaction of high-energy particles with the solar atmosphere could occur in processes unrelated to flares: Particles might be accelerated by a shock travelling through interplanetary space and then move back to the sun along magnetic field lines (Vestrand and Forrest 1993). Also, cosmic rays from outside the solar system bombarding the solar surface might produce γradiation (Seckel et al. 1991). To date only upper limits could be derived for quiet sun emission. For the 2.2 MeV line flux a limit of $5.1 \times 10^{-5}\,\mathrm{cm}^{-2}\,\mathrm{s}^{-1}$ was derived from GRS/SMM data (Harris et al. 1992) and $4.1 \times 10^{-5}\,\mathrm{cm}^{-2}\,\mathrm{s}^{-1}$ from COMPTEL/CGRO data (McConnell et al. 1997). Also for radioactive decay products, various upper limits on line fluxes have been determined by COMPTEL/CGRO (McConnell et al. 1997).

Flares are not a phenomenon unique to the sun. It is known that early-type stars (spectral types O, B, A) have strong magnetic fields. However, these fields are thought to be simple dipole fields, since the outer layers of these stars are not convective. Late-type stars (F, G, K, M), on the other hand, have a convection zone underlying the photosphere which could drive a dynamo similar to that operational at the bottom of the convection zone in our sun and hence provide the complex magnetic configurations necessary to generate flares. The so-called Wilson Ca II index allows us to monitor line emission from hot regions typical for flare activity in a stellar atmosphere, and it is a general indicator of stellar flare activity. It has been used to search for stellar activity variations in a number of late-type stars. Periodic variations with periods in the range of some days up to a few months have been found. At X-ray energies flare-like outbursts of stars have been observed repeatedly, some exceeding the usual X-ray fluxes measured in solar events by some orders of magnitude.

In the following Section we describe, after a short outline of how a flare occurs, the production of γ-rays and neutrons. In Sect. 5.3 the solar flare γ-ray spectrum and the analytical methods are explained. In Sect. 5.4 the temporal history of flares is discussed. In Sect. 5.5 we concentrate on particle acceleration mechanisms, and Sect. 5.6 concludes this chapter with a summary and expectations for the future.

5.2 Production of Gamma-Rays and Neutrons

5.2.1 Flare Scenario

Although a clear and precise definition of the flare phenomenon is not easy to state we will define a flare as follows: A Solar flare is defined as the sum of all phenomena that occur when a sudden release of magnetic energy takes place in a comparatively small volume of the Solar atmosphere.

In this context, the term "sudden" refers roughly to time scales of seconds for the bulk energy release, while more elementary energy release processes within a flare may be as rapid as a few milliseconds, as indicated by the fine structure of hard X-ray bursts. The duration of the whole flare is usually on the order of minutes but can also extend to some hours (see Sect. 5.4.3). Thermal emission which originates from the flare can develop very slowly and persists for even longer times.

The physical processes in a flare take place in a "comparatively small volume", in contrast to coronal mass ejections (CMEs), which are much more extended phenomena that can span a whole hemisphere of the sun and are characterized by bulk mass motions. Although the flare energy release may be rather localized, the typical geometry of flare sites involves magnetic loop structures which are anchored in active regions and extend up into the corona. These loop structures set the stage for the acceleration and transport of charged particles.

The energy released in a single flare can be as large as 10^{33} erg. However, most flares are much less energetic, and since the number of flares increases with decreasing energy the small flares might have important effects (see the discussion on microflares in Sect. 5.3.1). Only the nonpotential portion of the total magnetic energy – representing the free energy contained in the field – is available for liberation in a flare. Since this portion is quite small compared to the total field energy, the large-scale magnetic field configuration in an active region usually changes little. However, large-scale restructuring of magnetic loops has also been observed (Šimberová et al. 1993).

The flare energy is converted into different channels, mostly into mass motion and flows, into thermal heating of plasma (leading e.g. to H_α and

Ca II brightenings), and into nonthermal particle acceleration. These accelerated particles can make the sun an occasional source of X-rays and γ-rays when they dump their energy into the solar atmosphere, where it is in part converted into radiation. The mechanisms involved will be described in Sect. 5.2.2. Some fraction of the accelerated particles can also leave the sun and can then be measured by particle detectors in space and on the ground.

The interplay between plasma heating, mass motion, acceleration and transport of particles can be quite complex and is not understood in detail. Also the question of how the particles are accelerated to high energies has not been answered unambiguously and will be discussed further in Sect. 5.5.

Although details of the energy release and the energy conversion have never been directly observed on the sun, magnetic reconnection is widely accepted to be the energy release mechanism. Reconnection is well known and experimentally verified on both the dayside and nightside of the earth's magnetosphere. Although the densities and magnetic field strengths are different from coronal conditions, the same mechanism is expected to work in the solar corona.

5.2.2 Conditions for Gamma-Ray Production in the Solar Flare Environment

Solar flare X and γradiation as well as neutrons originate from the interaction of accelerated charged particles (called the *parent particles*) with nuclei of the ambient medium of the solar atmosphere. According to the species of parent particles the emission can be classified as

- electron-induced emission, or
- ion-induced (or nucleonic) emission.

Accelerated electrons manifest themselves predominantly in a bremsstrahlung continuum at X/γ-ray energies. For accelerated ions there exists a large variety of reaction channels, and their relative importance depends on the particle species and the energies involved. The principal mechanisms for γ-ray production are nuclear deexcitation, neutron capture, positron annihilation and the decay of neutral and charged pions. In the following we describe the principal properties of reactions that produce important features in the γ-ray flare spectra. Most interactions of the ion component are dominated by reactions that involve protons (and to a minor degree α particles), since they are much more abundant than heavier ions.

Electron Bremsstrahlung

From the interaction of electrons with matter a continuous spectrum results, extending from the energy of the most energetic electrons down to almost zero. The physical mechanism for bremsstrahlung is well known (Koch and

Motz 1959; Haug 1975), and the interpretation mostly depends on the plasma environment at the interaction site where the kinetic energy of the electrons is converted into radiation.

Nuclear Deexcitation Lines

Inelastic scattering of protons, α particles and ions and to a lesser degree spallation reactions of certain elements lead to the generation of excited nuclei. Since the lifetimes of the excited states are on the order of 10^{-12} s, the deexcitation lines are emitted promptly. The cross sections for the most abundant elements in the solar atmosphere peak around 10 MeV for inelastic scattering, while spallation reactions dominate at higher energies.

If accelerated protons and α particles interact with heavy ions of the ambient medium, the emitted lines are narrow. The full width at half-maximum (FWHM) of the 4.4 MeV carbon line, for instance, is about 90 keV. If, however, accelerated heavy ions interact with ambient hydrogen and helium, broad lines (called the *inverse component*) result due to Doppler broadening. These inverse lines are about 10 times broader than the narrow lines.

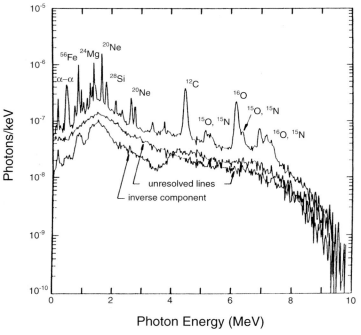

Fig. 5.2. Calculated γ-ray spectrum showing prompt nuclear deexcitation lines. The different components are explained in the text. (Reproduced with permission from Murphy et al. 1990)

Shown in Fig. 5.2 is a theoretical γ-ray spectrum (Murphy et al. 1990, based on calculations of Ramaty et al. 1979) of prompt deexcitation lines which was calculated under the assumption that the accelerated particles lose all their energy in nuclear reactions (thick target interactions). The lines are shown with a width corresponding to the resolution of a germanium solid-state detector. Besides the narrow and broad lines the figure indicates a third component, called unresolved lines. These originate from a superposition of many densely spaced weak lines that cannot be resolved by the detector. Also evident in the figure is a marked drop off above about 7.5 MeV. This is due to the fact that intense nuclear lines are absent above this energy (Crannell et al. 1979). The emission complex around 0.45 MeV (α–α line) is a blend of prompt lines at 0.429 MeV from ^{7}Be and at 0.478 MeV from ^{7}Li. These elements are formed in flares by nonthermal fusion reactions of α particles with ambient ^{4}He (Kozlovsky and Ramaty 1974; Share and Murphy 1997).

Neutrons and the 2.2 MeV Line

In addition to excitation of nuclei the interaction of accelerated ions with the solar atmosphere also leads to the production of neutrons (Wang and Ramaty 1974; Ramaty et al. 1975; Murphy et al. 1987). The most prolific neutron production reaction is that of protons on ^{4}He with a threshold of about 30 MeV. The produced neutrons can have different fates, as illustrated in Fig. 5.3.

- Free neutrons are not stable but decay spontaneously into a proton and an electron–antineutrino pair with a decay time of about 15 min.

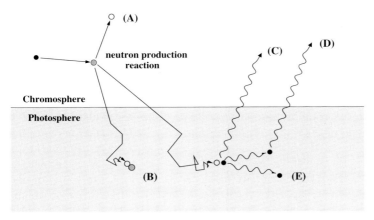

Fig. 5.3. The neutron capture process taking place in the solar photosphere. The neutrons generated in a nuclear reaction can have different possible fates (see Sect. 5.2.2 for explanation)

- Some fraction of the neutrons (about 60–70%) escapes from the sun and travels out into interplanetary space (neutron A in Fig. 5.3). Those neutrons that reach the earth before decaying can be detected with instruments in space and on the ground (Chupp 1987; Debrunner et al. 1983; Ryan et al. 1994). Also, the neutrons which decayed during travel can be identified by their energetic decay protons, which are trapped by the interplanetary field (Evenson et al. 1983, 1985).
- The remaining neutrons enter the photosphere. With increasing photospheric density, the scatter probability increases, and the neutrons are thermalized quickly. The thermal neutrons can be captured on ambient H in the reaction $^1\text{H}(n,\gamma)^2\text{H}$ and radiate the binding energy of the newly formed deuterium nucleus as a 2.223 MeV γ-ray photon (C). Another reaction channel (B), $^3\text{He}(n,p)^3\text{H}$, occurs with about the same probability; although the abundance ratio $^3\text{He}/\text{H}$ in the photosphere is only on the order of 10^{-5}, the neutron capture cross section of ^3He of about $\sigma_n(^3\text{He})/\sigma_n(\text{H}) = 10^5$ is uncommonly high. The capture on ^3He takes place without the emission of radiation, but it has the effect of decreasing the neutron capture time.

For the thermalized neutrons in the photosphere the probability of elastic scattering is some 500 times larger than that of a capture reaction, and it takes on the order of one minute for a neutron to finally be captured. This time delay is dependent on the density in the capture region (and therefore on its photospheric depth) and also on the abundance of ^3He. A study of the capture time delay can yield information about the capture depth and the ^3He abundance (Kanbach et al. 1981; Prince et al. 1983; Hua and Lingenfelter 1987; Trottet et al. 1993). It was found that the majority of the neutrons are captured at a photospheric depth of roughly 100 km, corresponding to a density of about $1.3 \times 10^{17}\,\text{H cm}^{-3}$ (Hua and Lingenfelter 1987).

Some important characteristics of the neutron capture should be noted: because the neutrons are captured at rest, the line is intrinsically very narrow (FWHM < 100 eV). The neutron capture region is deep enough in the photosphere that the generated γ-ray photons can be Compton-scattered on their way back up (D). This effect leads to an energy degradation of the photons, and in addition to the sharp 2.2 MeV line itself, a continuum to lower energies emerges (Vestrand 1990). However, this continuum has not yet been clearly identified in measurements. Also, significant attenuation of the 2.2 MeV line can occur if some fraction of the photons is absorbed within the photosphere (E). Both effects are most pronounced for flares at the solar limb, where the geometric path of the γ-ray photons through the photosphere becomes very long. For this reason the 2.2 MeV line can be drastically weakened in limb flares when compared to nuclear deexcitation lines which are formed at higher altitudes in the lower chromosphere.

The Positron Annihilation Line at 0.511 MeV

The positron annihilation line at 0.511 MeV is formed when positrons annihilate with ambient electrons. Since positrons have the highest probability for reactions at lowest velocities and hence have to be stopped before being annihilated, this line (like the 2.2 MeV line) is also emitted with a time delay. Positrons result from accelerated particle interactions which lead to β^+-emitting radioactive nuclei. The most prolific positron emitters are ^{11}C, ^{14}O and ^{15}O with half-lives of 20.5 min, 71 s and 2.06 min, respectively (Ramaty et al. 1975; Kozlovsky et al. 1987). Another production channel for positrons is the decay of positively charged pions created by the interaction of very energetic (several hundred MeV nucleon^{-1}) ions with ambient matter (Murphy et al. 1987). The production energies of the positrons from β^+ and pion decay are about 1 MeV and 10–100 MeV, respectively. The time for slowing down and annihilating is inversely proportional to the ambient density. For reactions occurring in the chromosphere, however, this time is short compared to the decay times of the radioactive nuclei. For this reason the observed delays (Share et al. 1983) reflect the half-lives of the β^+ emitters (Murphy and Ramaty 1984).

After slowing down the positrons can annihilate directly with ambient electrons to produce two 0.511 MeV photons per annihilation, or they form positronium, which is similar to a hydrogen atom but with the proton replaced by a positron (Crannell et al. 1976). According to the relative spin orientation, 25% of the positronium are in the singlet state, while 75% are in the triplet state. The singlet state decays into two 0.511 MeV photons. The triplet state decays into three photons which have less than 0.511 MeV photon^{-1} and the photon energy is distributed continuously, forming a continuum below the 0.511 MeV line (Murphy and Ramaty 1984; Ramaty 1986; Share and Murphy 1997).

Pion-Related Continuum

Very high-energy reactions lead to the production of charged as well as neutral pions (Murphy et al. 1987). Neutral pions decay after 10^{-15} s into two γ-ray photons of about 68 MeV each in the center-of-mass system. Due to the energy of the primary particle the pion decay spectrum is Doppler-broadened around 68 MeV.

Charged pions produce electrons and positrons which interact with the ambient matter and emit bremsstrahlung and a line at 0.511 MeV (see above). This bremsstrahlung is called *secondary*, in contrast to the *primary* bremsstrahlung which originates directly from accelerated electrons.

Since the energy threshold for the production of π^- mesons is substantially higher than for π^+ mesons, mostly positrons are generated and electrons are a minor constituent. Under realistic production conditions the spectrum resulting from pion decay is always a superposition of the broad emission feature centered at 68 MeV and the secondary bremsstrahlung continuum extending to lower energies.

5.3 The Solar Flare Gamma-Ray Spectrum

5.3.1 The Observed Gamma-Ray Spectrum

The observed solar flare γ-ray spectrum is a complex superposition of electronic and nucleonic emission components, namely electron bremsstrahlung continuum, narrow and broad line emissions and pion decay continuum.

Separation of Spectral Components

To obtain spectroscopic information about the γ-ray-producing particles, the different components of a flare spectrum have to be separated first. The separation is facilitated by the fact that below about 1 MeV electron bremsstrahlung is the major constituent, while above 1 MeV nuclear line radiation dominates the spectrum.

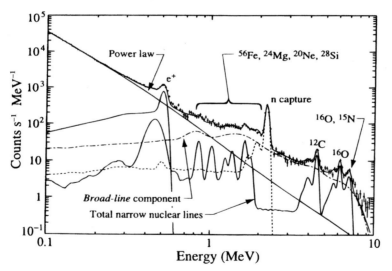

Fig. 5.4. The count rate spectrum during an observation interval lasting two minutes late in the June 4, 1991 flare as measured by OSSE. The different components which contribute to this flare spectrum give an impression of the complexity of the physical processes involved. While the power law originates from accelerated electrons, the broad and narrow nuclear line emission is caused by ion-induced reactions on the target nuclei which are indicated. Additionally the α–α complex, the 0.511 MeV annihilation line and the 2.2 MeV neutron capture line contribute to the spectrum. (Reproduced with permission from Murphy et al. 1997)

As an illustration of a measured spectrum, the time integrated and background subtracted count rate spectrum of the powerful flare on June 4, 1991 is shown in Fig. 5.4 for photon energies from 0.1 to 10 MeV, as observed by OSSE/CGRO (Murphy et al. 1997). Below 1 MeV the Li–Be (α–α) complex and the 0.511 MeV line are superimposed upon the bremsstrahlung spectrum. Towards higher energies nuclear lines and the neutron capture line dominate the spectrum. Although the flare occured relatively close to the solar limb at a heliocentric angle of 74.5°, the 2.223 MeV neutron capture line is the strongest feature in the spectrum.

A single power law fit is ascribed to the electron bremsstrahlung component. As Fig. 5.4 shows, the power law is determined by the data below the 0.511 MeV annihilation line, but is assumed to remain unchanged in the MeV range in order to separate the nucleonic from the electronic component. It has to be pointed out, however, that the application of an unbroken power law may be an oversimplification. For some flares, in particlular for impulsive flares, a broken power law is often a better representation of the electronic bremsstrahlung continuum.

The γ-ray spectrum at energies above about 8 MeV, where strong emission lines are absent, may be composed either of electron bremsstrahlung or pion decay continuum or a combination of both. For some flares the extrapolation of an unbroken power law from energies below 1 MeV agrees well with the measurements above 8 MeV. For other flares a flattening above 40 MeV is observed which can be explained either by a broken power law spectrum of the primary electrons or by additional emission from pion decay. In some cases this ambiguity between electronic and pionic emission introduces some uncertainty into the spectral analysis.

Flare Statistics

The first database that allowed statistical analysis of flares was provided by the Gamma-Ray Spectrometer (GRS) on SMM. A compilation of this database is published by Vestrand et al. (1999), hereafter called the "SMM flare atlas". Altogether, more than 250 flares were recorded that showed emission above 300 keV. Of these events 185 flares were intense enough for spectral analysis, and a subset of 67 flares shows γ-ray line emission. It is important to note that γ-ray continuum emission and even γ-ray line emission are not properties of giant flares. Although there is some correlation between γ-ray fluence and optical emission from flares, substantial γ-ray emission can result from optical subflares as well. In Fig. 5.5 the narrow nuclear line emission is plotted versus the >0.3 MeV continuum emission. Within relatively narrow limits these quantities are correlated. This implies that in flares high-energy electrons and ions are generally present in a more or less fixed proportion. An exception to this finding are the electron-dominated events, which are explained in Sect. 5.3.2.

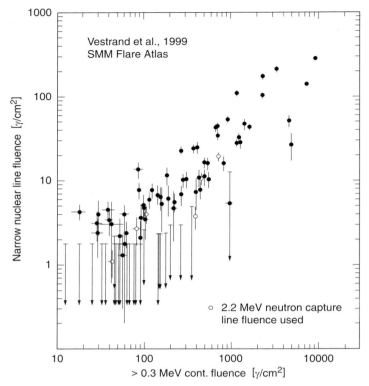

Fig. 5.5. Narrow nuclear line emission versus >0.3 MeV continuum emission for flares with continuum fluence $> 10 \, \gamma \, \text{cm}^{-2}$. *Arrows* denote flares which are either too weak to determine line emission fluxes or which show an intense electronic continuum at MeV energies so that line emission is obscured

Important global information about flare energetics can be gained from the so-called *flare size distribution*, which determines how many flares occur with a certain peak flux or integrated intensity. A number of measured flare size distributions exist using soft and hard X-ray data as well as radio observations (see the summary report by Crosby et al. 1993). Figure 5.6 (taken from Biesecker et al. 1994) is an example of a measured flare size distribution. It utilizes hard X-ray flare observations from BATSE and demonstrates how the number of flares increases with decreasing peak flux. Since the largest flares emit up to about 10^{33} erg total energy, flares that reach total energies in the ranges 10^{27} erg and 10^{24} erg are called *microflares* and *nanoflares*, accordingly. Such events have been observed in hard X-rays (Lin et al. 1984) and at ultraviolet wavelengths (Porter et al. 1987). Microflares are thought to be a viable explanation for the heating of the corona. It can be shown that if the power law of the flare size distribution continues down to about 10^{23} erg

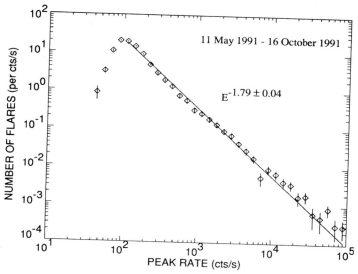

Fig. 5.6. Example of a measured flare size distribution. The number of flares with a given peak rate is plotted for all flares detected by BATSE during the indicated time interval. A power law fit to the distribution is included. At high peak rates the number of flares goes towards zero, while the turnover at low peak rates is caused by the sensitivity limit of the detector. [Reproduced with permission from Biesecker et al. 1994]

with a slope steeper than two there would be a sufficiently large number of small events to contribute a substantial fraction to the energy required for the coronal heating. On the other hand, for a distribution flatter than two the bulk flare energy would be liberated in only a few very large events. As Fig. 5.6 shows that from the measurements the slope has been mostly determined to be between about 1.4 and 1.8. However, nanoflares seem to be different in nature. They are not restricted to active regions but occur mainly on boundaries between supergranular cells, and are also called *network flares* for this reason. Interestingly, Krucker and Benz (1998) have found a power law index of about 2.3 for the size distribution of network flares. At present, the question of whether microflares can heat the corona cannot be answered conclusively, but it seems more likely that the newly observed population of network flares could fulfil the requirements rather than the "classical" (active region) flares.

5.3.2 Spectral Analysis

The Primary Electron Spectrum

The spectrum of the accelerated electrons can be deduced from the measured spectrum of the γ-ray continuum under some assumptions about the energy loss conditions. In general, the measured photon spectra follow very well a power law, indicating that also the electron spectrum itself follows a power law. In the case that the electrons lose all their energy in the interactions (thick target interactions), the spectral slope of the primary electron spectrum in the energy range 0.3–30 MeV is steeper by about 1.2 when compared to the slope of the γ-ray continuum (Bai 1977).

The slope of 185 flare spectra in the SMM flare atlas ranges from 1.7 to 4.5 with an average of 2.9±0.5. Therefore, the average power law index of the primary electron spectrum is about 4.1. As the spectral index of the bremsstrahlung component increases at lower photon energies, this implies that the bulk of the energy is contained at electron energies below about 100 keV. If this spectrum continued to very small energies, the energy content would become infinite. To avoid this "low-energy catastrophe" the electron spectrum is assumed to level off at energies of about 20 keV (Lin and Hudson 1976).

At even lower photon energies the spectrum is dominated by thermal emission, emitted by hot plasma at temperatures of several million degrees. In contrast to the nonthermal bremsstrahlung radiation, which is characterized by a power law spectrum, the thermal radiation exhibits an exponential spectrum. During flares a so-called *superhot component* can be measured in addition to the usual thermal emission (Lin et al. 1981). Since this exhibits an exponential spectrum it must be thermal in nature. It is thought to be caused by electrons of a few 10 keV which predominantly dump their energy at the loop footpoints and heat the plasma to high temperatures. Observations with Yohkoh show that its morphological development differs markedly from hard X-rays (see, e.g., Hudson and Ryan 1995 and references therein).

Although a discussion is beyond the scope of this article, it should be mentioned that electrons in the energy range of a few MeV not only radiate hard X-rays and γ-ray bremsstrahlung at the loop footpoints, but the same electrons also emit gyrosynchrotron radiation at microwave frequencies of a few hundred MHz up to a few ten GHz in the corona. These signals provide an additional source of information about the electron component, but since additional assumptions about thin or thick target interactions are necessary the interpretation is difficult (McLean and Labrum 1985).

Electron-Dominated Events

In general, the flare spectrum at MeV energies is dominated by nuclear line radiation. However, there are flares or episodes within flares where a continuum stretches over the whole γ-ray spectrum, extending up to several 10 MeV for some events.

This high-energy photon emission above 8 MeV, historically, pointed to the exceptional nature of these events (Rieger and Marschhäuser 1990). Since the continuum originates from electron bremsstrahlung these flares are called *electron-dominated* or *electron-rich* events.

In Fig. 5.7 the temporal history of a short episode of the large γ-ray line flare on March 10, 1989, is shown from X-rays to high-energy γ-rays. A burst of short duration appears simultaneously within the time resolution of the instrument at all energies up to photon energies of about 50 MeV. This is almost reminiscent of hard cosmic γ-ray bursts. That this event was indeed solar in origin could be verified by light travel time measurements of SMM in earth orbit and an experiment on one of the Phobos satellites which was at that time close to Mars. H_α images obtained at the National Solar Observatory at Sacramento Peak indicated that the event occurred in the same active region but well apart from the main flare (Wülser et al. 1990).

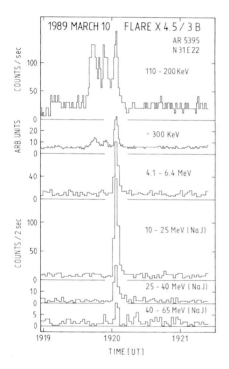
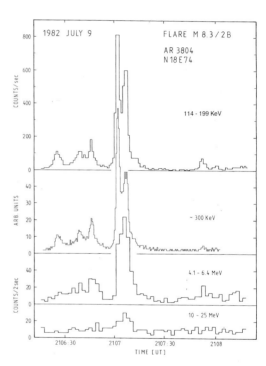

Fig. 5.7. An electron-dominated episode during the March 10, 1989, flare. The signal in the energy range 4.1–6.4 MeV of the third burst is continuum from electron bremsstrahlung

Fig. 5.8. Example of impulsive flare emission in different energy bands. The event on July 9, 1982, is not electron dominated (see Fig. 5.7); hence the 4.1–6.4 MeV band has its origin in nuclear interactions

Spectral analysis of this and similar events show that for comparable continuum fluences (the fluence is the total time-integrated flux of a flare) the nuclear contribution, if detectable, is smaller by an order of magnitude than for ordinary γ-ray line flares (Marschhäuser et al. 1994; Rieger et al. 1998).

These events have also been detected from other instruments in space, for instance, by PHEBUS on GRANAT (Pelaez et al. 1992; Vilmer and Trottet 1997), by GAMMA-1 (Akimov et al. 1994), by CGRO (Ryan et al. 1993; Dingus et al. 1994), by Yohkoh (Yoshimori et al. 1994), and possibly by Hinotori (Yoshimori et al. 1986).

Most of these events have short durations of a few tens of seconds. But also electron dominated flares with durations of more than 100 seconds can be found in the SMM flare atlas. Especially the short duration and impulsive events like the one presented in Fig. 5.7 are a major challenge for theoretical understanding.

The Primary Proton Spectrum

The relative contributions of different components of the nucleonic emission are dependent on the energy spectrum of the primary ions or protons. This allows us to evaluate the spectral hardness by comparing the flux ratios of different components with theoretical models. Murphy and Ramaty (1984) calculated the yields of the ^{12}C and ^{16}O emission lines, neutrons, pions and positrons as a function of the primary particle spectrum. It was assumed that the primary particles have an isotropic distribution and that they lose all their energy in the ambient medium (thick target interaction). The results are shown in Fig. 5.9, normalized to the interaction of 1 proton of energy >30 MeV. The calculations have been performed for spectra of different shapes and hardness. The right panel is valid for ion spectra represented by a power law function of the form E^{-s}. The left panel shows the results for ion spectra represented by Bessel functions. Here the characteristic parameter is αT (see Sect. 5.5), where a higher value of αT denotes a harder spectrum. A Bessel function is flatter at low energies and steeper at high energies when compared to a power law. Its slope increases with increasing energy. The dashed areas in Fig. 5.9 give the parameter ranges obtained by measurements. As can be seen from the figure, the yields for the given spectral features, in particular that for pion production, are different from each other and hence allow us to deduce the type and the hardness of a measured spectrum.

Further calculations by Ramaty et al. (1993) took into account different particle beam geometries and the influence of the composition of the ambient medium and accelerated particles. It was found that the ratio of the 2.223 MeV line fluence to the 4–7 MeV deexcitation line fluence does not depend much on the abundances of the target and beam. There is also no pronounced influence of the particle geometry, such as isotropic downward versus horizontally moving particles.

A number of flux or fluence ratios have been used to study accelerated proton spectra, each being sensitive to a different spectral range of the accelerated particles: the neutron yield as represented by the 2.2 MeV neutron capture line can be compared to the prompt nuclear deexcitation line yield

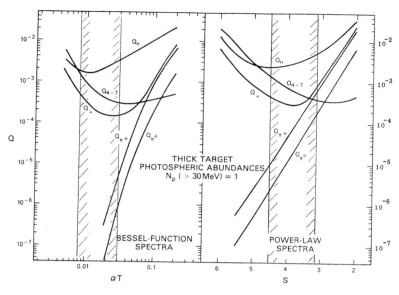

Fig. 5.9. Prompt line, neutron, positron, and pion yield Q per interacting ion calculated for ion spectra represented by a Bessel function and a power law. The range of the spectral parameters deduced from observations is indicated by the dashed area. (Reproduced with permission from Murphy and Ramaty 1984)

which dominates the 4–7 MeV range. This ratio gives information about the ion spectrum at energies of about 10–100 MeV nucleon^{-1}. Although the production processes for prompt lines have similar cross sections, the 1.6 MeV line of ^{20}Ne and the 6.1 MeV line of ^{16}O differ sufficiently so that their ratio is sensitive to the spectral hardness at about 5–20 MeV nucleon^{-1}. The flux ratio between prompt lines at 4–7 MeV and pion-related emission above 8 MeV provides information about higher proton energies of some 100 MeV nucleon^{-1}. Also the spectrum of directly measured neutrons can be used to determine the accelerated proton and ion spectrum.

The most commonly used measure of the hardness of the proton spectrum at energies of about 10 to 100 MeV is the flux ratio of the 2.223 MeV line and the 4–7 MeV band. However, some peculiarities due to the neutron capture process have to be kept in mind (see Sect. 5.2.2): The 2.223 MeV line photons are scattered and attenuated in the photosphere. Therefore, the calculations have to include the dependence of the 2.2 MeV to 4–7 MeV ratio on the heliocentric distance of the flare site. As a further complication the 2.223 MeV line is emitted with a time delay and cannot be compared directly to the promptly emitted deexcitation lines. For this reason the fluences rather than the fluxes themselves are often compared, leading to a single value of the spectral hardness for the whole flare. A more detailed analysis is possible by including the capture delay time directly as a parameter and modelling the measured 2.2 MeV flux. The prompt emission in other energy bands is used as a template for the time profile of the proton interactions (Prince et al. 1983).

At higher particle energies information can be gained by studying the photon flux from the decay of charged and neutral pions. Since the energy threshold for pion production is a few hundred MeV, flares which exhibit pion decay emission are rare and are found among the most energetic events. One event which showed strong pion decay emission was the June 15, 1991, flare. Kocharov et al. (1998) employed several methods to analyze a time interval late in the event: a comparison was made of the 4–7 MeV flux with both the 2.223 MeV line flux (both measured by COMPTEL/CGRO) and the pionic emission above 30 MeV (measured by GAMMA-1). Additionally, the neutron spectrum was recorded by COMPTEL and yielded another measurement of the proton and ion spectrum at energies of about 10–200 MeV nucleon^{-1}. It was found that a power law with an index of 3.3 is a good description for the proton spectrum from about 10 to a few hundred MeV. Above this energy the spectrum was found to become steeper.

Whereas the shape of the spectrum of primary particles at high energies can set constraints on the mechanism of how the particles are accelerated, the extension of the spectrum to lower energies (<10 MeV) is important, because it allows the overall number of accelerated particles and the total energy carried by them to be estimated. The Ne/O line intensity ratio was already mentioned and the cross section for excitation of the ^{20}Ne line at 1.63 MeV by protons extends further down to lower energies than any other cross section of the most abundant elements on the sun. This makes it an indicator for the study of the accelerated particle spectrum at energies lower than 10 MeV. From the analysis of SMM flares, Share and Murphy (1995) used the ratios between Ne and O to determine the spectral hardness. The derived values agree well with the results from the ratio between the 2.2 MeV line and the prompt 4.4 MeV carbon line. However, from the study of elemental abundances they found abnormally high abundance values for ^{20}Ne. To reconcile these values with those obtained from energetic particle measurements, a steep accelerated particle spectrum, namely a power law instead of a Bessel function, extending down to about 1 MeV nucleon^{-1} is assumed. In this case, the energy contained in flare-accelerated particles is larger by about an order of magnitude, and the energy contained in >1 MeV protons and ions is substantially larger than the energy content of the >20 keV electrons for a number of energetic flares.

Directionality of the Primary Particles

Although the present instrumentation does not permit us to resolve the sun spatially at γ-ray energies, it is possible to locate the flare sites on the solar disk with the aid of coinciding H_α images. This enables us to study the directionality of the γ-ray producing energetic particles. In this respect the sun is really a unique object in the sky. Bremsstrahlung photons are emitted anisotropically in the direction of motion of the electron, and the degree of anisotropy increases with energy (Elwert and Haug 1971). If the electrons move predominantly straight downwards (called a *pencil beam distribution*) or parallel to the solar surface due to the gyromotion at magnetic mirror points (called a *fan beam distribution*), a flare observed near the limb appears brighter than a flare observed close to disk center. Therefore, weak flares are detectable predominantly close to the limb. However, all efforts to show this limb brightening from X-ray flare observations did not lead to a positive result (Kane 1974; Datlove et al. 1977). This does not rule out a directionality of the primary electrons because the emitted photons may be scattered by electrons in the solar atmosphere (Compton backscattering), and thus any anisotropy of the emitted photon distribution may be smeared out (Bai and Ramaty 1978). However, if all SMM γ-ray flares of cycle 21 with emission above 0.3 MeV and known position on the disk are taken into account (Vestrand et al. 1987), a limb brightening begins to emerge.

The directionality of the electrons can best be investigated at higher energies (above the nuclear line emission, or about 8 MeV) where the Compton backscatter is negligible and the bremsstrahlung cross section is highly anisotropic. Figure 5.10 shows the H$_\alpha$ position on the sun for flares with photon emission above 8 MeV. During cycle 21 only SMM/GRS measurements are available (Rieger 1989) while during cycle 22 data from other spacecrafts

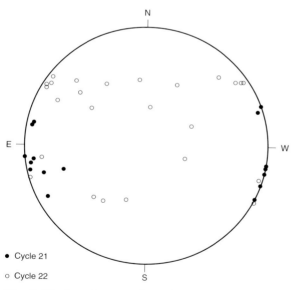

Fig. 5.10. Location on the solar disk of flares exhibiting emission above 8 MeV. The positions are determined from simultaneous H$_\alpha$ observations

were also used (Vilmer 1994). In cycle 21 the majority of the >8 MeV flares are concentrated towards the limb, thus calling for an anisotropic distribution of the radiating electrons, possibly a fan beam distribution (Kocharov et al. 1987; Ramaty et al. 1988), whereas in cycle 22 the >8 MeV flares are more or less evenly distributed over the solar disk and are in agreement with an isotropic angular distribution of the reacting electrons. Mandzhavidze and Ramaty (1993) show that the length of the flaring loop can influence the angular distribution of the >10 MeV γ-rays. Therefore, the difference between cycles 21 and 22 could result from changes in the magnetic field structure. In any case it will be of interest to see what distribution will prevail during the upcoming cycle 23. The angular distribution of the accelerated protons or ions cannot be obtained in the same way as for electrons because the prompt narrow line emission is only weakly dependent on angular effects. The 4.4 MeV carbon line itself, however, is of special interest because it is sensitive to the direction of the energetic particles; if γ radiation is observed at 90 degrees to a proton beam, the line has a minimum at its center and two symmetrical maxima, whose separation and height depend on the particle energy (Lang et al. 1987). Another feature sensitive to the beam geometry is the α–α line complex at about 0.45 MeV (see Sect. 5.2.2).

Line profiles for the 4.4 MeV carbon line and the α–α line have been calculated for different flare positions and particle distributions. The expected line shape variations are at the limit of the spectral resolution of current instruments. Up to now no conclusive results exist, but observations made by SMM/GRS (Share and Murphy 1997) and Yohkoh (Yoshimori et al. 1994) suggest that the possibility of downward beaming can be excluded.

Elemental Abundances

The sun is one of the primary sources of information about elemental abundances in the cosmos. The prompt nuclear lines are the fingerprints of the excited elements and allow us to evaluate the abundances for the most abundant elements in the solar atmosphere. The paper by Ramaty et al. (1979) laid the theoretical foundations for the computation of elemental abundances from the intensities of γ-ray lines. Since all narrow nuclear lines originate from the interaction of protons and α particles with the target medium (see Sect. 5.2.2), the intensity of the narrow lines allows us to study elemental abundances in the target, which in this case is the chromosphere.

Nuclear line fluences have been analyzed by Share and Murphy (1995) for 19 flares observed by SMM and by Murphy et al. (1997) for the June 4, 1991, flare observed by OSSE/CGRO (see Fig. 5.4). They found from flare-to-flare variations in line fluxes that the abundances of elements in the flare plasma can be grouped with respect to their first ionization potential (FIP). Line fluxes of elements with similar FIP correlate well from flare to flare. In contrast, the low-FIP to high-FIP line ratios vary from flare to flare.

It is well known from extreme ultraviolet (EUV) observations and from direct particle measurements in space that so-called *low-FIP* elements with an FIP $< 10\,\text{eV}$ (such as Mg, Si, or Fe) behave differently from *high-FIP* elements with an FIP $> 10\,\text{eV}$ (like C, N, O). In the solar corona, in the slow solar wind, in regions where magnetic flux emerges from below the photosphere and in solar energetic particle events, the low-FIP elements are about four times more abundant relative to high-FIP elements than in the photosphere. Since the dividing energy of about $10\,\text{eV}$ corresponds to the energy of the strong Lyman-α line, the photoionization of neutrals by Ly-α photons and the subsequent separation of neutrals and ions is thought to be responsible for the *FIP effect*.

For the abundances of elements in the flare plasma this means that low-FIP elements are enhanced relative to high-FIP elements when compared to photospheric abundances. Such abundances are more typical for coronal values. Flare-to-flare variations of the line ratios could be explained accordingly by variations of the respective coronal values. For the flare on June 4, 1991, it was possible to study changes in the elemental abundances during the flare, and it was found that the low-FIP to high-FIP ratio increased during the event.

Neon, another high-FIP element, is overabundant relative to C, N, O when compared to coronal abundance ratios deduced from particle measurements. Since the cross section for nuclear excitation of Ne reaches farther down to lower energies, this discrepancy could be solved by assuming a particle spectrum that is very rich in low-energy protons and ions (see Sect. 5.3.2).

5.4 Temporal History of Gamma-Ray Emission During Flares

In addition to the energy spectra of the primary particles the time history of a flare in different energy intervals is a major observational ingredient in the analysis of a flare and in setting constraints to acceleration models.

5.4.1 Evolution of Flares: Impulsive and Gradual Emission

Concerning the temporal evolution of solar flares, there is a large variety in their appearance at hard X-ray and γ-ray energies. The SMM flare atlas can offer a good impression of this. Some flares show emission contained in one single spike, but for the majority of flares the time structure is more complicated, including multiple spikes or a combination of spikes and slowly varying emission. Figures 5.8 and 5.11 show examples for different types of flare time profiles. In Fig. 5.8 a sharp and impulsive burst of short duration is shown. It can be seen that all energy ranges peak at the same time, within the instrumental time resolution of $2\,\text{s}$. Time profiles like these call for very prompt acceleration of particles, both ions and electrons, and the particles

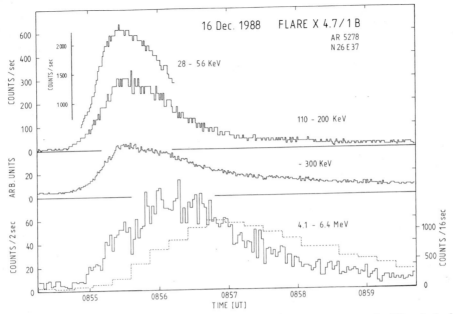

Fig. 5.11. Example of gradual flare emission in different energy bands. The *dashed* histogram in the lowest panel gives the profile for the 2.2 MeV neutron capture line

have to lose their energy in very dense layers of the solar atmosphere (Forrest and Chupp 1983).

Another type of emission is shown in Fig. 5.11. This gradual burst episode within a flare, which lasted for more than an hour, was observed by SMM. Although the burst commences more or less at the same time in all energy channels, the maximum is delayed as we go to higher energies. This suggests that the density of the medium where the energy is deposited is lower for this event, or that mirroring of particles plays a role.

Differences between flare types called *impulsive* and *gradual* have been known for some time, mainly from particle measurements in space (see e.g. Reames 1996). Gradual events exhibit long-lasting and structureless X-ray emission and are often accompanied by type II and IV radio bursts, which are indicative of bulk acceleration of electrons or shock fronts moving through the corona (see McLean and Labrum 1985 for an explanation of the solar radio burst nomenclature). The particle detectors find proton-rich emission, and the elemental abundances reflect coronal properties. Impulsive events are accompanied by strong but short duration X-ray emission and type I and III radio bursts, which are indicative of more beam-like particle populations. The electron-to-proton ratio is high, and high-Z elements (like Fe) are strongly enhanced. Of special interest is the ^{3}He/^{4}He ratio (measured in interplanetary space), which can be enhanced by a factor of up to 100 and cannot be explained by the FIP effect. The temperatures of the acceleration sites can be inferred from the charge states of the measured iron ions. Ionization states of +20 and +14, indicative of temperatures of about 10 and 2 million K, have been found for impulsive and gradual events, respectively. From these observations it appears that impulsive events are more localized events and that the particles are confined in small, hot flare loops. Gradual events are more related to large and open magnetic structures with lower density and temperature. Although there is consensus that the differences are caused by different conditions regarding where and how the flare occurs, questions on details of the acceleration mechanisms and the particle transport have not yet been answered conclusively.

5.4.2 Timing of Electron and Ion Interactions

Before γ-ray measurements were available, Wild et al. (1963) used radio data in their search for the origin of the flare energy release. From their observation that impulsive meter-wave type III bursts were followed after several minutes by meter-wave type II emission, indicative of a shockwave moving through the corona, they developed a concept of a two-phase acceleration. According to this idea, during the first phase electrons would be accelerated impulsively up to energies of about 100 keV, emitting the radiation seen in the type III bursts. The second, more gradual phase, which needs the first phase as a trigger, then accelerates the particles to relativistic energies. These particles would generate the type II emission and also would produce γ-ray lines and ground-level events.

This concept of a two-step particle acceleration was basically accepted until the early 1980s, when flare measurements with high time resolution in the hard X-ray and γ-ray regime became available through SMM and Hinotori. Among the many flares with γ radiation events were observed with impulsive γ-ray radiation occurring simultaneously in X- and γ-rays within the 1 to 2 s time resolution of the instruments. This is evident in Fig. 5.7 and 5.8.

If we take into account that emission versus time is a complicated superposition of acceleration, transport and energy-loss processes and that the last two have a tendency to smooth out fast temporal changes, the acceleration by itself may be even more impulsive. These events also suggest that the particles dump their energy in very dense material ($> 10^{12}\,\mathrm{cm}^{-3}$), which locates the position in the low chromosphere, and also that small magnetic loops are involved. This kind of acceleration has to happen in one single step, which puts constraints on the nature of the acceleration mechanism itself.

In addition to constraining the time scales of the acceleration process, timing analysis can also provide an idea of the geometry of the flare site. The time profiles of flares at hard X-ray energies show a wide variation. The interpretation is difficult because it is not clear how far the acceleration and the transport of electrons are related. Aschwanden et al. (1996) used hard X-ray data from 42 flares recorded by BATSE; they all showed a slowly varying envelope of X-ray emission with a number of rapid spikes superimposed. By studying the relative timing of different energies in the spikes, it was found that the low energies peak later when compared to the high energies. Most likely the time delay is caused by time-of-flight differences of the electrons, which allows us to calculate the distance between acceleration and precipitation sites. It is found that for most of these flares the acceleration region is not located within the loop itself, but above the top of the flare loop at a distance of about half the loop radius.

5.4.3 Extended Gamma-Ray Emission

With the instruments on CGRO a new class of flares was discovered which is characterized by prolonged γ-ray emission (see Hudson and Ryan 1995 for a review). In June 1991 a series of large solar flares occurred and could be observed by several instruments. The flares exhibited γ-ray and neutron emission which lasted for several hours after the impulsive phase. The long-term emission of these flares was dominated by nucleonic processes. Also strong pion decay emission occurred, indicating that particles have to be accelerated to high energies, a few hundred MeV nucleon^{-1}. Measurements were reported for the June 11 flare by EGRET/CGRO (Kanbach et al. 1993; Schneid et al. 1994), for the June 9, 11 and 15 flares by COMPTEL/CGRO (Ryan et al. 1994; Rank et al. 1996), for the June 4 flare by OSSE/CGRO (Murphy et al. 1997) and for the June 15 flare by GAMMA-1 (Akimov et al. 1994). It is thought that basically the same phenomenon occurred also in the June 3, 1982, flare, observed by SMM (Chupp et al. 1987).

Figure 5.12 shows the time profiles of the 2.223 MeV line emission for the flares on June 9, 11 and 15 (Rank et al. 1997). The emission could be detected for several hours, and the shape of the time profiles can be represented by a two-fold exponential decay with a time constant of about 10 min for the fast-decaying component and several hundred minutes for the slow component. It is interesting to note that despite the flares having quite different impulsive phase profiles the long-term time profiles of all three flares show a very similar overall structure. Since these events are associated with large coronal loops and the long-term time profiles are thought to be linked to the topology of these loops, the observations suggest that the large-scale magnetic structure of the active region did not change during the whole period of flare activity. Starting with an impulsive phase as a trigger, each flare could then evolve into the same overall long-term emission pattern.

To understand the nature of the long-lived γ-ray emission two types of models have been proposed, the trapping of highly energetic particles and the prolonged acceleration of particles.

Particles that are accelerated during the impulsive phase can subsequently be magnetically trapped in large flare loops due to reflection above the loop

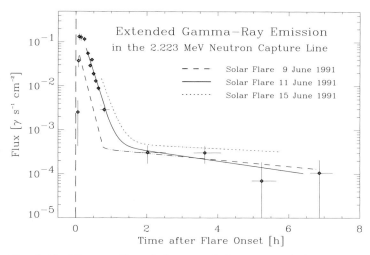

Fig. 5.12. Time profiles of the extended emission of the 2.223 MeV line for three flares in June 1991 as observed by COMPTEL/CGRO. The emission profile can be represented by a two-fold exponential decay. For the flare on 11 June both data points and fit function are included, for the flares on 9 and 15 June only the fit function is shown.

footpoints. The slow leakage of particles at the footpoints causes the prolonged γ-ray emission. The level of plasma turbulence must be very low so that only few particles are removed from the loop and long trapping times can be accomplished. This type of model is governed by transport rather than acceleration processes. The trapping model by Ramaty and Mandzhavidze (1994) was originally calculated for the EGRET >50 MeV measurements of the June 11 flare. The long precipitation time can be explained by a loop with a high mirror ratio, and a very low energy density. To avoid having particles removed by drift, the loops have to be either very large or twisted.

Particles in a turbulent flare loop will scatter frequently with plasma waves and can be diffusively confined to the loop. The energy contained in the turbulence, e.g. in Alfvén waves, can be transferred to the particles and lead to acceleration by stochastic acceleration processes (see Sect. 5.5.1). If the turbulence level is high enough, the particles can be accelerated efficiently enough to overcome energy losses. This prolonged particle acceleration can be continued as long as the energy content in the turbulence is maintained at a sufficiently high level. This model was calculated by Ryan and Lee (1991) originally to explain the June 3, 1982, event. If applied to the June 1991 flares very large loop sizes have to be assumed.

A scenario where many short episodes of particle acceleration are followed by subsequent short-term trapping has also been suggested (Ramaty and Mandzhavidze 1995). On the time scales accessible to observation this cannot be distinguished from a continuously operating process. This mechanism, however, requires high plasma turbulence to accelerate the particles, but on the other hand sufficiently low turbulence to enable trapping.

A possibility to discriminate between these models is to study γ-ray time profiles at different photon energies which are produced by particles of different energies. The energy-loss processes in trapping models are energy dependent, and therefore different precipitation times are expected for different spectral regimes. For instance, the production of nuclear lines and pion decay radiation originates from parent particles with $\sim$20 MeV nucleon^{-1} and >400 MeV nucleon^{-1}, respectively. For such particle energies the trapping model by Ramaty and Mandzhavidze (1994) predicts precipitation times of 5.5 min and 140 min, respectively. Under such conditions a 400 MeV proton would experience about 32 000 reflections within the flare loop before it is removed by a collision, a 20 MeV proton only 450. However, all measured datasets show similar decay times of the γ-ray emission in all observed energy bands. Also the decay times measured for electron- and proton-induced emission do not differ substantially. This result indicates that the observed extended γ-ray emission cannot be explained by long-term magnetic trapping but favors a continuously operating acceleration process.

5.5 Acceleration Mechanisms

5.5.1 Theoretical Considerations of Particle Acceleration

The storage and sudden explosive release of energy in sheared magnetic fields including the acceleration of particles is a common process in plasmas throughout the universe, from a region as close to the earth as its own magnetosphere to the most distant objects at cosmological distances such as quasars (see Chap. 12). The physical understanding of these processes is of basic importance to astrophysics and the solar–terrestrial relationship.

Any viable acceleration mechanism should be able to account for the observational facts: the measured time scales of electron and proton acceleration as well as the overall time structure, the observed energy spectra, the total number and ratio of accelerated particles. Further, it must be capable of imparting a large fraction of the available flare energy to the energetic particles. The fundamental processes which are most widely investigated are shock acceleration, stochastic acceleration, and acceleration by DC electric fields. Figure 5.13 sketches the basic mechanism for each of these processes in a qualitative manner. In the following we will explain the processes at a basic level. For a more detailed description see, for example, the review by Miller et al. (1997).

Shock Acceleration

Particles crossing a shock front can be accelerated by a first-order Fermi mechanism. The energy gain is proportional to v/c, where v is the velocity of

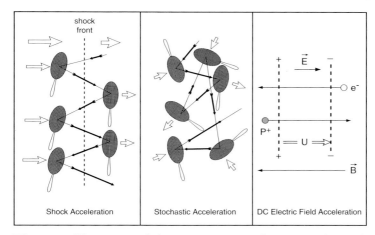

Fig. 5.13. Illustration of the most promising basic particle acceleration processes expected to be at work in solar flares

the shock front and c is the speed of light. To accelerate efficiently enough it is necessary that the particles are scattered many times at both sides of the shock at inhomogenities so that they stay close to the shock front and cross the shock many times (see the left panel in Fig. 5.13).

If losses only occur by particles diffusing away from the shock and if the shock front is much larger than the particle diffusion length (*planar shock*), the resulting particle spectrum can be shown to follow a power law in momentum space. The exponent is determined by the compression ratio r of the shock. The resulting particle spectrum is given by

$$\left(\frac{\mathrm{d}J}{\mathrm{d}E}\right)_0 \sim (E^2 + 2Em_0c^2)^{-\gamma}, \tag{5.1}$$

where m_0c^2 is the particle rest mass and $\gamma = 1/2(r+2)/(r-1)$.

Several effects can hinder the acceleration and thereby limit the spectrum at high energies, for instance, if the shock lifetime becomes comparable to the particle acceleration times or if the shock dimension becomes comparable to the particle diffusion length. If these effects are incorporated in the acceleration model, (5.1) is replaced by

$$\frac{\mathrm{d}J}{\mathrm{d}E} \sim \left(\frac{\mathrm{d}J}{\mathrm{d}E}\right)_0 \exp(-E/E_0). \tag{5.2}$$

Above the energy E_0 (called the *turnover kinetic energy*) the spectrum steepens considerably, while it follows a power law function below E_0 (Ellison 1984; Ellison and Ramaty 1985).

Stochastic Acceleration

In this scenario particles are immersed in a turbulent medium and change their energy randomly due to many interactions between the particles and plasma waves (see the middle panel in Fig. 5.13). Since the moving particles on average have more head-on than trailing collisions, their energy increases with time. However, since to the first order the particles lose the same fraction of energy as they gain, the net energy gain is of second order, i.e., proportional to v^2/c^2. This process is essentially second-order Fermi acceleration. In the solar corona, interactions with Alfvén waves and whistler waves are most important for protons and electrons, respectively. The resulting particle spectra can be described by two parameters, the acceleration efficiency, α, and average acceleration time, T. Assuming that both parameters are energy independent it has been shown that the spectra are represented by a modified Bessel function of second order in the nonrelativistic regime and by a power law in the ultra-relativistic regime. In both cases the product αT characterizes the shape of the spectra in the sense that a larger αT corresponds to a harder spectrum. For the nonrelativistic case ($E \ll mc^2$) and at energies $E \gg 3.26(\alpha T)^2$ MeV nucleon^{-1}, the equation

$$\frac{dJ}{dE} \sim E^{3/8} \exp\left[-\left(\frac{E}{3.26\,(\alpha T)^2}\right)^{1/4}\right], \tag{5.3}$$

where dJ/dE is the differential particle energy spectrum, is a good approximation for the modified Bessel function (Forman et al. 1986).

DC Electric Field Acceleration

The most direct way to accelerate charged particles is by large-scale quasistatic DC electric fields (see the right panel in Fig. 5.13). Such fields are associated with magnetic reconnection in the vicinity of magnetic neutral lines and current sheets and appear perpendicular to the magnetic field. Particle acceleration is also possible in electric fields which are parallel to the magnetic field. Parallel electric fields originate from the interruption (due to plasma instabilities) of the parallel currents associated with twisted magnetic flux tubes and from the formation of electric double layers.

One problem in explaining acceleration by electric fields is that oppositely charged particles are accelerated in opposite directions and a large-scale charge separation occurs. While particles gain energy from the electric field, they also experience an oppositely directed Coulomb drag force from the particles of opposite sign. It is the interplay between these two forces that governs whether or not electrons or ions can be accelerated out of a bulk particle distribution.

5.5.2 Acceleration Models for Solar Flares

We have shown several candidates for particle acceleration, but do they work under the conditions of a realistic solar environment?

The flaring sun is the brightest object in the γ-ray sky and as explained in Sect. 5.3.1 the "fingerprints" in the form of high-energy particles and γ-radiation can be used to investigate the acceleration processes of these particles. For this reason the acceleration models can be confronted with a wealth of observational results. These are, apart from particle measurements in space which have not been discussed here, the photon spectrum of the flare and the temporal history in different energy bands.

The measured photon spectra can be used to obtain information about the parent particle spectra. This was already described in Sect. 5.3.2 for electrons and in Sect. 5.3.2 for protons and ions.

The SMM/GRS database was used to deduce information on proton spectra from γ-ray line ratios. It is remarkable that in all flares observed by GRS, except a single burst during one flare, the primary particle spectrum softens at high energies. This behavior can be well described by a Bessel function. Also the neutron time profiles recorded by GRS after two flares are in agreement with a Bessel function spectrum of the primary protons (Murphy and Ramaty 1984). A spectrum of this type favors stochastic acceleration as the underlying mechanism.

These conclusions were drawn from the analysis of high-energy photons and neutron observations and thus pertain to primary particles with energies above about 100 MeV nucleon^{-1}. However, the situation becomes more complicated at lower energies.

Spectral investigations of major γ-ray line flares carried out recently suggest that the particle spectra at energies below about 10 MeV nucleon^{-1} are more likely power laws than Bessel functions. These findings would then argue for shock acceleration instead. The spectral cut-off at energies above 100–200 MeV nucleon^{-1} which is necessary to explain the absence of a high-energy pion decay signal in the measurements during the main bursts can be achieved with a suitable turnover kinetic energy (see Sect. 5.5.1).

An argument for the occurrence of stochastic acceleration processes comes from the fact that some elements such as Fe and ^{3}He are overabundant in flares, as deduced from spectral analysis and particle measurements. The only mechanism that can explain these abundances is gyroresonant stochastic acceleration. The term "gyroresonant" means that the gyrofrequency of certain ions is a multiple of the prevailing wave frequency. Hence these elements are accelerated more efficiently than others.

From measurements of the flare continuum above some 100 keV we know that the photon spectrum has a power law slope of typically $E^{-2.9}$, which, under the assumption that no pions are generated in the flare, translates into a spectral slope of about -4 for the primary electrons. Calculations carried out by Miller et al. (1996) assuming a stochastic type of acceleration mechanism yield electron spectra with about the correct slope at a few MeV, but the spectra are curved as expected from a stochastic acceleration process. To reconcile theory and measurements, the energy dependence of the escape of the electrons from the acceleration region also needs to be taken into account. Shock acceleration up to MeV energies is considered a less likely process for electrons, because they need to be pre-accelerated by some other process in order to obtain the high energies that are observed.

This discussion suggests that electrons and protons, if they are accelerated in the same volume, may derive their energy from different processes.

From the analysis of the temporal history of impulsive events such as the one shown in Fig. 5.8, we can deduce that protons are accelerated to energies of a few ten MeV within a few tenths of a second. This should be within the reach of all acceleration mechanisms discussed in Sect. 5.5.1. Therefore, a two-stage scenario to generate highly energetic particles is not needed.

On the other hand there are some very intense flares which show a gradual peak after the main phase at energies above 10 MeV, and the spectrum of this peak is harder than that of the main burst. This second peak can be interpreted as a second-phase acceleration by a shockwave with a high compression ratio moving through the corona.

Also the flares which show extended emission (see Sect. 5.4.3) suggest that two different mechanisms are at work. The impulsive phases of the June 1991 flares are rather different from each other and exhibit a quite soft accelerated particle spectrum without a pronounced pion decay signal. The emission during the extended phase appears to be smoother and more gradual with a harder spectrum and a strong pion decay emission. The exact mechanisms behind this type of emission, however, are not completely understood.

As far as electrons are concerned, stochastic mechanisms should be able to accelerate them to MeV energies within a second. A problem may arise for electron-dominated events where >10 MeV electrons have to be produced within sub-second time scales, which may strain stochastic mechanisms. DC electric field acceleration may be an alternative for these cases. Then, electrons and protons are accelerated to high energies with about the same efficiency. This would also quite naturally explain the dominance of the electron continuum compared to the nucleonic spectral features. Potential differences on the order of 50 MV are necessary to accelerate the particles within the time scales suggested by observation. Barium ion jet experiments carried out in the earth's auroral zone have confirmed that field-parallel electric potential differences of a few kV can build up (first measured by Haerendel et al. 1976). The electrons and protons accelerated in these electric fields interact with the earth's atmosphere and are responsible for the aurora. In this sense, electron-dominated episodes during flares might be a phenomenon similar to auroral processes but on a larger scale.

The conventional theoretical approach to modelling solar flares is to study the basic mechanisms involved – energy build-up and storage, energy release, different types of acceleration processes (as explained in Sect. 5.5.1), and particle transport processes – and then to put these mechanisms together to obtain a more global picture. To enable a mathematical description of the processes, simple loop topologies are studied. In reality, however, flares preferentially happen in complex magnetic configurations, and it has been found that no single acceleration process is able to explain the wealth of observational facts.

A different approach is tried in statistical flare models (see Vlahos et al. 1995 for an overview). For the cells of a three-dimensional grid, relevant parameters for magnetic and plasma conditions are randomly chosen and then allowed to dynamically evolve under coronal conditions. A set of "rules" governs the interaction processes and leads to random formation of discontinuities, the release of energy and the acceleration of particles. In avalanche models (introduced by Lu and Hamilton 1991) so-called *statistical flares* can be generated, where small random energy release triggers further release processes on different energy scales.

The statistical approach is a promising tool to gain understanding of the overall characteristics of flare time profiles and the flare size distribution. However, since it is a global approach, it does not tell explicitly what physical processes are relevant in an individual event.

5.6 Conclusions

5.6.1 Our Current Understanding of the Flare Phenomenon

Measurements of solar flares which are carried out at γ-ray energies are ideally suited to answer questions about the processes of particle acceleration and transport in the solar atmosphere. From the SMM/GRS detector, whose lifetime of nearly 10 years covered the latter half of cycle 21 and the onset of cycle 22, we now know that γ-ray emission during solar flares is not a unique property characterizing the largest events, but can be detected rather frequently.

From the temporal history of impulsive flares in different energy bands we deduce that protons are promptly accelerated to a few 10 MeV or more within about a second. The concept that a second phase which further accelerates particles from a previous phase is not necessary. The same is true for MeV electrons.

Spectral analysis of the flares shows that high-energy protons and electrons are produced in about the same proportion in all flares. Exceptions to this result are the electron-dominated events, where nuclear radiation seems absent or is buried within the intense electronic continuum. Some of these events are very short lived, with durations of a few seconds, and the radiation is emitted simultaneously from X-rays to high-energy γ-rays. The primary proton spectrum as deduced from γ-ray line ratios in most flares investigated so far cuts off at energies above 100–200 MeV nucleon^{-1}. At energies below

~10 MeV a power law particle spectrum is suggested by the measurements. This points to shock acceleration rather than stochastic acceleration. To accelerate electrons to high energies, stochastic acceleration mechanisms are invoked. But to explain the power law distribution of the electrons deduced from the continuum spectrum, modifications to the acceleration mechanism are necessary.

Evidence for a directionality of the accelerated particles was obtained for the first time by GRS/SMM measurements. From the analysis of the line shapes and the temporal history of the flux of the escaping neutrons it is concluded that the interacting high-energy ions were either distributed in a fan beam parallel to the solar surface or distributed isotropically. A hint of directionality for high-energy electrons was given by the near-limb distribution of the flares which occurred in cycle 21. During cycle 22, however, the events were distributed uniformly over the disk. This difference in the distribution of flares could reflect changes in the magnetic field structure from cycle to cycle.

5.6.2 Expectations for the Future

Theoretical investigations of the existing flare database will undoubtedly improve our understanding of the flare process, but a big step forward will only be possible by obtaining new data with advanced detectors.

The main issues are to improve the temporal and spectral resolution from X-rays to high-energy γ-rays, and ultimately also to enable spatial resolution up into the MeV range.

A better temporal resolution of milliseconds over a wide energy range will further help constrain the acceleration mechanisms. It will be of special interest to find out if even at sub-second time scales the electronic emission in electron-dominated events is still simultaneous from X-rays to high-energy γ-rays. It will also be of interest to search for temporal fine structure in flares. This would suggest that the basic driver for a flare is a superposition of many elementary processes rather than one large-scale mechanism.

The most progress in the understanding of solar flares will probably be achieved by using high-purity Ge detectors. This, however, requires some technical effort since the Ge crystals have to be cooled to low temperatures, around 80 K. Improved spectral resolution enables more accurate elemental abundance determinations and expands the studies to less intense emission lines. Especially the investigation of important abundance ratios like ^{3}He/^{4}He, Ne/O or Fe/(C,N,O) will profit from such measurements. Directionality effects can be investigated by resolving line shapes, particularly the α–α line complex and the 4.438 MeV carbon line. In addition, breaks in the electron spectrum or deviations from a strict power law can be investigated in more detail.

Spatial resolution in the γ-ray domain will be an important achievement. Up to now γ-ray emission has only been associated with a flare site by other spectral ranges, usually to optical H_α localizations. It is of great interest to image the flare sites directly in γ-rays and see how localized or spreadout the production region of the emission is. In particular from observations of 2.2 MeV emission from a flare that occurred behind the solar limb (Vestrand and Forrest 1994), it is expected that the γ-ray brightenings will be rather delocalized for some flares. The expected improvement can be judged by comparing the insight which came from the Yohkoh soft and hard X-ray images with former experiments yielding only spectra and light curves of the integrated sun.

Scheduled for launch in 2001 is the High-Energy Solar Spectroscopic Imager (HESSI; Lin et al. 1998). It combines high spectral resolution with spatial resolution and is based on a rotating collimator technique. It provides angular resolution varying from 2 arc seconds below 35 keV to about 1 arc minute at MeV energies, depending on the total high-energy emission of the flare. The spectral resolution of about 2 keV at 1 MeV is a clear improvement compared to typical resolutions of scintillators of about 70 keV at 1 MeV, as used by GRS/SMM and the CGRO instruments. It is expected to detect a large number of microflares but also to study large events during the next solar cycle in great detail.

References

Akimov, V.V., Leikov, N.G., Belov, A.V. et al., 1994, High-Energy Solar Phenomena, ed. Ryan, J.M., Vestrand, W.T., AIP Conf. Proc., **294**, 106
Aschwanden, M.J., Kosugi, T., Hudson, H.S. et al., 1996, Astrophys. J., **470**, 1198
Bai, T., Ph.D. Thesis, Maryland Univ., GSFC, Greenbelt, MD, 1977
Bai, T., Ramaty, R., 1978, Astrophys. J., **219**, 705
Biermann, L., Haxel, O., Schlüter, A., 1951, Z. Naturforsch., **6A**, 47
Biesecker, D.A., Ryan J.M., Fishman, G.J., 1994, High-Energy Solar Phenomena, ed. Ryan, J.M., Vestrand, W.T., AIP Conf. Proc., **294**, 183
Chambon, G., Hurley, K., Niel, M. et al., 1981, Sol. Phys., **69**, 147
Chupp, E.L., 1964, in: The Physics of Solar Flares, ed.: Hess, W.N., NASA SP **50**, 445
Chupp, E.L., 1987, Phys. Scripta **T18**, 5
Chupp, E.L., 1996, High Energy Solar Physics, ed.: Ramaty, R., Mandzhavidze, N., Hua, X.-M., AIP Conf. Proc. **374**, 3
Chupp, E.L., Forrest, D.J., Higbie, P.R. et al., 1973, Nature, **241**, 333
Chupp, E.L., Debrunner, H., Flückiger, E. et al., 1987, Astrophys. J., **318**, 913
Crannell, C.J., Joyce, G., Ramaty, R., Werntz, C., 1976, Astrophys. J., **210**, 582
Crannell, C.J., Ramaty, R., Crannell, H., 1979, Astrophys. J., **229**, 762
Crosby, N.B., Aschwanden, M.J., Dennis B.R., 1993, Sol. Phys., **143**, 275
Datlove, D.W., O'Dell, S.L., Peterson, L.E., Elcan, M.J., 1977, Astrophys. J. **212**, 561
Debrunner, H., Flückiger, E. ,Chupp, E.L., Forrest, D.J., 1983, Proc. 18th Internat. Cosmic Ray Conf. Papers, **4**, 75
Dingus, B.L., Sreekumar, P., Bertsch, D.L. et al., 1994, High-Energy Solar Phenomena, ed. Ryan, J.M., Vestrand, W.T., AIP Conf. Proc., **294**, 177

Dolan, J.F., Fazio, G.G., 1965, Rev. Geophys., **3**, 319
Ellison, D.C., 1984, J. Geophys. Res., **90**, 29
Ellison, D.C., Ramaty, R., 1985, Astrophys. J., **298**, 400
Elwert, G., Haug, E., 1971, Sol. Phys., **20**, 413
Evenson, P., Meyer, P., Pyle, K.R., 1983, Astrophys. J., **274**, 875
Evenson, P., Meyer, P., Yanagita, S., Forrest, D.J., 1985, Astrophys. J., **283**, 439
Forman, M.A., Ramaty, R., Zweibel, E.G., 1986, in: The Physics of the Sun, ed.: Sturrock, P.A., Holzer, T.E., Mihalas, D., Ulrich, R.K., Dordrecht, D. Reidel Publishing Co., Vol. II, p. 249
Forrest, D.J., Chupp, E.L., 1983, Nature, **305**, 291
Haerendel, G., Rieger, E., Valenzuela, A. et al., 1976, ESA **SP-115**, 203
Harris, M.J., Share, G.H., Beall, J.H., Murphy, R.J., 1992, Sol. Phys. **142**, 171
Haug, E., 1975, Sol. Phys., **45**, 453
Hua, X.-M., Lingenfelter, R.E., 1987, Astrophys. J., **323**, 779
Hudson, H.S., Bai, T., Gruber, D.E. et al., 1980, Astrophys. J. Lett., **236**, L91
Hudson, H.S., Ryan, J.M., 1995, Annu. Rev. Astron. Astrophys., **33**, 239
Kanbach, G., Pinkau, K., Reppin, C. et al., 1981, Proc. 17th Internat. Cosmic Ray Conf. Papers, **10**, 9
Kanbach, G., Bertsch, D.L., Fichtel, C.E. et al., 1993, Astron. Astrophys. Suppl. **97**, 349
Kane, S.R., 1974, in: Coronal Disturbances, ed.: Newkirk, G. Jr., IAU Symp. **57**, 105
Koch, H.W., Motz, J.W., 1959, Rev. Mod. Phys., **31**, 920
Kocharov, G.E., Kovaltsov, G.A., Mandzhavidzhe, N.Z., Semukhin, P.E., 1987, Proc. 20th Internat. Cosmic Ray Conf. Papers, **3**, 74
Kocharov, L., Debrunner, H., Kovaltsov, G. et al., 1998, Astron. Astrophys., **340**, 257
Kozlovsky, B., Ramaty, R., 1974, Astron. Astrophys., **34**, 477
Kozlovsky, B., Lingenfelter, R.E., Ramaty, R., 1987, Astrophys. J., **316**, 801
Krucker, A., Benz, A.O., 1998, Astrophys. J. Lett., **501**, L213
Lang, F.L., Werntz, C.W., Crannell, C.J. et al., 1987, Phys. Rev. C, **35**, 1214
Lin, R.P., Hudson, H.S., 1976, Sol. Phys., **50**, 153
Lin, R.P., Schwartz, R.A., Pelling, R.M., Hurley, K.C., 1981, Astrophys. J. Lett., **251**, L109
Lin, R.P., Schwartz, R.A., Kane, S.R. et al., 1984, Astrophys. J., **283**, 421
Lin, R.P., Hurford, G.J., Madden, N.W. et al., 1998, Proc. SPIE, **3442**, 2
Lingenfelter, R.E., Ramaty, R., 1967, in: High-Energy Reactions in Astrophysics, ed.: B.S.P. Shen, Benjamin, New York, p. 99
Lu, E.T., Hamilton R.J., 1991, Astrophys. J. Lett., **380**, L89
Mandzhavidze, N., Ramaty, R., 1993, Nucl. Phys. B Suppl., **33 A, B**, 141
Marschhäuser, H., Rieger, E., Kanbach, G., 1994, High-Energy Solar Phenomena, ed. Ryan, J.M., Vestrand, W.T., AIP Conf. Proc., **294**, 171
McConnell, M.L., Bennett, K., MacKinnon, A. et al., 1997, Proc. 25th Internat. Cosmic Ray Conf. Papers, **1**, 13
McLean, D.J., Labrum, N.R., 1985, Solar Radiophysics, Cambridge Univ. Press, Cambridge
Miller, J.A., La Rosa, T.N., Moore, R.L., 1996, Astrophys. J., **461**, 445
Miller, J.A., Cargill, P.J., Emslie, A.G. et al., 1997, J. Geophys. Rev., **102**, A7, 14631

Morrison, P., 1958, Il Nuovo Cimento, **7**, 858
Murphy, R.J., Ramaty, R., 1984, Adv. Space Res., **4**, No. 7, 127
Murphy, R.J., Dermer, C.D., Ramaty, R., 1987, Astrophys. J. Suppl., **63**, 721
Murphy, R.J., Share, G.H., Letaw, J.R., Forrest, D.J., 1990, Astrophys. J., **358**, 298
Murphy, R.J., Share, G.H., Grove, J.E. et al., 1997, Astrophys. J., **490**, 883
Pelaez, F., Mandrou, P., Niel, M. et al., 1992, Sol. Phys., **140**, 121
Peterson, I.E., Winckler, J.R., 1958, Phys. Rev. Lett. **1**, 205
Porter, J.G.; Moore, R.L.; Reichmann, E. J et al., 1987, Astrophys. J., **323**, 380
Prince, T.A., Ling, J.C., Mahoney, W.A. et al., 1982, Astrophys. J. Lett., **255**, L81
Prince, T.A., Forrest, D.J., Chupp, E.L. et al., 1983, Proc. 18th Internat. Cosmic Ray Conf. Papers, **4**, 79
Ramaty, R., 1986, in: The Physics of the Sun, ed.: Sturrock, P.A., Holzer, T.E., Mihalas, D., Ulrich, R.K., Dordrecht, D. Reidel Publishing Co., Vol. II, p.291
Ramaty, R., Mandzhavidze, N., 1994, High-Energy Solar Phenomena, ed. Ryan, J.M., Vestrand, W.T., AIP Conf. Proc., **294**, 26
Ramaty, R., Mandzhavidze, N., 1995, Proc. of Kofu Symposium, NRO Report, **360**, 275
Ramaty, R., Kozlovsky, B., Lingenfelter, R.E., 1975, Space Science Rev., **18**, 341
Ramaty, R., Kozlovsky, B., Lingenfelter, R.E., 1979, Astrophys. J. Suppl. **40**, 487
Ramaty, R., Miller, J.A., Hua, X.-M., Lingenfelter, R.E., 1988, Nuclear Spectroscopy of Astrophysical Sources, ed: Share, G.H., Gehrels, N., AIP Conf. Proc. **170**, 217
Ramaty, R., Mandzhavidze, N., Kozlovsky, B et al., 1993, Adv. Space Science, **13**, (9) 275
Rank, G., Bennett, K., Bloemen, H. et al., 1996, High Energy Solar Physics, ed.: Ramaty, R., Mandzhavidze, N., Hua, X.-M., AIP Conf. Proc. **374**, 219
Rank, G., Debrunner, H., Lockwood, J. et al., 1997, Proc. 25th Internat. Cosmic Ray Conf. Papers, **1**, 5
Reames, D.V., 1996, High Energy Solar Physics, ed.: Ramaty, R., Mandzhavidze, N., Hua, X.-M., AIP Conf. Proc. **374**, 35
Rieger, E., 1989, Sol. Phys., **121**, 323
Rieger E., Marschhäuser, H., 1990, Proc. Max'91 Workshop Nr. 3, ed.: Winglee, R., Kiplinger, A., p. 68
Rieger, E., Gan, W.Q., Marschhäuser, H., 1998, Sol. Phys., **183**, 123
Ryan, J.M., Lee, M.A., 1991, Astrophys. J., **368**, 316
Ryan, J.M., Forrest, D.J., Lockwood, J. et al., 1993, Compton Gamma-Ray Observatory, ed.: Friedlander, M., Gehrels, N., Macomb, D., AIP Conf. Proc., **280**, 631
Ryan, J., Forrest, D., Lockwood, J., 1994, High-Energy Solar Phenomena, ed. Ryan, J.M., Vestrand, W.T., AIP Conf. Proc., **294**, 89
Schneid, E.J., Brazier, K.T. S., Kanbach, G. et al., 1994, High-Energy Solar Phenomena, ed.: Ryan, J.M., Vestrand, W.T., AIP Conf. Proc., **294**, 94
Seckel, D., Stanev, T., Gaisser, T.K., 1991, Astrophys. J., **382**, 652
Share, G.H., Murphy, R.J., 1995, Astrophys. J., **452**, 933
Share, G.H., Murphy, R.J., 1997, Astrophys. J., **485**, 409
Share, G.H., Chupp, E.L., Forrest, D.J., Rieger, E., 1983, Positron-Electron Pairs in Astrophysics, AIP Conf. Proc. **101**, 15
Šimberová, S., Karlický, M., Švestka, Z., 1993, Sol. Phys. **146**, 343

Trottet, G., Vilmer, N., Barat, C. et al., 1993, Astron. Astrophys. Suppl., **97**, 337
Vestrand, W.T., 1990, Astrophys. J., **352**, 353
Vestrand, W.T., Forrest, D.J., 1993, Astrophys. J. Lett., **409**, L69
Vestrand, W.T., Forrest, D.J., 1994, High-Energy Solar Phenomena, ed.: Ryan, J.M., Vestrand, W.T., AIP Conf. Proc., **294**, 143
Vestrand, W.T., Forrest, D.J., Chupp, E.L. et al., 1987, Astrophys. J., **322**, 1010
Vestrand, W.T., Share, G.H., Murphy, R.J. et al., 1999, Astrophys. J. Suppl., **120**, 409 ("SMM Solar Flare Atlas")
Vilmer, N., 1994, Astrophys. J. Suppl., **90**, 611
Vilmer, N., Trottet, G., 1997, in: Coronal Physics from Radio and Space Observations, ed.: Trottet, G., Lecture Notes on Physics, **483**, p. 28
Vlahos, L., Georgoulis, M., Kluiving, R., Paschos, P., 1995, Astron. Astrophys., **299**, 897
Wang, H.T., Ramaty, R., 1974, Sol. Phys., **36**, 129
Wild, J.P., Smerd, S.F., Weiss, A.A., 1963, Ann. Rev. Astron. Astrophys., **1**, 291
Wülser, J.-P., Canfield, R.C., Rieger E., 1990, Proc. Max'91 Workshop Nr. 3, ed.: Winglee, R., Kiplinger, A., p. 149
Yoshimori, M., Okudaira, K., Yanagimachi, T., 1986, J. Phys. Soc. Jpn., **55**, 3683
Yoshimori, M., Suga, K., Morimoto, K. et al., 1994, Astrophys. J. Suppl., **90**, 639

6 Gamma-Ray Pulsars

Gottfried Kanbach

6.1 Introduction

6.1.1 Radio Pulsars: Discovery of Rotating Neutron Stars

In 1967 a new radio telescope was commissioned in England. A group of astronomers under the direction of Anthony Hewish at the University of Cambridge had completed this receiver system to study the newly discovered quasi stellar radio sources and the rapid scintillations that result when radio waves from distant cosmic point sources pass through the turbulent interstellar and interplanetary plasma. One goal was to study the properties of these cosmic plasma clouds, and the sensitive receiver at 81.5 MHz was therefore fitted with a high-speed recorder. The intensity of celestial sources passing over the fixed field of radio receivers was recorded on strip charts and had to be analyzed by visual inspection. This task fell naturally to the graduate student of the group, Jocelyn Bell. On August 6, 1967, she first noticed a peculiar train of radio signals when the sky at right ascension 19h19min passed through the field of view. What could have caused such a transient periodic signal? The first suspicion was of course interference from some electric equipment, like the ignition of a passing car or a satellite. But to the surprise of Bell the signal appeared again at about the same time of day.

After a few months it was obvious that the regular pulses were coming from a celestial source beyond our solar system. Furthermore a recording of the source with sub-second time resolution on November 28, 1967, revealed pulses repeating at a regular period of 1.33 s. At that point – the discovery of the phenomenon was still kept secret – the thought that radio signals from an extraterrestrial civilization had been recorded was seriously considered, under the code "LGM" (little green men). The discovery of three more pulsating sources at different locations in the sky and with different periodicities did away with the LGM hypothesis, and it became clear that a fascinating new natural phenomenon, soon called pulsars, had been discovered.

The publication by Hewish et al. (1968), aside from presenting the observations of CP1919 (CP for Cambridge Pulsar), discussed two possible explanations: regular oscillations of white dwarfs and rotation of "compact" stars. Immediately a veritable hunt for these new astrophysical objects started at radio observatories around the world: several installations had the necessary equipment already in operation (rumor has it that a few had caught pulsar signals already before the Cambridge group but failed to notice the significance of these "interferences or malfunctions"). Soon dozens of new pulsars were known, with periods extending from fractions of a second to about 2 s. When the central source in the Crab Nebula, the remnant of a supernova in the year 1054 A.D., was discovered as a pulsar with a frequency of about 30 Hz, stellar oscillations had to be abandoned as a plausible hypothesis because the natural modes of white dwarfs have to be much slower. Pacini (1968) and Gold (1968) arrived at the still valid interpretation: the pulsar signals were generated by rotating, magnetized neutron stars, and the radiation luminosity derives ultimately from rotational energy. One essential prediction of this hypothesis was that the rotation must slow down due to the energy loss – after several months of observations this small change of frequency was found and the case was proven.

Now, 30 years after the discovery of pulsars, about 1500 radio pulsars are known and it can be expected that this number will continue to grow as more refined detection equipment is used. Figure 6.1 displays the distribution of currently observed radio pulsars in a period–period-derivative ($P\dot{P}$) diagram together with lines indicating the "rotational" age of the pulsars and their dipole field strength (see below). The range of rotational periods extends now from millisecond pulsars (the fastest PSR B1937+21 has a period of 1.56 ms) to the slowest pulsars with periods of several seconds. Catalogues of radio pulsars and their properties can be found in archives maintained at Princeton University[1], the Parkes Observatory in Australia[2], or at the European Pulsar Network[3]. Current population studies of pulsars suggest that our Galaxy contains $\sim 10^5$ active radio pulsars.

6.1.2 The Formation of Neutron Stars and Their Basic Properties

Neutron stars had already been speculated on in the 1930s as possible end products of stellar evolution. After stars of various masses evolve from the contraction of primordial gas clouds (probably controlled by rotation and magnetic fields), their individual stability is determined by a delicate balance of attractive forces (gravitation) and repulsive forces (thermal pressure from normal gas or quantum pressure from degenerate gas). The mass of a star determines the gravitational pressure in its center. During the life of normal main sequence stars, thermal pressure powered by nuclear fusion maintains the internal balance. More massive stars need a higher thermal pressure to stay in equilibrium; this is provided by a higher rate of nuclear processes in their cores ($\propto T^{5.3}$ [pp cycle], $\propto T^{15.6}$ [CNO cycle], T = core temperature),

---

[1] http://pulsar.princeton.edu/pulsar/catalog.shtml
[2] http://www.atnf.atnf.csiro.au/research/pulsar/psr/archive
[3] http://www.mpifr-bonn.mpg.de/div/pulsar/epn/

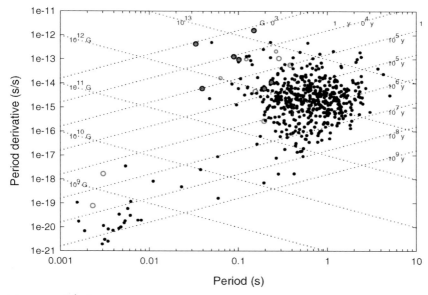

Fig. 6.1. $P\dot{P}$ distribution of presently known radio pulsars. The labeled lines indicate the "rotational" age (6.4) and the dipole field strength of the pulsars (6.3). *Colored symbols* indicate pulsars detected above 1 MeV (*filled* = high confidence detection, *open* = tentative pulsation or positional coincidence, see Table 6.1)

at the cost of much faster fuel consumption and a shorter life. After the nuclear fuel in the stellar core is exhausted and the available lighter nuclei have essentially all been fused into ^{56}Ni (which later decays to ^{56}Fe), the star has to transit into a new configuration. For a star of roughly solar dimensions, i.e., with mass less than 1.44 $M_\odot$ (the Chandrasekhar limit), the burnt-out core restabilizes by the pressure of a degenerate electron gas in its interior. These "white dwarf" stars have typically the size of Earth (radius $\sim 10^4$ km) and show strong magnetic fields of the order of 10^7 Gauss.

More massive stars have to follow a different path to a final, stable configuration: according to our current understanding these stars will form either neutron stars or black holes depending on the mass that collapses in their interior. Formation of a compact core is accompanied by the release of gravitational binding energy ($\sim 10^{52}$ erg) and the emission of neutrinos and gravity waves from the core. Together with a nuclear detonation front reaching the outer layers of the star the input of this central energy leads to the spectacular explosions we observe as supernovae (see Chap. 8). The outer layers of the massive star will be ejected with velocities of more than 10^4 km s^{-1} while a massive central core contracts and cools to form a sphere with a radius of about 10 km containing several solar masses of fusion-processed material. The central density of such a stellar remnant is of the order of 10^{15} g cm^{-3}. Under these conditions the nuclei originally present in the core disintegrate into nucleons and electrons; protons and electrons recombine in inverse β-decays to form a sea of neutrons, which as fermions fill their available phase space. The pressure exerted by this degenerate Fermi gas of neutrons is sufficient to stabilize a stellar core up to masses of $\sim 3.6\,\mathrm{M_\odot}$, which constitutes a fundamental limit. The equation of state of neutron-star matter, i.e., the relation between pressure and density, which determines the actual mass of newly formed neutron stars is not yet exactly known. From the study of X-ray binary systems one finds that most neutron stars have masses of $\sim 1.8-2.0\,\mathrm{M_\odot}$. If the compact stellar core contains more than the limiting mass, it is currently assumed that nothing can prevent further collapse, and a black hole is formed.

Although the detailed processes during the formation of a neutron star are complex and not completely understood (core magnetic forces, differential rotation and cooling behavior), basic conservation laws already provide estimates of the general characteristics of young neutron stars.

Conservation of angular momentum ($= I\omega$, where $I \propto Mr^2$ is the moment of inertia and ω the angular frequency) leads to the conservation of $r^2\omega$. A stellar core that collapses from a radius of $r_\star \approx 10^{11}$ cm to $r_{NS} \approx 10^6$ cm will therefore increase its rate of rotation by $\sim 10^{10}$, and a solar period of rotation of about 27 days speeds up to periods of milliseconds. The interior of a star is a fully conducting medium. Magnetic flux (Br^2) will therefore also be conserved during core collapse. This will increase the typical field of a normal star ($\sim 10^2$ Gauss) to values of the order of 10^{12} Gauss. Stellar core collapse is therefore expected to lead to a rapidly spinning highly magnetized neutron star. The total rotational energy content of a young neutron star ($\frac{1}{2}I\omega^2$, with the canonical moment of inertia $I \approx 10^{45}$ g cm^2) is of the order of 10^{51} erg.

Another estimate for the magnetic field strength of neutron stars can be obtained from the observed slowing down of the rotation. A spinning magnetic dipole radiates electromagnetic energy at a rate of $\frac{2}{3}\mu_\perp^2\omega^4 c^{-3}$, where $\mu_\perp = \mu\sin\alpha$ is the perpendicular component of the magnetic moment $\mu \sim B_0 r$. The angle α is the inclination of μ to the axis of rotation. Equating the dipole radiation loss with the slowing down of rotation, i.e.,

$$\frac{d}{dt}\left(\frac{1}{2}I\omega^2\right) = I\omega\dot\omega = -\frac{2}{3}\mu_\perp^2\omega^4 c^{-3}, \tag{6.1}$$

leads to

$$\mu_\perp^2 = -\frac{3Ic^3}{2}\frac{\dot\omega}{\omega^3} = \frac{3Ic^3}{2}P\dot P. \tag{6.2}$$

Inserting $I = 10^{45}$ g cm^2, $\alpha \sim 90°$, and P in seconds, one obtains the following estimate for the polar field strength:

$$B_0 \approx 3 \times 10^{19}\sqrt{P\dot P} \quad \text{Gauss.} \tag{6.3}$$

The lines of constant magnetic field strength in Fig. 6.1 follow from this estimate.

Another important pulsar characteristic is the "rotation age" τ. It can be derived from the dipole radiation energy-loss rate under the assumption that $\mu_\perp$ remains constant. It follows that $\dot\omega \propto -\omega^3$, which can be integrated to yield

$$\tau = \frac{\omega}{2\dot\omega} \times \left[1 - \left(\frac{\omega}{\omega_0}\right)^2\right], \tag{6.4}$$

where ω_0 is the initial spin frequency. Since for most pulsars we expect $\omega_0 \gg \omega$, the age reduces to $\tau \sim \omega/2\dot\omega$. Lines of constant age are also shown in Fig. 6.1.

6.1.3 Expectation of High-Energy Phenomena Around Young Pulsars

A rotating, magnetized, conducting neutron star forms a so-called unipolar inductor. The positive and negative charges inside the star experience Lorentz forces in opposite directions under the influence of the moving magnetic field $[F \sim \boldsymbol{v}\times\boldsymbol{B} \sim (\boldsymbol{\omega}\times\boldsymbol{r})\times\boldsymbol{B}]$. The charges have to redistribute themselves in such a way that electric fields from charge separation counterbalance the magnetic forces and no permanent currents flow in the star ($\boldsymbol{E}\cdot\boldsymbol{B} = 0$). Outside the star the electric fields can then be calculated by Laplace's equation from the interior charge distribution, requiring continuity for the potentials and tangential electric fields at the stellar surface. The radial components of the electric field will in general have a discontinuity at the stellar surface that implies a surface charge layer. If the magnetic moment is assumed to be aligned with the axis of rotation (Goldreich and Julian 1969) the potential outside the neutron star has a quadrupole form with maxima over the poles and the equator (of opposite signs) and zero values at mid-latitudes:

$$\Phi = \frac{B_0 \omega r_{ns}^5}{6cr^3}(3\cos^2\theta - 1). \tag{6.5}$$

and the electric field derived from Φ at the stellar surface has components along the magnetic field lines

$$\boldsymbol{E}\cdot\boldsymbol{B} = \frac{\omega r_{ns}}{c} B_0^2 \cos^3\theta. \tag{6.6}$$

These fields and potentials are illustrated in Fig. 6.2. The parallel electric field is of the order of $6\times 10^{10} P_s^{-1}$ V cm^{-1}. It overwhelms by far the gravitational force on the surface charges. The charges are therefore ejected from the surface and fill the pulsar magnetosphere, where they rearrange themselves in the same way as the internal charges of the star until the electric and

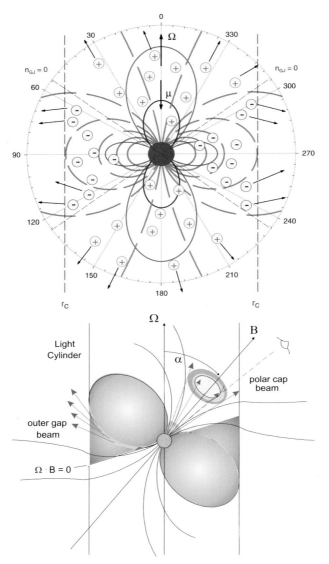

Fig. 6.2. *Top*: the magnetosphere of an aligned pulsar. Shown are the dipolar field lines $\boldsymbol{B}$ (*green*), the electrostatic potential Φ (*red*), and the parallel electric field ($\boldsymbol{E} \cdot \boldsymbol{B}$) on the stellar surface (*blue*). Following Goldreich and Julian (1969) the magnetosphere is filled with a charge-separated plasma. Regions that are connected to the outside by field lines crossing the speed of light cylinder ($r_c = c/\omega$, indicated by *dashed lines*) can lose their plasma by outflow. *Bottom*: Schematic for an inclined pulsar. The potential acceleration regions above the poles ("polar cap") and in the outer magnetosphere ("outer gap") are indicated

magnetic forces are in equilibrium. The required charge density, often called the Goldreich–Julian density, is given by

$$n_{GJ} \simeq -\frac{\boldsymbol{\omega} \times \boldsymbol{B}}{2\pi c} = 7 \times 10^{10} B_{\parallel,12} P_s^{-1} \text{ cm}^{-3}, \quad (6.7)$$

where $B_{\parallel,12}$ (units of 10^{12} G) is the field component parallel to $\boldsymbol{\omega}$ and P_s the rotation period in s.

However there can be no static equilibrium in the co-rotating magnetosphere of a pulsar. At a distance of $r_c = c/\omega \sim 5 \times 10^4 P_s$ km from the axis of rotation (the "light cylinder"), particles and fields would have to co-rotate with the speed of light. Relativistic effects of retardation and plasma mass loading will severely distort the pulsar magnetosphere close to the light cylinder. Field lines trying to extend across r_c are forced open to the outside and release their charges into the pulsar wind zone. In the inner magnetosphere this outflow leads to a deficit of charges in the so-called gap regions above the magnetic poles ("polar cap") and between the zero charge density surfaces ($n_{GJ} = 0$) and the closed magnetosphere close to r_c (the "outer gap" regions). In these gap regions the electrostatic potential of the rotating dipole is not balanced by charges and is available to accelerate particles to very high energies. Another source of accelerating potentials is given by relativistic effects close to the neutron-star surface. As Muslimov and Tsygan (1990, 1992) have shown, frame dragging leads to a sizeable electric field parallel to $\boldsymbol{B}$ in the inner polar cap region. As a consequence of the described dynamics, in a rotating pulsar magnetosphere high-energy particles with Lorentz factors $\sim 10^6$–10^7 are accelerated in regions with open field lines and propagate under the influence of the magnetic field and the radiation environment until the energy losses due to magnetic bremsstrahlung (synchro-curvature radiation) and inverse Compton scattering balance the acceleration. The size of the acceleration region is probably also limited by the screening of potentials due to the formation of a dense electron–positron plasma from pair creation of primary high-energy photons and the magnetic field (γ–γ and γ–$\boldsymbol{B}$). The current of relativistic particles flowing from the polar cap of area $A_{PC} \sim \pi r_{NS}^2 (\frac{r_{NS}}{r_c})$ (called the Goldreich–Julian current) can be estimated as

$$\dot{N} = A_{PC} \cdot n_{GJ} \cdot c \sim 1.7 \times 10^{38} \dot{P}^{\frac{1}{2}} P^{-\frac{3}{2}} \text{ s}^{-1}. \quad (6.8)$$

The power carried by these particles with Lorentz factor $\gamma \sim 10^6$–10^7 corresponds to about $(2-4) \times 10^{38} \dot{P}^{\frac{1}{2}} P^{-\frac{3}{2}}$ erg s^{-1}. It should be noted that the product $\dot{P}^{\frac{1}{2}} P^{-\frac{3}{2}}$ is proportional to $\sqrt{\dot{E}_{rot}}$ as given in 6.1. The particle flux and the possible initiation of particle/photon cascades with pair creation are further discussed in Sect. 6.3.

A direct consequence of this theoretical framework is the prediction of high-energy radiation phenomena around rotating magnetized neutron stars. In fact, if the nonthermal polarized radiation from the Crab Nebula is generated by synchrotron emission of electrons in the nebular magnetic fields, then the electrons must have typical energies of 10^8 eV for radio emission and 10^{15} eV for γ-rays. Pacini (1967) concluded before the actual discovery of pulsars that these short-lived energetic electrons could be provided by a rotating neutron star in the center of the nebula. The predictions for high-energy emission from pulsars were soon widely published and the chase for pulsars outside the radio range began. Detections were reported first for the Crab pulsar at X-ray (from balloon-borne detectors) and optical wavelengths. Today more than 30 pulsars are known at X-ray energies. In the optical range, four to five pulsating counterparts and a similar number of candidates based on positional coincidence in deep Hubble Space Telescope images have been found.

6.1.4 Early History of the Observation of Gamma-Ray Pulsars

Before the launch of the Compton Gamma-Ray Observatory (CGRO) in 1991 only two pulsars, Crab (PSR B0531+21) and Vela (PSR B0833-45), were known at γ-ray energies. Crab, the youngest known pulsar, was a prime candidate for early searches with instruments on balloon platforms. The pulsar's γ-ray light curve was first resolved at energies above 0.6 MeV in 90 min of balloon flight data using a large unshielded plastic scintillator (Hillier et al. 1970). A hard X-ray balloon telescope had recorded photons with accurate timing from Crab up to 200 keV in 1967 (Fishman et al. 1969). The analysis and the detection of pulses in these early Crab observations was achieved after the existence of the Crab pulsar was discovered in 1968. The familiar two-peaked light curve known from radio and optical observations was also seen at X- and γ-ray energies.

In 1972 the SAS-2 mission provided the first significant measurements above 30 MeV. The Crab pulsar was readily detected by its well-known pulsation (Kniffen et al. 1974). Two other sources (which are now known as Geminga and Vela) were at first not identified. However, applying a contemporary radio ephemeris (i.e., $P, \dot{P}$) for Vela, the 89 ms pulsation appeared clearly, making Vela the brightest identified γ-ray source in the sky (Thompson et al. 1975). The COS-B mission (1975–1982) allowed for very detailed temporal and spectral investigations of Crab and Vela (e.g., Clear et al. (1987), Grenier et al. (1988)) and established many facts about these sources, which were corroborated by later measurements with the CGRO instruments. Geminga defied all attempts at identification, although extensive counterpart searches were conducted. These resulted in finding a plausible X-ray source and a very weak, blue optical object compatible with the Geminga position. Finally joint detections of pulsations at X-ray energies (ROSAT, German-US-UK X-ray mission, 1990-1999) and γ-rays (Energetic Gamma-Ray-Experiment Telescope, EGRET on CGRO) in 1992 established Geminga as a true high-energy pulsar, i.e., an object that emits orders of magnitude more power at X- and γ-ray energies than anywhere else in the electromagnetic spectrum. Despite several claims for other γ-ray pulsars in the SAS-2 and COS-B data, no more solid detections could be found.

6.2 Current Observational Results of High-Energy Emission from Pulsars

6.2.1 Detections of High-Energy Pulsars

Ranking Pulsars by $(\eta \cdot \dot{E}_{rot})/4\pi f_\Omega d^2$, Instrumental Sensitivity Limits, and Data Analysis Methods

Expectations for the discovery of more pulsars with γ-ray emission were high when the CGRO was launched in 1991. In Fig. 6.3 the known radio pulsars are ranked according to a $F_E = (\eta \cdot \dot{E}_{rot})/4\pi f_\Omega d^2$ scale, which corresponds to apparent brightness. Included are factors for the emission efficiency of γ-radiation, $\eta = \mathcal{L}_\gamma/\dot{E}_{rot}$, and a beaming factor f_Ω, which describes the fraction of sky illuminated by the pulsar's γ-ray beam. From population studies of radio pulsars and the detected γ-ray pulsars, Helfand (1994) suggested that for the bulk of pulsars $\eta > 0.1$ and $0.1 < f_\Omega < 0.6$. It is often assumed, and I follow that assumption in the present estimates, that the beaming factor is $f_\Omega = \frac{1}{4\pi}$, i.e., the beam illuminates 1 sr. For the efficiency η we use either a constant value of 10% or a functional dependence proportional to the Goldreich–Julian current ($\approx \sqrt{\dot{E}}$) as described in Sect. 6.3. A coarse

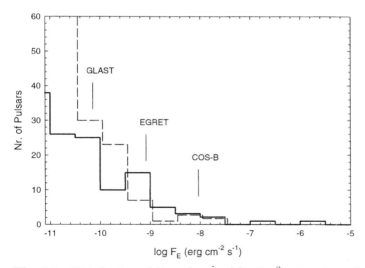

Fig. 6.3. Distribution of $F_E = (\eta \cdot \dot{E}_{rot})/4\pi f_\Omega d^2$ values for radio pulsars. For the γ-ray efficiency η two assumptions are used: $\eta = $ const. $= 0.01$ (*solid line*) and $\eta \approx \sqrt{\dot{E}}$ (*dashed line*, see Sect. 6.3). The beaming factor is assumed to be $f_\Omega = \frac{1}{4\pi}$. Coarse sensitivity limits for COS-B, EGRET and GLAST are shown

sensitivity limit (not accounting for various background conditions etc.) for the COS-B, EGRET and GLAST (Gamma-ray Large-Area Space Telescope) experiments is indicated in Fig. 6.3. It is apparent that COS-B (essentially the same sensitivity as SAS-2) could have detected pulsars aside from Crab and Vela. The COS-B catalogue sources 2CG195+04 and 2CG342-02 were indeed identified more than 15 years after COS-B in EGRET observations as γ-ray pulsars "Geminga" and PSR B1706-44. This was only possible because their pulsational ephemeris had been discovered in the meantime in the X-ray and radio ranges respectively.

Detection and Analysis Methods of Periodicity in Gamma-Ray Data

Here it is appropriate to define the methods commonly used in the analysis of pulsar γ-ray data and discuss the meaning of "detecting" a γ-ray pulsar. The basic property that clearly confirms the identification of a γ-ray source with a pulsar is the detection of periodic modulation in the rate of photons from the source. For most high-energy sources the rate of photon detections is extremely low. For EGRET, a source with a flux above 100 MeV of $10^{-6}\,\gamma\,\mathrm{cm}^{-2}\,\mathrm{s}^{-1}$ leads typically to a count rate of $10^{-3}\,\mathrm{s}^{-1}$ (~ 0.3 Crab), i.e., one photon in ~ 15 min. Therefore long observations, spanning days to weeks, are required to collect enough photons from a source. Correspondingly, of the order of 10^7 pulsar rotations have to be searched and analyzed for a periodic signal, which is done by folding the arrival times of pulsar photons with the pulsar period. To maintain the pulsar rotational phase accurately throughout the observation one not only needs extremely accurate pulsar parameters (the precision on the rotational frequency ν must be about $1/T_{obs}$, where T_{obs} is the duration of the data set) but must also correct for all geometrical and relativistic time delays and time-dependent Doppler shifts of ν resulting from the motion of the detector in Earth orbit and around the Sun. The topocentric (at the moving observatory) arrival time of every detected photon t_topo is therefore first transferred to a frame of reference that is at "rest" (more precisely "not in accelerated motion") with respect to the pulsar. The generally accepted origin of this rest frame is the "solar-system barycenter (SSB)" , which is defined for each point in time by the masses and positions of all major bodies in the solar system. The solar system ephemeris most widely used in pulsar research comes from the Jet Propulsion Laboratory[4] and a freely available procedure for pulsar timing correction and analysis is the TEMPO[5] program, maintained by several radio observatories. Let the vector from the observer to the SSB be called $r_\mathbf{SSB}$ and the vector to the pulsar $r_\mathbf{PSR}$. The arrival time t_SSB at the SSB is given by

$$t_\mathrm{SSB} = t_\mathrm{topo} - \frac{r_\mathbf{SSB} \cdot r_\mathbf{PSR}}{c \cdot |r_\mathbf{PSR}|}. \tag{6.9}$$

---

[4] ftp://navigator.jpl.nasa.gov/pub/ephem/
[5] http://pulsar.princeton.edu/tempo/index.html

After further corrections, as mentioned above, are applied t_{SSB} (called t below) can be converted to "phase of pulsar rotation" by the expansion:

$$\varphi = \varphi_0 + \nu(t - t_0) + \frac{1}{2}\dot{\nu}(t - t_0)^2 + \frac{1}{6}\ddot{\nu}(t - t_0)^2 + \cdots, \quad (6.10)$$

where the pulsar rotational phase φ_0 occurs at epoch t_0 and $\nu, \dot{\nu}$, and $\ddot{\nu}$ are the frequency of rotation and its derivatives at t_0. The data set of the epoch, rotational parameters, and position is called the pulsar ephemeris. Since pulsars are not perfect clocks (glitches and phase drifts occur) the ephemerides have to be monitored regularly and are valid only for specified time intervals. The decimal fractions of the phase values for all events φ_i are binned in histograms (values x_i) and display the so-called light curve of pulsar emission. Several cases can now be distinguished:

- The contemporary ephemeris of pulsation is known from other observations (e.g., radio or X-ray): γ-ray phases and light curves can be derived directly. Tests for modulation in the data can be performed with various methods. One of the most basic approaches uses a (binned) χ^2 test on the light curve values x_i:

$$\chi^2 = \frac{1}{\bar{x}} \sum_{i=1}^{n}(x_i - \bar{x})^2 \text{ , where } \bar{x} = \frac{1}{n}\sum_{i=1}^{n} x_i. \quad (6.11)$$

Pulses or other deviations from a uniform distribution in the light curve will result in large χ^2 values.

Another popular test is the Z_m^2 test which uses the phase values of the individual photons (φ_i) and sums the Fourier power over the first m harmonics (Buccheri et al. 1983):

$$Z_m^2 = \frac{2}{n} \sum_{k=1}^{m} \left[\left(\sum_{i=1}^{n} \cos 2\pi k \varphi_i\right)^2 + \left(\sum_{i=1}^{n} \sin 2\pi k \varphi_i\right)^2 \right]. \quad (6.12)$$

The H-statistic test (De Jager et al. 1989) is an extension of the previous test:

$$H \equiv \max_{1 \leq m \leq 20}(Z_m^2 - 4m + 4). \quad (6.13)$$

H tests the uniformity of the density of phase values on a circle. The sensitivity of these tests for weak periodic signals depends somewhat on the shape of the light curves and has to be investigated for individual cases. The significance of detection or the limits for a periodic signal are often difficult to assess because hidden degrees of freedom (e.g., event-selection criteria, binning, or assumptions on the pulse duty cycles) should be taken into account. Therefore it is generally accepted that high thresholds on significance must be used to obtain reliable results.

- If the pulsar period is not measured simultaneously it has to be extrapolated from a more or less distant epoch. This will result in a range of search periods, reflecting the errors of measurement and extrapolation. The γ-ray data then have to be scanned over ranges of ν and $\dot{\nu}$, and tests for modulation as described above are used to search for pulsar signals. In this case the threshold for a significant detection has to be even higher because of the additional degrees of freedom in the search.

- If the γ-ray source is strong (above ~ 100 detected photons per day) and is detected with a high signal-to-noise ratio, it is possible to perform periodicity searches in the γ-ray data alone. In the EGRET data only the Geminga source is strong enough for such a search. Mattox et al. (1996) used a FFT method and Brazier and Kanbach (1996) an evolutionary period search algorithm to find the (already known) correct period for Geminga. Similar procedures will become important to investigate the nature of unidentified γ-ray sources when the highly significant data of the next generation of space γ-ray telescopes becomes available.

Current Detections

The instruments on CGRO have detected a total of 7 pulsars with high significance. Aside from the already mentioned pulsars Crab, Vela, Geminga and PSR B1706-44, two more pulsars have been discovered above 30 MeV: PSR B1055-52 and PSR B1951+32. PSR B1509-58 was detected by the low-energy instruments on CGRO up to 30 MeV and therefore is counted as the seventh γ-ray pulsar. Some other "candidate" γ-ray pulsars, among them PSR B0656+14, have been reported as potential detections above 100 MeV. Because the lower significance in these cases requires confirmation, we will not review these sources further. Table 6.1 gives an overview of the pulsars emitting high-energy γ-rays and their multi-wavelength detections. It is remarkable that, in terms of the $\dot{E}/d^2$ rank, the six top ranked radio pulsars have all been detected at γ-ray energies. This argues in favor of beaming models where the radio beam overlaps with the γ-ray emission. Geminga, which is a very weak radio source, however, also indicates that the possibility of radio-quiet γ-ray pulsars exists. Objects of this type could explain some of the yet unidentified galactic γ-ray sources.

6.2.2 Gamma-Ray Searches and Limits for Other Radio Pulsars

A detailed search for pulsed γ-ray emission from all known radio pulsars in the first 3.5 years of EGRET data was performed by Nel et al. (1996). Using contemporaneous radio timing ephemerides, only upper limits could be determined for about 350 pulsars and no new detections could be made. However there is evidence for localized γ-ray sources at the position of some of the investigated pulsars. Although most cases appear to be chance coincidences because of the pulsar's low ranking $\dot{E}/d^2$ values, Table 6.1 contains 3 pulsars

6 Gamma-Ray Pulsars   139

Table 6.1. High-energy pulsars: multi-wavelength detections, candidates, and positional coincidences

PSR	P (ms)	$\dot{E}/d^2$ rank	Radio	Optical	EUV	X_{low}	X_{hi}	γ_{low}	γ_{hi}
High-confidence γ-ray detections									
B0531+21 (Crab)	33.4	1	P	P		P	P	P	P[1]
B0833-45 (Vela)	89.3	2	P	P	P			P	P[3]
J0633+1746 (Geminga)	237.1	3	?	?	P	P		?[9]	P[7]
B1706-44	102.5	4	P			D			P[4]
B1509-58	150.7	5	P	D		P	P	P[2]	
B1951+32	39.5	6	P			P		P[10]	P[5]
B1055-52	197.1	33	P	D		P		P[11]	P[8]
Candidate γ-ray detections									
B0656+14	384.9	18	P	?	D	P		?[10]	?[6]
B0355+54	156.4	36	P			D			?[12]
B0631+10	287.7	53	P			D			?[13]
B0144+59	196.3	120	P						?[14]
Candidate ms-pulsars γ-ray detections									
J0218+4232	2.32	43	P			P			?[15]
B1821-24	3.05	14	P			P			?[16]
Likely pulsar–γ-ray source coincidences									
B1046-58	123.7	8	P			D			
J1105-6107	63.2	21	P[17]			D			
B1853+01	267.4	27	P						

Notes: P: pulsed emission; D: positional coincidence; ?: low significance pulsed detection

Refs: [1] Nolan et al.(1993); [2] Matz et al.(1994), Kuiper et al.(1999a); [3] Kanbach et al.(1994); [4] Thompson et al.(1992, 1996); [5] Ramanamurthy et al.(1995); [6] Ramanamurthy et al.(1996); [7] Mayer-Hasselwander et al.(1994); [8] Fierro et al.(1993); [9] Kuiper et al.(1996); [10] Hermsen et al.(1997); [11] Thompson et al.(1999); [12] Thompson et al.(1994); [13] Zepka et al.(1996); [14] Ulmer et al.(1996); [15] Kuiper et al.(1999b); [16] Thompson et al.(1997); [17] Kaspi et al.(1997)

that rank above PSR B1055-52, coincide with a γ-ray source, but show no significant high-energy pulsations.

6.2.3 Light Curves of Gamma-Ray Pulsars

The high-energy light curve of a pulsar can be regarded as a direct projection of the sites and beaming directions of radiation sources in the pulsar's magnetosphere onto the celestial sphere. Since γ-ray photons are created in direct

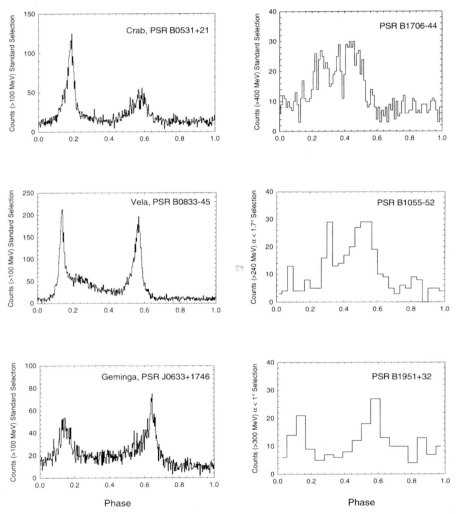

Fig. 6.4. High-energy light curves of γ-ray pulsars (> 100 MeV, unless indicated differently)

interactions between particles, photons and fields (Sect. 6.3) the investigation of light curves at different energies allows the most direct insight into the radiation processes inside a pulsar's magnetosphere. A fully self-consistent model for the acceleration and radiation of particles in the magnetosphere of a γ-ray pulsar presents a tremendously complex problem and has not been achieved so far. Such a model would have to account for a range of effects, including the following:

- the source particle densities, energy spectra and trajectories along the (possibly distorted) magnetic field lines;
- photon source functions due to different processes;
- photon propagation through the magnetosphere and around the neutron star (gravitational lensing);
- light travel delays across the magnetosphere; and
- general relativistic effects (e.g., beaming, aberration) arising from source co-rotation velocities close to the speed of light in the outer magnetosphere.

In spite of this highly complex scenario the observational data show that γ-ray pulsars are a rather homogeneous group and that the possibility exists to deduce fundamental information on the radiation processes and energetics of these objects.

Figure 6.4 shows γ-ray ($> 100\,\mathrm{MeV}$) light curves for the detected γ-ray pulsars. At high energies four pulsars basically show γ-ray light curves with two peaks separated by about 0.4 to 0.5 in phase ($140° - 180°$ of rotation). The emission peaks (if enough counting statistics are available) are very narrow: their cusp-shaped profiles display significant peaks of only a few degrees in phase. Two pulsars show a more complex emission pattern: PSR B1706-44 has possibly 3 peaks extending over about $120°$ of phase, and PSR B1055-52 shows two peaks at a phase distance of $\sim 70°$ which partly overlap.

6.2.4 Multi-Wavelength Light Curves

The rich phenomenology of multi-wavelength light curves for high-energy pulsars is shown in Figs. 6.5 to 6.7. The Crab pulsar (Fig. 6.5), basically emits a phase-aligned double-peaked emission profile over all available wavelengths, from radio to γ-rays. However we also notice some extra features: at low radio frequencies a precursor to the first main peak is detected. Its spectral and polarization properties indicate that this precursor is very similar to the single light curve peaks seen in other radio pulsars. Above 2.7 GHz the second main peak changes phase to a slightly earlier position and two peaks trailing the second main peak appear. At X- and γ-ray energies the region between the main peaks is filled with emission. Presently we have no theoretical concept that would explain these extra features in the Crab light curve. The other multi-wavelength pulsars show completely different light curves in

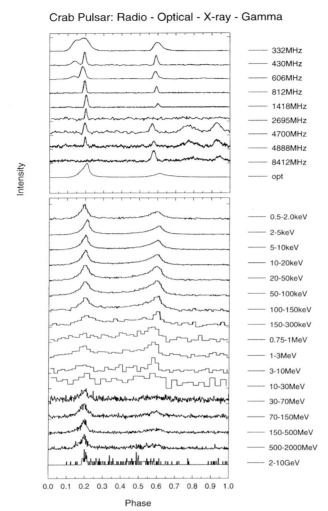

Fig. 6.5. Multi-wavelength light curves of the Crab pulsar

different bands. Figure 6.6 displays the next brightest γ-ray pulsars, Vela and Geminga. Vela emits a single radio pulse, followed by double emission peaks at optical and γ-ray energies. The peak separation changes from optical to γ-rays. A detailed analysis in the γ-ray range alone has revealed that the light curve peaks of Vela move together at the highest recorded energies. In Geminga the radio and optical pulsations have been detected only recently and their significance is still not fully established. It is, however, quite apparent that Geminga has different light curves at all energies, even from the low X-ray range, where thermal emission might dominate, to the emission above

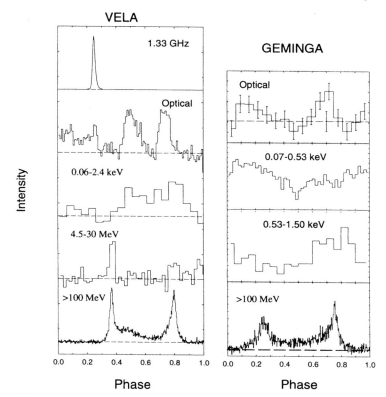

Fig. 6.6. Multi-wavelength light curves of the Vela and Geminga pulsars

1 keV, which is believed to originate from nonthermal processes in the magnetosphere. The other γ-ray pulsars, shown in Fig. 6.7, have different light curves over the detected spectral ranges. Assuming a random orientation of pulsars with respect to the observer it is clear that only a larger sample of detected light curves will help to reveal the full emission pattern of young pulsars.

6.2.5 Spectra of Electromagnetic Radiation from Pulsars

Thompson et al. (1999) compiled the multi-wavelength spectra for γ-ray pulsars shown in Fig. 6.8. The spectral intensities are displayed in terms of νF_ν, which is equivalent to the power emitted per unit of $\ln(\nu)$. These spectra clearly emphasize that emission in the X- and γ-ray region dominates the radiation budget of these pulsars. The spectra of the pulsed emissions are generally harder than a power law with index -2, which would give a horizontal line in the spectral plot. In several cases the spectra are curved in the γ-ray range and lead up to a cut-off or turnover at a few GeV.

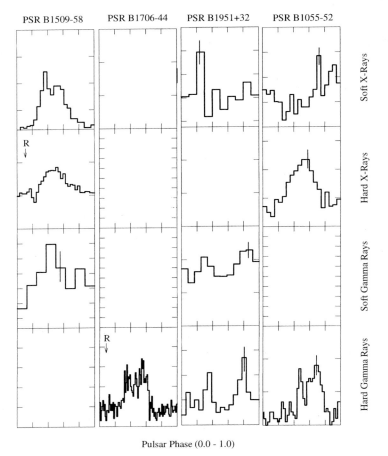

Fig. 6.7. Multi-wavelength light curves of the pulsars PSR B1706-44, PSR B1509-58, PSR B1951+32, PSR B1055-52

The Crab pulsar emits the maximum power at about 100 keV and the spectrum below and above that maximum can be described well by power laws extending from the optical ($\sim 1\,\mathrm{eV}$) to about 10 GeV. In addition to the pulsed emission, the Crab shows a strong unpulsed component at high energies. This component has been interpreted to come from the inner Crab Nebula as synchrotron radiation below a few GeV and as inverse Compton radiation, upscattered from optical photons, up to TeV energies.

PSR B1509-58 has been detected in the hard X-ray range by Ginga and in low γ-ray energies by BATSE (Burst and Transient Source Experiment), OSSE (Oriented Scintillation-Spectrometer Experiment), and COMPTEL (Compton Telescope). The spectrum shows a break at around 100 keV. EGRET may show a marginal detection of this pulsar between 30 and 70 MeV

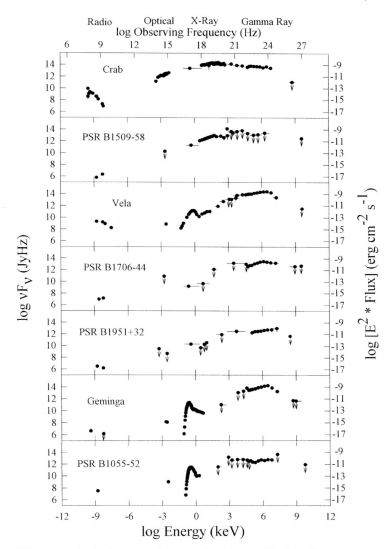

Fig. 6.8. Pulsed spectral power (units of νF_ν describe the power radiated per decade of frequency) observed from γ-ray pulsars over the complete electromagnetic spectrum. (Reproduced with permission from Thompson et al. (1999), their Fig. 4)

but places significant upper limits on the flux above this energy range, which indicates that this spectrum must turn over again in the MeV range.

The Vela pulsar has a pronounced spectral break at 2 GeV. Due to the strength of the source, very detailed spectra for the individual phase components, e.g., the peaks and the interpulse emission, are available. The spectra of the emission peaks are generally softer than the spectra of the interpeak regions. The softest components are observed in the leading and trailing wings of the peaks, which has been explained as a low-energy spill-out from the main γ-ray-producing cascades in the outer magnetosphere of the pulsar.

PSR B1706-44, a γ-ray source that had been known under the name 2CG342-02 since the observations with COS-B in the 1970s, was discovered in 1992 as a radio pulsar with a 102 ms pulsation period. Although an X-ray source is found coincident with the pulsar, no pulsation is found at keV energies. This source has also been detected as a steady emitter at TeV energies (Kifune et al. 1995), similar to the finding for the Crab. The maximum power of PSR1706-44 is emitted at around 1 GeV in the EGRET energy range.

PSR B1951+32 is detectable as a weak γ-ray DC source only at energies above 300 MeV. Pulsational analysis has only recently been successful, after extended EGRET observations became available, to confirm the identification of the source as PSR B1951+32. The spectrum of this pulsar does not show a characteristic fall off at energies above several GeV, although the statistics for this weak source are very limited.

Geminga, the enigmatic γ-ray source known in the Galactic Anticenter since its detection with SAS-2 in 1972, was identified as a pulsar in 1992. Observations with ROSAT at X-ray energies revealed 237 ms pulsations in Geminga, and the corresponding γ-ray signal was found in EGRET, COS-B, and SAS-2 data (Mayer-Hasselwander et al. 1994). The spectrum of the pulsed emission is generally very hard and shows marked variations over the rotation period. The maximum power of Geminga is emitted at about 1 GeV, and above a few GeV the spectrum breaks off sharply. Geminga is the first identified specimen of a true high-energy pulsar. The power in optical ($m_v \sim 25.5$, pulsation marginally detected) and radio emissions (recent marginal detections of pulsed emission at 102 MHz of about 100 mJy [Kuz'min and Losovskii 1997, Malofeev and Malov 1997]) is lower than the γ-ray emission by more than 6 orders of magnitude. Speculations that many of the other unidentified galactic γ-ray sources are similar "Geminga-like" pulsars have been discussed widely. Geminga itself however is a very close, low-luminosity pulsar. The other γ-ray sources indicate by their galactic distribution that they are at least 10 times more distant. If the above hypothesis is maintained, one has to assume that much younger, high-luminosity pulsars can also operate in a "Geminga-like" mode, where the emission patterns at different wavelengths do not coincide.

PSR B1055-52 has a very hard energy spectrum that seems to extend from the X-ray range into the γ-ray range. Again the maximum power is emitted around 1 GeV, and no break in the spectrum is visible up to 4 GeV. Above that energy a break is required by upper limits for PSR B1055-52 at TeV energies.

6.2.6 Phase-Dependent Gamma-Ray Spectra of Pulsars

Summing data from all of the presently available EGRET pulsar observations has enabled the derivation of very detailed light curves and spectra, not only for the total emission but also for shorter phase intervals of the pulsars rotation. For the 4 brightest pulsars a spectral hardness parameter, i.e., the intensity ratio > 300 MeV/(100–300 MeV), was derived as a function of phase, and the result is shown graphically in Fig. 6.9. There are some clear trends: pulsars with a double-peak structure tend to have a harder, more energetic spectrum at later phases of emission. This effect is particularily noticeable in an older pulsar like Geminga. Model calculations for polar cap cascades have successfully reproduced the phase-dependent spectra of Vela (Daugherty and Harding 1996). The outer-gap modelling of Romani (1996) was also able to reproduce the observed trends in the phase-resolved spectra.

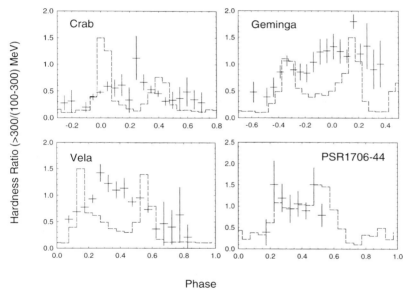

Fig. 6.9. Spectral hardness of the pulsar emission as a function of rotational phase

6.2.7 Pulsar Luminosities and Gamma-Ray Efficiencies

The multi-wavelength spectra and light curves presented in the previous sections show the detected high-energy pulsars as individual objects. One can also attempt to describe these pulsars as a class of sources, and derive some global characteristics, e.g., the energy budget of high-energy emission from optical to γ-ray wavelengths. The smooth and continuous photon spectra,

Table 6.2. Properties of high-energy pulsars (Thompson et al. 1999)

Name	P (ms)	τ (10^3 yr)	$\dot{E}$ (10^{36} erg s^{-1})	F_E (erg cm^{-2} s^{-1})	d (kpc)	L_{HE} (erg s^{-1})	η ($E > 1$ eV)
Crab	33	1.3	450	1.3×10^{-8}	2.0	5.0×10^{35}	0.001
B1509-58	150	1.5	18	8.8×10^{-10}	4.4	1.6×10^{35}	0.009
Vela	89	11	7.0	9.9×10^{-9}	0.5	2.4×10^{34}	0.003
B1706-44	102	17	3.4	1.3×10^{-9}	2.4	6.9×10^{34}	0.020
B1951+32	40	110	3.7	4.3×10^{-10}	2.5	2.5×10^{34}	0.007
Geminga	237	340	0.033	3.9×10^{-9}	0.16	9.6×10^{32}	0.029
B1055-52	197	530	0.030	2.9×10^{-10}	1.5	6.2×10^{33}	0.207

except for the thermal emission peaks in three pulsars, suggest that the radiation sources and processes are based on a rather homogeneous population of energetic particles. The energy flux F_E carried in high-energy photons is derived by integrating the wideband spectra in Fig. 6.8. The result is given in Table 6.2. For the calculation of the efficiency η, with which the pulsar generates energetic photons from the loss of rotational energy, it is assumed that the emission is beamed into 1 sr of solid angle. This is a problematic assumption especially for the wide beaming patterns generated in outer gap

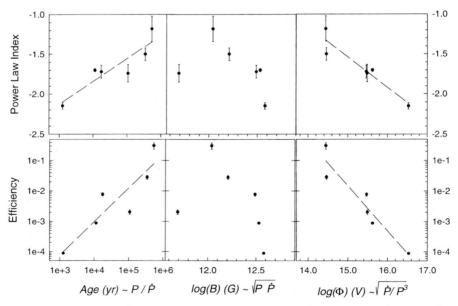

Fig. 6.10. Efficiency/spectral index of γ-ray emission (E > 100 MeV) versus the rotational energy loss, age, polar magnetic field and open field line potential of pulsars

models. We consider the following derived pulsar parameters: the rotational age ($\tau \propto P/\dot{P}$), the magnetic field ($B \propto \sqrt{P\dot{P}}$) and the total voltage available over the pulsars open field lines ($\Phi \propto \sqrt{\dot{P}/P^3}$). Figure 6.10 shows the resulting correlations. As many previous investigations have shown, both the efficiencies and the spectral indices correlate well with the apparent age and the open field line potentials of the pulsars. Pulsars become more efficient in their conversion of rotational energy into $E > 100$ MeV radiation and their spectra become harder with increasing age or decreasing potential values (Arons 1996). The correlation with the inferred magnetic field [see also Goldoni et al. (1995)] is not at all as compelling. Indeed, there seems to be an indication of a decreasing efficiency for both the highest B fields and for the objects with low B. Similarily the spectra appear softer at the marginal magnetic field values.

6.2.8 Search for Identification of Other Gamma-Ray Sources

The true nature of the yet unidentified galactic γ-ray sources poses both a challenge and a puzzle to present astronomical research (see Chap. 13). Let us assume here that several of these objects are pulsars that have not yet been discovered, for lack of emission at other wavelengths. There is some justification for this assumption in the fact that spatial, temporal, and spectral studies of the galactic γ-ray sources (Kanbach et al. (1996), Merck et al. (1996), Grenier (1997), Mereghetti and Pellizzoni (1997)) have resulted in global characteristics of the source population that is very compatible with the results from the detected pulsars. If Geminga is taken as a rôle model for this population of pure γ-ray pulsars, possibly with some faint X-ray and optical emission, then the approach which successfully led to the identification of Geminga (Bignami and Caraveo 1996) should be followed. Presently the search for X-ray counterparts is conducted with the intention to locate the γ-ray source sufficiently accurately for an optical search. The first results from such counterpart searches are available. Brazier et al. (1996) have found that the γ-ray source 2EG J2020+4026, a source that had been known as 2CG 078+2 for nearly 20 years, coincides with a weak ROSAT X-ray source located in the center of the supernova remnant G78.2+2.1, also called γ-Cyg. The X-ray source is still too weak to yield a significant periodicity. In this case, as in several others, further observations, in X-rays as well as in the optical, followed by pulsation studies will have to be done to secure an identification of the high-energy source. Another potential γ-ray pulsar identified in this search for X-ray counterparts is located inside the supernova remnant G119.5+10.2 (CTA 1, Brazier et al., 1998); the EGRET designation of this object is 2EG J0008+7307, and it has a γ-ray spectrum very similar to a typical pulsar spectrum.

The contribution of unresolved, distant pulsars to the diffuse galactic emission at high energies could be in the range of 5–30%, as recent work by

Sturner and Dermer (1996), Kanbach et al. (1996), Pohl et al. (1997) and Grenier (1997) has indicated.

The next generation of γ-ray telescopes will be much more sensitive and will be able to obtain the counting rates needed for an independent period search in γ-rays. For Geminga it was shown by Brazier and Kanbach (1996) and Mattox et al. (1996) that a direct search for γ-ray pulsations can be successful if about 100 counts day^{-1} are available from a source. This rate should be well attainable with the next high-energy space telescope, GLAST (Michelson 1996).

6.3 Theory of Pulsar High-Energy Emissions

6.3.1 Acceleration Zones and Production of Relativistic Particles

A spinning, highly magnetized neutron star generates enormous potential differences between different parts of its surface and its magnetosphere. As was described in Sect. 6.1 it was recognized early on that the pulsar magnetosphere cannot remain a vacuum but will be filled with a charge-separated plasma that is either extracted from the neutron-star surface (Goldreich and Julian 1969) or generated by pair-creation cascades directly in the magnetosphere (Sturrock 1970, 1971). In the closed regions of the pulsar magnetosphere, i.e., regions not connected by field lines to space beyond the speed-of-light cylinder (r_c), the charge-separated plasma settles in an equilibrium configuration to completely balance the magnetically induced forces on the charges by electrostatic forces, such that no net currents are induced. The open field lines, originating from the polar caps of the neutron star and extending beyond r_c, form an open conduit for the plasma of the polar regions and cannot retain the charges in the vicinity of the neutron star. It is therefore widely accepted that in these open regions of the magnetosphere accelerating potentials become available, driven by the imbalance generated by outflow from the otherwise static magnetosphere. Two "gap" regions are identified (see Fig. 6.2): close to the stellar surface the "polar gap" and further out, close to r_c above the zero-charge density surface where $\boldsymbol{\omega} \cdot \boldsymbol{B} = 0$, the "outer gap" regions. The perturbations and electric fields in the open magnetosphere are not easily calculated in a self-consistent model but their existence follows directly from the assumption that the particle currents from the polar regions carry away the star's angular momentum as it slows down.

From the observations of γ-ray pulsars it is also clear that ultrarelativistic particles up to $\gamma \sim 10^6 - 10^7$ must be generated in these acceleration zones. They propagate along the local fieldlines and are subject to energy loss and cascading processes as described in the next section. The particle photon cascade is expected to exceed the threshold for pair creation ($\gamma + B_\perp \Rightarrow e^+ \, e^-$ and $\gamma + \gamma \Rightarrow e^+ \, e^-$) in energetic young pulsars after a certain pathlength H in the accelerator. At this point, called the pair-formation front (PFF), the exponentially increasing density of the pair plasma will quench the accelerating fields (screening), and the accelerated beam will coast through the remaining magnetosphere and create more γ-rays. The saturation energy and the particle spectra that can be reached depend on the strength and spatial extent of the accelerating electric field and on the energy-loss processes that the particles encounter. A number of authors, e.g., Arons (1981), have investigated the existence of such steady electrically driven flows above the polar cap. For typical pulsar parameters estimates of the accelerating potential from the surface up to a PFF of the order of 10^{12} V and a height of the PFF of about a kilometer were derived. Later studies pointed out that the situation is probably more complicated: additional accelerating fields are generated in fully relativistic treatments of the electrodynamics close to the rotating neutron star ("frame dragging"), inverse Compton scattering of particles occurs with photons from the hot surface, and positrons returning to the surface can set up a lower PFF (Harding and Muslimov 1998). The consequence of these studies is that the acceleration region is higher above the polar cap ($H \sim r_{ns}$) and that the energy the particles can reach ($5 \times 10^{12} - 5 \times 10^{13}$ eV) is not only higher than in the earlier estimates but also rather insensitive to the pulsar parameters (self-regulating system). An important aspect influencing the formation of PFFs in high-field pulsars might be the onset of photon splitting ($\gamma \Rightarrow \gamma + \gamma$, a third-order QED process in fields $B \sim B_{cr}$), which competes with pair formation and could lead to radio-quiet high-energy pulsars.

The outer gap regions and their possible particle acceleration zones have been evaluated theoretically to a lesser extent than the polar-cap scenarios. The basic scheme of particle acceleration as a result of the development of vacuum gaps by outflow from the magnetosphere is similar in both, outer gap and polar cap, models. In contrast to the polar regions with their distance of a few stellar radii, the outer magnetosphere provides a different environment. Relativistic effects from stellar gravity are of minor importance and the magnetic field strengths are lower by many orders of magnitude. In the outer gaps the relativistic particles will therefore mostly lose their energy by synchrotron and curvature radiation. Processes of pair production are also expected to form cascades and PFFs that will terminate the accelerating regions. These occur via photon–photon interactions between energetic curvature radiation photons with thermal X-rays from the stellar surface. The radiation emerges from the outer gaps along the local field line directions; it has to be transformed from the rotating frame of reference of the magnetosphere to inertial space (beaming, aberration).

6.3.2 Radiation and Propagation Processes

The theoretical treatment of electromagnetic conversion processes for relativistic particles in intense magnetic fields has to go beyond the classical description of electrodynamics ("magnetic bremsstrahlung" or "synchro-curvature radiation"). It also has to account for relativistic and quantum effects like pair production and photon splitting. Erber (1966) introduced a compact formalism to describe the energy states and transition probabilities of charged relativistic particles of energy E in fields with perpendicular components of $B_\perp$ mostly in terms of a single dimensionless parameter,

$$\Upsilon = \frac{E}{mc^2} \frac{B_\perp}{B_{cr}}, \qquad (6.14)$$

where $B_{cr} = m^2 c^3/e\hbar = 4.414 \times 10^{13}$ G is the natural quantum-mechanical measure of the field strength. In the typical magnetosphere of a young pulsar, particle energies and field strengths vary considerably from the polar cap surface to the outer reaches close to r_c. The parameter Υ therefore assumes a wide range of values. Close to the polar cap $\Upsilon \gg 1$ leads to dominant rates of curvature radiation, photon-field pair production or photon splitting. In the outer magnetosphere, where $\Upsilon \ll 1$ due to the much reduced field strengths, classical formulae for relativistic synchro-curvature radiation can be applied. Everywhere in the pulsar's magnetosphere acelerated relativistic particles and photons encounter a bath of radiation with various origins. The thermal emission from the hot neutron star surface ($T \sim 10^6$ K) or synchrotron photons from the particle cascades themselves can lead to inverse-Compton interactions and to photon–photon pair creation events, which also add to the formation of strong particle/photon cascades and eventually to the quenching of the accelerating fields by a massive PFF.

The result derived by Erber (1966) for the spectral distribution of magnetic bremsstrahlung with photon energy ϵ_γ from electrons of energy E under the assumption that $\epsilon_\gamma \ll E$ is given by

$$I(E, \epsilon_\gamma, B_\perp) = \left(\frac{\sqrt{3}\alpha mc^2}{\lambda_c}\right) \left(\frac{\Upsilon}{E}\right) \left(1 - \frac{\epsilon_\gamma}{E}\right) \kappa(2\zeta), \qquad (6.15)$$

where α is the fine structure constant, λ_c the Compton wavelength, and $\kappa(z)$ the incomplete Bessel function integral, which is also known in classical synchrotron theory:

$$\kappa(z) = z \int_z^\infty K_{5/3}(x)\,\mathrm{d}x \quad \text{and} \quad \zeta = \frac{\epsilon_\gamma}{(E - \epsilon_\gamma)3\Upsilon}. \qquad (6.16)$$

In the inner pulsar magnetosphere charged particles are tightly constrained to move along the field lines because any nonzero pitch angle would be damped on time scales of 10^{-16} s. Since the dipole field lines are curved [polar regions have $\rho_c \sim \left(\frac{r_{ns}c}{\omega}\right) \sim 10^8 P_s^{1/2}$ cm, but close to the star up to a height h multipole fields could lead to even smaller values of ρ_c] the charges have to move on curved trajectories and emit radiation. This "curvature" emission is equivalent to synchrotron radiation, only the role of $B_\perp$ is replaced by $\rho_c \propto B_\perp^{-1}$.

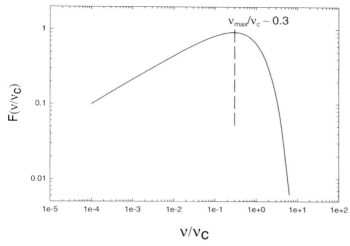

Fig. 6.11. Spectral distribution of magnetic bremsstrahlung from a relativistic electron in a magnetic field

In Fig. 6.11 a typical magnetic bremsstrahlung spectrum for monoenergetic electrons in a homogeneous perpendicular magnetic field is shown. In a real environment this spectrum is modified by the following mechanisms:

- the spectrum of electrons and any inhomogeneity of the field have to be properly folded into the magnetic bremsstrahlung spectrum.
- photons are absorbed as they travel in the magnetosphere by conversion to electron–positron pairs either directly on the field or by γ–γ collisions with the intense X-ray density from the hot neutron-star surface. On the other hand the inverse-Compton scattering of primary or secondary electrons will energize ambient photons into the γ-ray regime.

In summary, pulsar γ-ray spectra are thought to be composed of several emission components and are shaped by corresponding absorption processes, especially if a polar cap origin is considered.

6.3.3 Field Line Geometries, Beaming and Light Curves

In polar cap emission models the origin of the observed radiation is placed at a height of a few neutron star radii. The electron–photon cascades emanating from the outer PFF should have a distinct hollow-cone-like profile, due to the better development of the cascades at the edges of the polar cap, where the field lines are more curved. We also expect the γ-ray beam to have such a profile. An observer that happens to be in the path of the rotating cone will see one (tangential crossing) or two (full crossing) peaks of emission with a filled interval between the peaks. Depending on the angle the observer forms with the axis of rotation and the opening angle of the emission cone, various peak separations can be realized to account for the observations. If the emission should come from very close to the neutron-star surface, the opening angle of the beam and the corresponding solid angle would be very small. Then it is required that the angle between the spin and magnetic axes is also small in order to generate peak separations of close to 180°. In this case only a small fraction of all pulsars would be observed. A potential problem for simple polar cap models could lie in the explanation of light curves at different wavelengths which are all different and have various phases.

The latter problem is addressed by outer gap emission models. The zones of high-energy emission in these models are located in the open outer magnetosphere just above the closed co-rotating plasma regions. Towards the neutron star the outer gap is limited by the boundary layer between regions of positive and negative space charge. The field-line geometry close to the speed-of-light cylinder is shaped by relativistic effects of retardation, and any photon emitted along the field lines in the rotating pulsar's frame of reference must be properly transformed into the observer's frame (aberration, light travel time, Doppler shifts). Outer gap models are therefore characterized by a complex beaming pattern generally extending over large solid angles ("fan beam"). A recent calculation of the outer gap geometry by Romani and Yadigaroglu (1995) resulted in a good fit to the observed Vela light curve where the sharp emission peaks are explained as the result of caustic magnetospheric surfaces that bundle their emission into one direction. In such models the variety of multi-wavelength light curves could also be explained by, for example, placing the radio emission in a pencil beam on the magnetic axis and the γ-ray beam at widely different angles of emission along the field-lines of the outer magnetosphere. The possibility of finding pulsars where we only intercept the wide high-energy beam and not the radio pulse (Geminga types) is inherent to the outer gap models.

6.3.4 Spectra, Luminosities and Efficiencies

Pulsar spectra are shaped by the superposition of several emission mechanisms, which in turn will each be shaped locally by the convolution of particle spectra, fields and photon environments. Furthermore the pulsar magnetosphere is an "optically thick" medium for high-energy photons coming from the inner regions of the polar cap and travelling at an angle to the magnetic field. Theoretical γ-ray spectra for pulsars have been derived in polar cap and outer gap models and generally resulted in acceptable fits to the data. As an example the outer gap model spectrum of Romani (1996) is shown in Fig. 6.12 for the case of the Vela pulsar. An additional component to the magnetospheric nonthermal emissions in the form of a thermal spectrum from the neutron star surface could be added at low X-ray energies.

In the range from MeV to GeV energies we presently seem to have no clear spectral signature to distinguish the competing theoretical models. There is

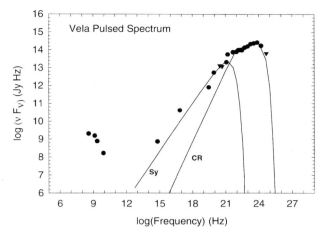

Fig. 6.12. The theoretical outer gap γ-ray spectrum as calculated by Romani (1996) fitted to the observed data of the Vela pulsar

however a prediction from the outer gap models that inverse-Compton interactions of relativistic electrons and hard X-ray photons should lead to a pulsed spectral component around 1 TeV. Similar photons from a polar cap scenario would be absorbed in the magnetosphere. Observations by current air Cherenkov telescopes have not been able to detect pulsed emission at TeV energies, and the upper limits are beginning to restrict the theoretically expected inverse-Compton component. The next generation of γ-ray telescopes in space and on the ground should provide decisive data for this problem.

The luminosities of high-energy pulsars, derived from spectra integrated over all energies above 1 eV, assuming a beaming fraction of $\frac{1}{4\pi}$, i.e., a beam size of 1 sr, and source distances derived from dispersion measures, are shown in Fig. 6.13 (Thompson et al. 1999). A clear trend is visible: the younger pulsars are more luminous. In contrast to the luminosities of soft X-ray pulsars (Becker and Trümper 1997), where $L_X \sim 10^{-3}\dot{E}$, the γ-ray emission varies proportionally to $\sqrt{\dot{E}} \sim \dot{N}$, the Goldreich–Julian current flowing out of the open magnetosphere [see (6.8)]. As was noted by Arons (1996) and Rudak and Dyks (1998) $\sqrt{\dot{E}}$ is also proportional to the voltage across the open field lines of the polar cap ($\Phi \sim 4 \times 10^{20} \dot{P}^{\frac{1}{2}} P^{-\frac{3}{2}}$ V). The luminosity trend in Fig. 6.13 could therefore be caused either by a proportionality to the number of accelerated particles, which would all be accelerated to the same energy, or by the same particle flow from different pulsars with correspondingly different acceleration energies. An argument for the first explanation could be that the acceleration of primary particles is limited by radiation reaction. The second scenario would correspond to pulsars with a limited supply of primary particles. Figure 6.13 also shows that the γ-ray efficiency of older pulsars increases (see also Fig. 6.10) to values around 20% for PSR B1055-52. It is

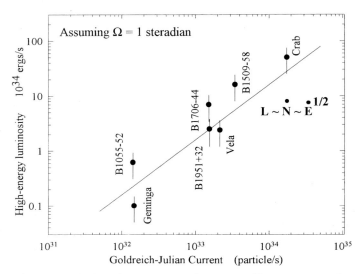

Fig. 6.13. Pulsar luminosities above $\sim 1\,\text{eV}$, assuming a beaming pattern size of 1 sr, versus $\sqrt{\dot{E}}$, which is proportional to the Goldreich–Julian current flowing from the open magnetosphere. [Reproduced with permission from Thompson et al. (1999), their Fig. 5]

clear that the luminosity trend must discontinue for pulsars with lower values of $\dot{N}$, because efficiencies of more than 100% would be implied for these objects.

It will be one of the foremost scientific challenges for the next generation of γ-ray telescopes to find more γ-ray pulsars. This should further advance our understanding of these fascinating natural laboratories, where relativistic particles, photons and fields generate observable radiation with the highest energies and allow us to probe physics in a realm which is unique in the universe.

References

Arons, J., 1996, Space Sci. Rev. **75**, 235
Arons, J., 1981, Astrophys. J. **248**, 1099
Becker, W., Trümper, J., 1997, Astron. Astrophys., **326**, 682
Bignami, G.F., Caraveo, P.A., 1996, Ann. Rev. Astron. Astrophys., **34**, 331
Brazier, K.T.S., Kanbach, G., 1996, Astron. Astrophys. Suppl. Ser., **120**, 85
Brazier, K.T.S., Kanbach, G., Carraminana, A., et al., 1996, MNRAS, **281**, 1033
Brazier, K.T.S., Reimer, O., Kanbach, G. et al., 1998, MNRAS, **295**, 819
Buccheri, R., Bennett, K., Bignami, G.F. et al., 1983, Astron. Astrophys., **128**, 245
Clear, J., Bennett, K., Buccheri, R. et al., 1987, Astron. Astrophys., **174**, 85
Daugherty, J.K., Harding, A.K., 1996, Astrophys. J., **458**, 278

De Jager, O.C., Swanepoel, J.W.H., Raubenheimer, B.C., 1989, Astron. Astrophys., **221**, 180
Erber, T., 1966, Rev. Mod. Phys. **38**, 626
Fierro, J.M., Bertsch, D.L., Brazier, K.T.S. et al., 1993, Astrophys. J., **413**, L27
Fishman, G.J., Harnden, F.R., Haymes, R.C., 1969, Astrophys. J. Lett., **156**, L107
Gold, T., 1968, Nature **218**, 731
Goldoni, P., Musso, C., Caraveo, P.A., et al., 1995, Astron. Astrophys., **298**, 535
Goldreich, P., Julian, W., 1969, Astrophys. J., **157**, 869
Grenier, I., 1997, in *The Transparent Universe*, Proc. 2nd INTEGRAL Workshop, ESA SP-382, 187
Grenier, I.A., Hermsen, W., Clear, J., 1988, Astron. Astrophys., **204**, 117
Harding, A.K., Muslimov, A.G., 1998, Astrophys. J., **508**, 328
Helfand, D.J., 1994, MNRAS **267**, 490
Hermsen, W., Kuiper, L., Schönfelder, V. et al., 1997, in *The Transparent Universe*, Proc. 2nd INTEGRAL Workshop, St. Malo, France, ESA SP-382, 287
Hewish A., Bell, S.J., Pilkington, J.D., Scott, P.F., Collins, R.A., 1968, Nature **217**, 709
Hillier R.R., Jackson, W.R., Murray, A., Redfern, R.M., Sale, R.G., 1970, Astrophys. J., **162**, L177
Kanbach, G., Arzoumanian, Z., Bertsch, D.L. et al., 1994, Astron. Astrophys., **289**, 855
Kanbach, G., Bertsch, D.L., Dingus. B.L. et al., 1996, Astron. Astrophys. Suppl. Ser. **120**, 461
Kaspi, V.M., Bailes, M., Manchester, R.N. et al., 1997, Astrophys. J., **485**, 820
Kifune, T., Tanimori, T., Ogio, S. et al., 1995, Astrophys. J. Lett., **438**, L91
Kniffen, D.A., Hartman, R.C., Thompson, D.J. et al., 1974, Nature **251**, 397
Kuiper, L., Hermsen, W., Schönfelder, V. et al., 1996, NATO-ASI No.515, p. 211
Kuiper, L., Hermsen, W., Krijger, J.M. et al., 1999a, Astron. Astrophys., **351**, 119
Kuiper, L., Hermsen, W., Verbunt, F. et al., 1999b, Astrophys. Lett. Commun., **38**, 33
Kuz'min, A.D., Losovskii, B.Ya., 1997, Astron. Lett., **27**, 295
Malofeev, V.M., Malov, O.I., 1997, Nature **389**, 697
Mattox, J.R., Koh, R.C., Lamb, D.J. et al., 1996, Astron. Astrophys. Suppl. Ser., **120**, 95
Matz, S.M., Ulmer, M.P., Grabelsky, D.A., et al., 1994, Astrophys. J., **434**, 288
Mayer-Hasselwander, H.A., Bertsch, D.L., Brazier, K.T.S. et al., 1994, Astrophys. J., **421**, 276
Merck, M., Bertsch, D.L., Dingus. B.L., et al., 1996, Astron. Astrophys. Suppl. Ser., **120**, 465
Mereghetti, S., Pellizzoni, A., 1997, in *The Transparent Universe*, Proc. 2nd INTEGRAL Workshop, St. Malo, France, ESA SP-382, 283
Michelson, P.F., 1996, SPIE, **2806**, 31
Muslimov A.G., Tsygan A.I., 1990, Astron. Zh., **67**, 263
Muslimov A.G., Tsygan A.I., 1992, MNRAS **255**, 61
Nel, H.I., Arzoumanian, Z., Bailes, M., et al., 1996, Astrophys. J., **465**, 898
Nolan, P.L., Arzoumanian, Z., Bertsch, D.L. et al., 1993, Astrophys. J., **409**, 697
Pacini, F., 1967, Nature **216**, 567
Pacini, F., 1968, Nature **219**, 145
Pohl, M., Kanbach, G., Hunter, S.D., et al., 1997, Astrophys. J., **491**, 159

Ramanamurthy, P.V., Bertsch, D.L., Dingus. B.L. et al., 1995, Astrophys. J., **447**, L109

Ramanamurthy, P.V., Fichtel, C.E., Kniffen, D.A. et al., 1996, Astrophys. J., **458**, 755

Romani, R.W., 1996, Astrophys. J., **470**, 469

Romani, R.W., Yadigaroglu, I.-A., 1995, Astrophys. J., **438**, 314

Rudak, B., Dyks, J., 1998, MNRAS, **295**, 337

Sturner, S.J., Dermer, C.D., 1996, Astron. Astrophys. Suppl. Ser., **120**, 99

Sturrock, P.A., 1970, Nature **227**, 465

Sturrock, P.A., 1971, Astrophys. J., **164**, 529

Thompson, D.J., Fichtel, C.E., Kniffen, D.A., Ögelman, H.B., 1975, Astrophys. J., **200**, L79

Thompson, D.J., Arzoumanian, Z., Bertsch, D.L. et al., 1992, Nature **359**, 615

Thompson, D.J., Arzoumanian, Z., Bertsch, D.L. et al., 1994 Astrophys. J., **436**, 229

Thompson, D.J., Bailes, M., Bertsch, D.L. et al., 1996, Astrophys. J., **465**, 385

Thompson, D.J., Harding, A.K., Hermsen, W. et al., 1997, Proc. of the 4th Compton Symp., eds. Dermer, C.D. et al., Williamsburg, Va., AIP CP410, 39

Thompson, D.J., Bailes, M., Bertsch, D.L. et al., 1999, Astrophys. J., **516**, 297

Ulmer, M.P., Matz, S.M., Grabelsky, D.A., et al., 1996, Astrophys. J., **448**, 356

Zepka, A., Cordes. J.M., Wasserman, I., et al., 1996, Astrophys. J., **456**, 305

7 X-Ray Binaries as Gamma-Ray Sources

Werner Collmar

7.1 The Nature of X-Ray Binaries

X-ray astronomy, for which measurements above the Earth's atmosphere are necessary as in γ-ray astronomy, is also a relatively new astronomical field, which was opened roughly 40 years ago. A rocket flight in 1962, carrying an X-ray detector above the Earth's atmosphere, resulted in two important discoveries: the detection of an extrasolar X-ray source, and the detection of a bright diffuse X-ray sky, in contrast to a dark optical one (Giacconi et al. 1962). Roughly five years after the discovery of this X-ray source in the constellation Scorpius, which was subsequently called Scorpius X-1 (Sco X-1), it was suggested that this source is a mass-accreting compact object in a binary system. It took another five years, however, until this picture was experimentally proven. Optical spectroscopy of the supergiant HD 226868, the optical counterpart of Cygnus X-1 (the first X-ray source detected in the constellation Cygnus), revealed a sinusoidal 5.6 day variation in radial velocity (Webster and Murdin 1972), which was a convincing sign of orbital motion in a binary system. This discovery immediately revealed a surprise. The mass function derived from the orbital period and amplitude in radial velocity, together with a canonical mass estimate of ≥ 15 $M_\odot$ for the supergiant, implied a mass of the compact object of more than 3 $M_\odot$, the maximal mass limit of a neutron star, and therefore strongly suggested that the compact object was a black hole. So, surprisingly, the first X-ray source with a proven binary nature most likely contains a black hole as the compact object.

X-ray binaries (XRBs) consist generally of a binary-star system with (at least) one component being a compact object at the end of its stellar evolution: a white dwarf, a neutron star, or a black hole. A schematic of an XRB is shown in Fig. 7.1. Their X-ray luminosity is powered by accretion of matter from the companion star onto the compact object. The released gravitational energy is thereby converted to radiation via different physical processes (e.g., heating of matter). The mass flow from the companion can be provided either by a strong stellar wind, which is typical for massive stars ($\gtrsim 10$ $M_\odot$) or by Roche-lobe overflow through the inner Lagrange point, which is typical for low-mass stars. The "Roche surface" is defined as the innermost common equipotential surface of the two binary star components, separating uniquely the volumes of their gravitational influence. Assuming

Fig. 7.1. Artist's conception of an X-ray binary. Matter is transferred from a "normal" companion star via a gas stream and an accretion disk onto a compact object. Due to the angular momentum of the accreted matter the accretions disk is flat. It becomes hotter towards its center, because the energy release per unit infall length increases towards the compact object, which is located in the center of the disk. (From http://universe.gsfc.nasa.gov/videos/old_faithful.html; source: NASA/GSFC; W. Feimer/Allied Signal)

that the gravitational energy of the accreted matter can be totally converted to radiation, the luminosity derived from steady accretion ($L_{\rm acc}$) is given by

$$L_{\rm acc} \approx \left(\frac{GM_{\rm c}}{R_{\rm c}}\right) \dot{M} \; , \tag{7.1}$$

where G is the gravitational constant, M_c and R_c the mass and the radius of the compact object, and $\dot{M}$ the mass accretion rate. By assuming a mass of 1.4 $M_\odot$ and a radius of 10 km, which are typical values for a neutron star, we derive

$$L_{acc} \approx 10^{20} \dot{M}_{\rm g/s} \; ({\rm erg \; s^{-1}}) \; , \tag{7.2}$$

where $\dot{M}_{\rm g/s}$ is the accretion rate measured in units of g s^{-1}. To derive luminosities of the order of 10^{36} to 10^{38} erg s^{-1}, as typically measured in X-rays, an $\dot{M}$ of 10^{16} to 10^{18} g s^{-1} or roughly 10^{-10} to 10^{-8} $M_\odot$ yr^{-1} is necessary.

XRBs are classified in various ways: (a) according to the type of the compact object in black-hole binaries, neutron-star binaries, or so-called cataclysmic variables, in which the compact object is a white dwarf; (b) according to the mass of the donor star in so-called high- and low-mass systems;

(c) into X-ray pulsars or non–pulsing sources, depending on whether they show regular X-ray pulsations or not, which physically reflects the strength of the magnetic field of the compact object; and (d) in transient or persistent sources, depending on whether they are seen episodically or permanently. Recently, a new group – called microquasars – was established: these sources occasionally expel bipolar radio jets with relativistic speeds, resembling the ones of active galaxies (see Chap. 12).

Prior to the Compton Gamma-Ray Observatory (CGRO) era, the soft γ-ray ($\geq$200 keV) and the γ-ray ($\geq$500 keV) emission from XRBs was a field of very limited knowledge. Only one source, Cyg X-1, was convincingly detected at these energies, because several experiments reported evidence for emission at $\sim$1 MeV and above. However, non–detections were reported as well, indicating a possible time-variable nature of the high-energy emission of Cyg X-1. Most notable are the reports of episodic levels of high flux between several hundred keV and $\sim$2 MeV, the so-called MeV bump (e.g., Ling et al. 1987; McConnell et al. 1989). Because Cyg X-1, a black-hole system, was an established γ-ray source, it was natural to search for γ-rays in similar sources as well. Spectral features near 500 keV for several black-hole candidates were observed by the French SIGMA experiment in the early 1990s. They were interpreted as broadened and redshifted annihilation radiation, emerging from the vicinity of the compact object. In general, from extrapolation of the well-defined X-ray spectra, black-hole candidates were considered to be promising objects for CGRO γ-ray observations.

In the very high-energy (VHE, $\sim$TeV photon energies) and ultrahigh-energy (UHE, $\sim$PeV photon energies) γ-ray bands several XRB detections were reported in the 1970s and 1980s, Cyg X-3 being the most prominent of them. However, because the experimental techniques at these high γ-ray energies still faced severe problems in these early years, skepticism was often associated with these results. If true, it was natural to assume that, if such energetic γ-ray photons are produced, lower-energy γ-rays (GeV and MeV photons) should be generated as well, making XRBs promising candidates for CGRO γ-ray observations also from the high-energy point of view.

For the rest of this chapter, I use the following definitions for the energy bands: the soft, medium and hard X-ray bands cover the ranges in photon energy between 0.1 and 2 keV, 2 and 20 keV, and 20 and 200 keV, respectively. The transition region between hard X-rays and γ-rays – 200 keV to 500 keV – is called the soft γ-ray band. Finally, at 500 keV the γ-ray band starts.

7.2 High-Energy Emission from Black-Hole Systems

As mentioned above, one subdivision of XRBs is made according to the nature of the compact object in these systems, i.e., a white dwarf, a neutron star or a black hole. The compact object is considered to be a black hole if its mass is determined or estimated to be in excess of 3 $M_\odot$, the theoretical upper limit for the mass of a stable neutron star according to current equations of state. The most reliable mass estimates for compact objects in binary systems result from measurements of the system dynamics in the optical and infrared bands. Spectroscopic radial velocity measurements of the (visible) companion star provide its orbital period, P_{orb}, its projected orbital velocity, V_{orb} (derived from its measured radial velocity curve), and subsequently the projected radius, $A_{\text{orb}} = P_{\text{orb}} V_{\text{orb}}/2\pi$, of its orbital motion. These values define the so-called *mass function*, $f(M)$, which is given by

$$f(M) \equiv \frac{M_c^3 \sin^3 i}{(M_c + M)^2} = \frac{4\pi^2 A_{\text{orb}}^3}{G P_{\text{orb}}^2} = \frac{V_{\text{orb}}^3 P_{\text{orb}}}{2\pi G} \, , \qquad (7.3)$$

where M_c is the mass of the compact object, M the mass of the companion, i is the inclination angle of the system, and G the gravitational constant. The mass function provides a lower limit on M_c, the mass of the (invisible) compact object, because $\sin^3 i \leq 1$. If M (for example, from spectral-type measurements) and i (for example, from modelling of the optical light curve) can be additionally determined, then, in principle, all system parameters are known, including the mass of the compact object.

To date such dynamical mass determinations have led to reliable masses of the compact object significantly in excess of 3 $M_\odot$ for 11 XRBs. These systems, listed in Table 7.1 together with some of their relevant binary parameters, are the most promising astrophysical objects to contain a stellar-mass-sized black hole. Because no physical reasons for the existence of neutron stars with masses larger than 3 $M_\odot$ exist, I will call these systems "black-hole systems", but keeping in mind that the final proof for their black-hole nature is still missing. In addition to these 11 prime black-hole candidates, there are several other systems which are suspected to host a black hole. Their tentative identification is based on similarities of their observational properties (e.g., time variability, X- and γ-ray spectra) to those of the prime black-hole systems. Examples are the sources GX 339-4, which is classified as a black-hole binary by its spectral behavior, and GRS 1915+105, by its relativistic jet. The presence (neutron stars) or absence (black holes) of a solid surface of the compact object may lead to such observational differences. However, since the emission characteristics and processes of XRBs are currently only partly understood, the black-hole nature of the compact objects in these systems is less secure than that for those with dynamic mass determinations. To avoid confusion I will also call them "black-hole systems" for the rest of the chapter.

The 11 black-hole systems listed in Table 7.1 are all significantly detected up to hard X-ray energies by current X-ray instruments, such as the American Rossi X-ray Timing Explorer (RXTE) or the Italian–Dutch *Beppo*SAX X-ray satellite, providing a wealth of new information (almost on a daily basis) on their emission characteristics in this energy range, resulting in a fast growth of this field. Important new results, with respect to γ-ray emission,

Table 7.1. This table provides a currently (July 1999) complete list of black-hole binary systems with dynamically established masses. Several parameters are given: $f(M)$ is the mass function, i the orbital inclination angle, M_{BH} the estimated black-hole mass, M_{opt} the companion mass, and *Source Type* the X-ray emission type, where P stands for persistent emission, N nova-like transients, and M multiple outbursts (see Sect. 7.2.1). The data in this table, except for Cyg X-1 (Herrero et al. 1995) and 1543-475 (Orosz et al. 1998) are taken from Table 1 of Zhang et al. (1997), which provides the references for the individual sources and source parameters

Black-hole system	Other name	$f(M)$ ($M_\odot$)	i (deg)	M_{BH} ($M_\odot$)	M_{opt} ($M_\odot$)	Source type
Cyg X-1	1956+350	0.252±0.010	~35	~10 (4.8-14.7)	~18	P
LMC X-1	0540-697	2.3±0.3	40–63	4–10	20–25	P
LMC X-3	0538-641	0.14	50–60	7–14	4–8	P
GRO J0422+32	XN Per 1992	1.21±0.06	48±3	3.57±0.34	0.4	N
A 0620-00	XN Mon 1975	2.91±0.08	66±4	4.9–10	0.4–0.7	N
N Muscae '91	GRS 1124-683	3.01±0.15	54–65	5.0–7.5	0.8	N
Nova '71,'83,'92	1543-47	0.22±0.02	20–40	2.7–7.5	2.3–2.6	N
N Sco '94	GRO J1655-40	3.16±0.15	69.5±0.08	7.02±0.22	1.2-1.5	M
N Oph '77	H 1705-250	4.0±0.8	70±10	4.9±1.3	0.7	N
N Vul '88	GS 2000+25	4.97±0.10	65±9	8.5±1.5	0.4–0.9	N
V404 Cyg	GS 2023+338	6.08±0.06	56±2	12.3±0.3	0.9	N

are the detections of different intensity states in soft and medium X-rays (up to ~20 keV), which are called the "very high," "high," "low" and "off" (or "quiescent") states according to their luminosities in this band (e.g., Tanaka 1989; Esin et al. 1997). These states show two characteristic spectral shapes in X-rays: in the very high and high states a soft X-ray spectrum is observed, and in the low state a hard X-ray spectrum is measured. It is currently believed that these states are connected to the mass-accretion rate, $\dot{M}$, onto the black hole; the sequence given above would correspond to a decreasing $\dot{M}$. The observed soft and medium X-ray continuum emission is simply generated by two components which overlap in energy and whose relative intensities can vary by large amounts. The so-called soft component is only visible at energies below ~10 keV and can be described by a blackbody-type emission with a temperature of $kT \sim 1$ keV. This component is generally considered to be the emission from an optically thick, geometrically thin accretion disk. If relativistic effects are neglected, the temperature distribution $T(r)$ of a standard disk is given by (e.g., Frank et al. 1992)

$$T^4(r) = \frac{3GM_c\dot{M}_d}{8\pi\sigma r^3} \, , \tag{7.4}$$

where G is the gravitational constant, M_c the mass of the compact object, $\dot{M}_d$ the mass accretion rate in the accretion disk, σ the Stefan–Boltzmann constant, and r the radial distance to the compact object. This disk component dominates the X-ray spectrum in the very-high and high states. The second component has a simple power-law shape in X-rays and is either visible throughout the whole X-ray band or only above $\sim$10 keV, depending on the absence or presence of the soft component. For further details on the X-ray properties of these sources, I refer the reader to recent reviews, by, for example, Tanaka and Lewin (1995) and van Paradijs (1998), and for multi-wavelength aspects the review by Liang (1998).

At γ-ray energies of about 1 MeV and above only two sources have been detected: the persistent source Cyg X-1 (McConnell et al. 1994) is significantly detected, and the transient X-ray nova GRO J0422+32 (van Dijk et al. 1995) was detected during an X-ray outburst with marginal significance. Cyg X-1 was regularly "seen" by COMPTEL (Compton Telescope) around 1 MeV when it was located favorably within the COMPTEL field of view with a reasonable exposure time ($\geq$2 weeks). Therefore, Cyg X-1 is the major black-hole system for which information on the MeV emission of these sources can be obtained. At energies above 100 MeV no black-hole system has been detected yet by EGRET (Energetic Gamma-Ray Experiment Telescope) or at higher energies by ground-based Cherenkov telescopes or air shower experiments.

7.2.1 Time Variability

The high-energy emission of the black-hole systems is generally found to be variable in flux on all accessible time scales. The long-term (months to years) light curves in hard X-rays (Fig. 7.2), predominantly observed by BATSE (Burst and Transient Source Experiment) via its all-sky monitoring capability, clearly show three different types (see, e.g., Zhang et al. 1997): The first (or sometimes so-called FRED) type shows a *fast* *r*ise (a few days) with a slower *e*xponential *d*ecay (several weeks). Smaller outbursts show the same profile and occur occasionally on top of such a pattern. This type of transient nova-like behavior is shown, for example, by the sources GRO J0422+32 and Nova Muscae '91, which belong to the class called soft X-ray novae or soft X-ray transients. The recurrence times of such outbursts is of the order of years to tens of years. The companion stars in these systems are low-mass stars (<1 $M_\odot$), which provide a low mass-transfer rate. In quiescent states a significant part of the mass is stored in the accretions disk, and the rest is accreted onto the compact object at a low level. An outburst is probably triggered by a thermal instability in the outer accretion disk (Lasota 1997), which leads to a significantly larger accretion rate, and subsequently to an outburst of radiation. In this process the disk is basically emptied and at some time will be refilled. The second type of light curve, which also displays a transient behavior, shows multiple outbursts with roughly equal rise and decay times

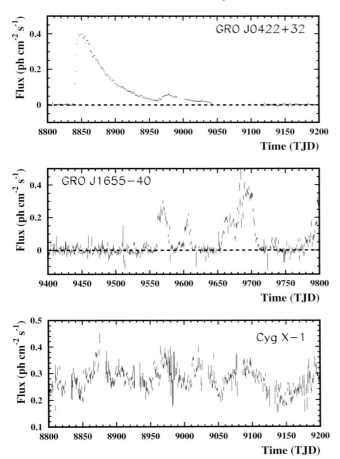

Fig. 7.2. Light-curve types of black-hole binary systems as observed by the BATSE instrument in the 20–100/200 keV hard X-ray band. The type "fast rise with slower exponential decay" (nova-like) is shown by GRO J0422+32 (*upper panel*), the "multiple outbursts" type is shown by GRO J1655-40 (middle panel), and the persistent type is shown by Cyg X-1 (lower panel). The time axis is in units of days (TJD: truncated Julian date). (Data obtained from BATSE Team)

on the order of several weeks to months. Such light curves are shown for example by the relativistic jet sources GRO J1655-40, GRS 1915+105 and GX 339-4. The outburst behavior in these sources is less well understood. These systems are rare and – apart from GRO J1655-40 – knowledge of their companion stars (e.g., masses, spectral types) is sparse. It is assumed that in these systems also the companion supplies matter at a low rate, resulting in a low or undetectable hard X-ray flux. An initial outburst triggers further outbursts via X-ray heating of the companion star, which responds via increased mass transfer (e.g., Hameury et al. 1986; Chen et al. 1993). Subsequently, this leads to another outburst at some later time, which can be considered as an "echo" of a previous one. Such a scenario is only viable for particular system configurations, which is consistent with the fact that these systems are rare. Persistent emission constitutes the third type of light curve. The flux varies significantly but always remains detectable for BATSE, for example Cyg X-1 shows this behavior. The companion stars in these systems are high-mass O or B stars, which permanently maintain a high mass-transfer rate via strong stellar winds or Roche-lobe overflow, which always keeps the sources "on".

In addition to these long-term flux variations (days, weeks, and longer) all black-hole systems show strong and rapid aperiodic variability from time scales of hundreds of seconds down to milliseconds. This variability is usually investigated by Fourier analysis techniques, where the Fourier power is a measure of the variance of a time series around its mean value and is usually expressed as fractional root-mean-square (rms) amplitude, which is the square root of the variance of the intensity variations divided by the average intensity. The power spectra are usually shown as power density versus frequency, where the power is normalized to the square of the fractional rms amplitude divided by the frequency. At X-ray energies such power spectra are structured and seem to depend on spectral states. The continuum can show three components: a plateau of low-frequency noise below about 0.1 Hz, a roughly $1/f$ power-law component at intermediate frequencies, and white (statistical) noise at higher frequencies. Some sources sometimes show quasi-periodic oscillations (QPOs) or "peaked noise" on top of this continuum. For a recent review of the aperiodic variability in XRBs and of the analysis techniques see van der Klis (1995).

Figure 7.3 shows a power spectrum from an OSSE (Oriented Scintillation Spectrometer Experiment) hard X-ray observation of GRO J0422+32 . It is physically very interesting because it shows the same shape for two different energy bands: breaks around 0.05 Hz and around 5 Hz with a peaked noise component at 0.23 Hz. The continuum, apart from the peaked noise component, is modelled by a superposition of randomly occurring accretion events ("shots") which each generate a burst of radiation with a particular duration. A shot model with two typical durations of individual radiation events of ~ 2.2 s and ~ 50 ms could reasonably fit the power spectrum (Fig. 7.3) apart from the peaked noise component (Grove et al. 1998b). The peaked noise component might be triggered by viscous instabilities in the accretion disk or by oscillations of a Comptonizing corona (Grove et al. 1998b). The frequency-dependent time-lags, derived by cross-correlation analysis techniques (e.g., Nowak et al. 1999), show that the hard emission (75–175 keV) lags the softer one (35–60 keV) by ~ 0.3 s at low frequencies and by $\sim 10^{-3}$ s at about 10 Hz, showing a roughly $1/f$ dependence. This "hard lag" can generally be interpreted in the framework of a Comptonized emission scenario, where a source

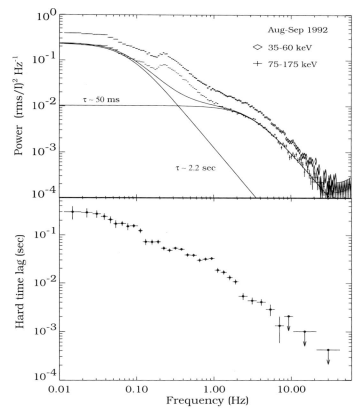

Fig. 7.3. Upper panel: Power density spectra of GRO J0422+32 in the ~35–60 keV and 75–175 keV bands observed by OSSE. Both energy bands show the same shape. The contribution to the power density spectrum of a two-component shot model with two time constants is also given. Lower panel: Time-lag spectrum between the two energy bands. The higher-energy photons lag those at lower energy in time showing, however, an obvious frequency dependence. [From Grove et al. (1994) reproduced with permission from The American Institute of Physics]

of soft photons is embedded in a cloud of hot electrons. If a photon population is generated in an accretion event, and the photons gain energy by scattering on energetic electrons, the emerging higher-energy photons should – on average – have experienced more scatterings than the lower-energy ones and therefore should lag them in time. However, the frequency dependence of the time-lags is in disagreement with standard Comptonization models, for which a uniform (frequency-independent) time shift is expected (e.g., Miyamoto et al. 1988). Recent modelling has shown that the frequency-dependent hard time-lags can still be explained by Comptonization; however, the picture of a uniform homogeneous Comptonizing region has to be replaced by a nonuniform one, for example, with a gradient in the density of the Comptonizing electrons (e.g., Kazanas et al. 1997; Böttcher and Liang 1998). The time-lag spectrum of Fig. 7.3 is interpreted in this sense (Grove et al. 1998b).

The physical importance of studying the rapid (≤ 100 s) aperiodic variability is that it reflects the dynamics of the accretion process, i.e., the intensity variations should somehow trace the matter infall onto the compact object. Because most of the radiation is generated in the vicinity of the compact object, the rapid aperiodic variability provides information on the accretion physics close to compact object, i.e., on the accretion flow in the innermost accretion disk. Thereby it might provide clues on the density structure of the accreted gas. Information on the properties of the compact object (e.g., spin, mass, magnetic moment) might be revealed as well, because this innermost accretion flow is interacting with the object. In particular it is hoped that the observed rapid aperiodic variability will become a tool for distinguishing between neutron-star and black-hole systems. However, currently no commonly accepted interpretation of the continua of power density spectra and their correlation with energy ("time-lags") exists. The long-term light curves trace global accretion process parameters like accretion stabilities/instabilities, the mass-transfer and mass-accretion rate (e.g., when no matter is accreted the sources are in the off state), and maybe the disk geometry (e.g., flat or warped disk).

Hardly anything is known about the time variability of black-hole systems at γ-ray energies, because at ~ 1 MeV and above only two such systems have been detected so far. While GRO J0422+32 was detected only once (between 1 and 2 MeV) during an X-ray outburst, when the X-ray flux was at its maximum (van Dijk et al. 1995), Cyg X-1 is generally detectable by COMPTEL in the energy band 0.75–2 MeV, with no significant evidence for variability. Clearly, more sensitive measurements are needed to improve our knowledge in this respect, and these must await next-generation instruments.

7.2.2 Energy Spectra

As mentioned above, X-ray measurements (≤ 10 keV) have revealed different intensity states of black-hole XRBs, which show two distinct major spectral shapes: a hard and a soft shape which are usually referred to as the X-ray hard and soft states. During recent years, the CGRO experiments BATSE and OSSE have found that these characteristic spectral shapes have their counterparts at hard X-ray and soft γ-ray energies (~ 20 to ~ 500 keV) (e.g., Grove et al. 1998a).

In the hard state, the X-ray hard power-law shape with a photon index of about 1.7 continues up to energies of ~ 50 keV and then cuts off exponentially with an e-folding of the order of 150 keV. In this spectral state the bulk of energy is emitted around 100 keV, i. e. the luminosity in the hard X-ray band is about a factor 5 larger than in the X-ray range (< 10 keV). In the soft state, the soft power-law shape, which becomes visible above ~ 10 keV,

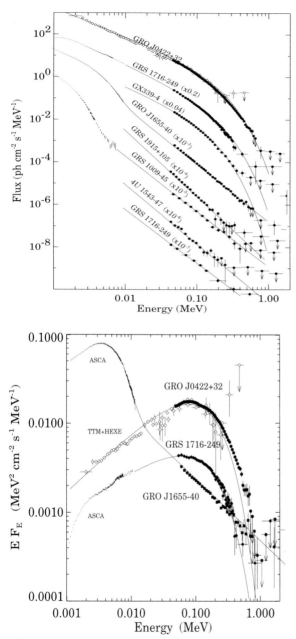

Fig. 7.4. *Upper panel*: Differential photon spectra as measured by OSSE for seven black-hole systems. The two different spectral states at X- and γ-ray energies are visible in various sources. The source GRS 1716-249 is shown in both states. The two sources GRO J1655-40 and GRS 1009-45 are in the X-ray soft state. They clearly show the soft X-ray excess below $\sim 10\,\mathrm{keV}$. *Lower panel*: ν-F_ν-type spectra ($E^2 \times$ differential photon flux) of three black-hole systems. In this representation the differences between the two spectral shapes are more obvious. [Reproduced with permission from Grove et al. (1998a)]

continues towards the γ-ray band without any obvious cut off below energies of about 0.5 MeV. Its shape is described well by a single photon index of typically ∼2.5 (e.g., Grove et al. 1998a) up to the sensitivity limits of the detectors. The spectrum below ∼10 keV is dominated by a strong soft excess above the high-energy power-law component, which mimics a soft spectrum at X-ray energies. Due to this soft excess, the X-rays dominate the system luminosity in this spectral state. Examples of X- to γ-ray spectra in the two different spectral states for several black-hole systems are shown in Fig. 7.4, and generic spectra for the two spectral states are given in Fig. 7.7.

Prior to CGRO, transient line – or line-like hump – features were reported from several sources by different experiments. For example, a broad "bump" around 1 MeV occurring occasionally in the spectrum of Cyg X-1 (e.g., Ling et al. 1987; McConnell et al. 1989) has been reported, as well as several line features near 511 keV (Goldwurm et al. 1992; Sunyaev et al. 1992) in various other systems (e.g., in Nova Muscae 1991). The latter features were interpreted as redshifted and split (due to disk rotation) annihilation lines. None of the CGRO experiments, however, despite their significantly improved sensitivities at these energies and their large exposure times on these sources have yet detected any line feature, broad or narrow, transient or steady, from any of these systems. In one case, simultaneous measurements are even contradictory. The SIGMA experiment reported a broad-line excess from the black-hole candidate 1E 1740.7-2942 in 1992 (Cordier et al. 1993). OSSE, however, observing simultaneously the same sky region by chance and being roughly a factor of 10 more sensitive at these energies, could not confirm this event (Jung et al. 1995). So the current status is that no transient line or bump feature is unambiguously detected in hard X-ray, soft γ-ray or γ-ray energies in black-hole systems, which is unfortunate, because spectral features are generally very helpful in revealing the underlying physics.

At MeV energies, only Cyg X-1 has been detected significantly. Its time-averaged spectrum shows significant emission out to several MeV (McConnell et al. 1999). If compared to contemporaneous BATSE and OSSE hard X-ray and soft γ-ray measurements and to their spectral model fits, it provides clear evidence for a high-energy (MeV) tail (Fig. 7.5). The MeV spectrum (0.75–5 MeV) is consistent with a power-law shape (photon index: ∼3), which shows no evidence for a cutoff at higher energies. Due to its significant MeV detection, Cyg X-1 is a promising candidate for providing physical insights on the emission processes at these energies. The single detection of GRO J0422+32 between 1 and 2 MeV was made near the COMPTEL detection threshold. Although no spectral details can be inferred convincingly, the MeV flux is above the extrapolation of the spectral models fitting well at hard X-ray and soft γ-ray energies, resembling the hard MeV tail of Cyg X-1. In particular, this measurement proves that other black-hole systems can also emit MeV photons at a detectable flux level. The relationship of these source detections to the spectral source states discussed above is still unexplored. It should be

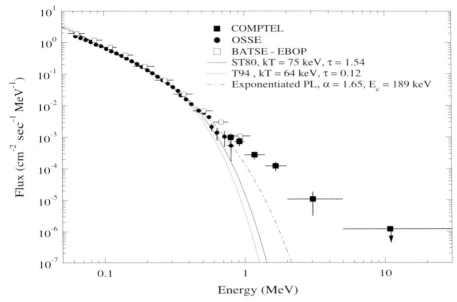

Fig. 7.5. Time-averaged hard X-ray to γ-ray spectrum of Cyg X-1. The spectrum was compiled for times when Cyg X-1 was in the hard X-ray state. Significant MeV emission up to ∼5 MeV is clearly seen and is well above the various Comptonization models which represent the spectrum well at lower energies. [From McConnell et al. (1999) reproduced with permission from Proc. 26th ICRC, Salt Lake City, Utah (1999) and M. McConnell]

noted, that the MeV spectrum of Cyg X-1 (Fig. 7.5) is determined completely from the X-ray hard state, where Cyg X-1 spends most (∼90%) of its time.

7.2.3 Interpretation

Recent CGRO hard X-ray and soft γ-ray observations of a number of black-hole X-ray binaries have revealed two distinct different spectral states in hard X- and soft γ-rays, which correspond to the hard and soft spectral states in X-rays. The generic shapes are sketched in Fig. 7.7. In general it is believed that the different states are the result of different accretion rates which trigger different emission geometries and mechanisms. Currently very popular and promising is a scenario which considers an interplay between a cold optically thick, geometrically thin outer accretion disk with a standard accretion flow where radiation cooling is efficient and a much hotter optically thin, geometrically thick inner accretion disk with an advection-dominated accretion flow (ADAF), where radiative cooling is inefficient and the bulk of the released gravitational energy is carried across the event horizon of the black hole (e.g., Narayan and Yi 1995). The transition point between these two types of accretion disks or accretion flows depends on the mass-transfer rate in the disk. For a high rate, the optically thick, geometrically thin disk reaches close to the black hole, and only a small (if any) inner hot disk exists. This configuration provides the abundant blackbody photons and therefore is connected to the X-ray soft state. At small accretion rates the accretion flow is different: the optically thick disk stops further away from the black hole, and further in the inward accretion flow becomes advection dominated, resulting in a "dim" inner disk. This configuration provides less blackbody photons and therefore is connected to the X-ray hard state.

The spectrum corresponding to the X-ray hard state is modelled above 10 keV with two components: a power-law component ($\alpha \sim$ 1.6–1.7) which cuts off at $\sim$150 keV and a Compton-reflection component on top of this which significantly alters the spectral shape in total above $\sim$10 keV (Fig. 7.6). The origin of this exponentially truncated power-law component, which is thought to be the primarily generated X-ray emission, is interpreted as thermal Comptonization of soft seed photons in a hot Comptonizing plasma. The reflection component is modelled with a solid angle Ω of the reflector, which is found to be only 30% to 40% of 2π (e.g., Gierlinski et al. 1997), and therefore rules out a simple, homogeneous slab-like accretion disk corona, for which 2π would be expected. Within the framework of the thermal Comptonization and reflection model, a geometry seems possible in which a geometrically thick, optically thin hot inner disk is surrounded by a cold outer disk. The irradiation of

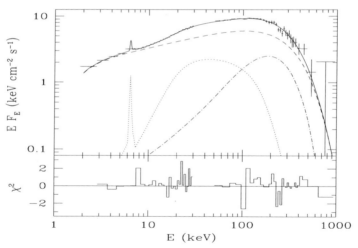

Fig. 7.6. Simultaneous broadband spectrum of Cyg X-1 observed by the Japanese X-ray satellite ASCA and OSSE together with the currently favored model. The individual model components (*dashed and dot-dashed lines*: primary and secondary Comptonization components, respectively; *dotted line*: iron line and Compton-reflection component of primary Comptonization) as well as their sum (*solid line*) are shown. [Reproduced with permission from Gierlinski et al. (1997)]

the cold outer disk by the hot inner one would generate a Compton-reflection component with a small Ω. A sketch of the anticipated emission scenario is shown in Fig. 7.7.

The X-ray soft state shows a strong soft component, displaying a blackbody shape with $kT_{bb} \sim 1$ keV, and above ~ 10 keV a soft power-law shape (index $\alpha \sim 2.5$), which reaches up to the detection thresholds of current instruments without any cutoff. There is a general consensus that the soft component is radiated from the innermost part of the optically thick and geometrically thin accretion disk, which – in this state with an assumed

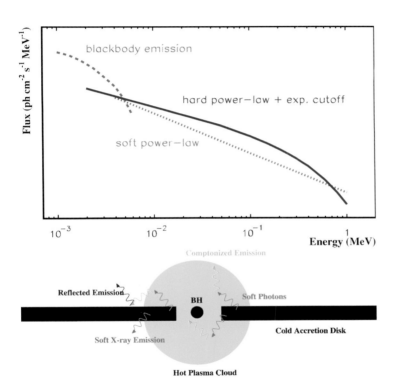

Fig. 7.7. *Upper panel*: Generic differential high-energy spectra of black-hole systems. The different X-ray states are indicated by color (soft state: *red*; hard state: *blue*), and their major emission components are shown. *Lower panel*: Sketch of a possible plasma geometry of the central region of a black-hole XRB, which shows the major emission processes and components. The plasma geometries in the different states are probably different and depend on the mass-accretion rate. The interpretation is that in the soft state (high accretion rate) the optically thick and cold accretion disk reaches much closer to the black hole than in the hard state. As a consequence of that the strong blackbody emission which presumably is radiated at the innermost accretion disk is visible in the soft state and is absent in the hard state

large $\dot{M}$ – reaches close to the black hole and therefore becomes very hot and luminous in its innermost region, where it subsequently radiates a blackbody spectrum. The power-law component is currently less well understood. Several explanations have been proposed but no consensus has yet been reached. The most viable models all assume some kind of Comptonization of the abundant soft blackbody accretion-disk photons by "hot" (high-energy) electrons. These models basically differ with respect to the energy distribution of the electrons and their geometry/location in the system. The following models are proposed: (a) standard thermal Comptonization; (b) bulk motion Comptonization in which a quasispherical accretion flow inside a minimum stable orbit is assumed, which in the vicinity of the black hole Comptonizes the blackbody photons (e.g., Ebisawa et al. 1996); and (c) nonthermal Comptonization, which assumes a population of nonthermal electrons (e.g., accelerated in an accretion disk corona to relativistic energies by magnetic reconnection events), which finally provide the soft power-law spectra without any cutoff (e.g., Poutanen and Coppi 1998). Sensitive spectral measurements at γ-ray energies, showing either a continuation of these power-law shapes well into the MeV band or a cutoff at some energy, have the potential to discriminate among these models, which make clear predictions for the high-energy part of the spectra. For example, the bulk Comptonization models predict a spectral cutoff around 500 keV, near the electron rest energy ($m_e c^2$), which could be proven or disproven by accurate γ-ray measurements.

From Fig. 7.5 it is obvious that the MeV emission of Cyg X-1 is not explicable by the thermal Comptonization models valid for the X-ray, hard X-ray, and soft γ-ray emissions. Those emissions predict a fall off which is too rapid at energies around ~ 1 MeV. No uniquely accepted explanation for this MeV tail exists yet. A nonthermal emission component, indicating particle acceleration somewhere in the system, or a second higher-temperature Comptonizing plasma, are currently favored as the origin. During its outburst in August 1992, GRO J0422+32 was detected up to ~ 2 MeV. Like Cyg X-1, its MeV flux was significantly above the extrapolation of the thermal Comptonization models which fit well at lower energies. The spectral similarities to Cyg X-1 suggest the same – yet unknown – emission mechanism. In any case both measurements point to the fact that more than simple thermal Comptonization is at work in these sources.

In summary, the CGRO measurements have provided real progress in our observational and physical understanding of these black-hole systems by extending their X-ray spectra towards significantly higher energies. For one source, Cyg X-1, the spectrum extends up to several MeV. Up to the soft γ-ray band, a two-state emission scheme has been found which in principle is physically understood; however, most details are still missing. At MeV energies we now have the same situation as below ~ 500 keV about 10 years ago. One source – Cyg X-1 – is significantly detected by COMPTEL, and there is evidence for (at least) one more with, however, marginal detection significance. This situation generates a "picture" which is dominated by an individual source. Only if Cyg X-1 is a "prototype" source is more physical insight achievable about the MeV emission of black-hole systems with the current instrumentation. Cyg X-1 may be such a prototype source, because it has many characteristics in common with other black-hole systems, and therefore it is reasonable to presume that this is the case for the MeV emission as well. This presumption is supported by the fact that the same high-energy tail seems to be present in GRO J0422+32.

7.3 High-Energy Emission from Neutron-Star Binaries and Cataclysmic Variables

Roughly 30 neutron-star XRBs have been detected in the hard X-ray range at energies above 20 keV. With the exception of some peculiar cases, which will be discussed in Sect. 7.4, no such source is detected at energies above 200 keV. This clearly shows that neutron-star XRBs are not γ-ray emitters, at least not for current instruments. Therefore I only briefly summarize their main soft and hard X-ray properties and briefly mention their emission scenarios. For more details I refer the reader to the relevant X-ray literature (e.g., Tavani and Barrett 1997; van Paradijs 1998, and references therein).

Two distinct spectral shapes are observed in neutron-star XRBs at X-ray energies, which are clearly related to two different types of objects. Typical spectra are shown in Fig. 7.8, where they are compared to the spectrum of a black-hole system. A soft spectrum, showing the signature of thermal bremsstrahlung with a plasma temperature of typically ∼5 keV, is observed from low-mass X-ray binaries (LMXRBs). Figure 7.8 shows a spectrum of Scorpius X-1, which is a member of this class, and which in fact was the first extrasolar X-ray source detected (see Sect. 7.1). LMXRBs have a low-mass (M < 1 $M_\odot$) companion star. These objects are usually not detected at energies above 20–30 keV. In contrast to the soft spectra of LMXRBs, X-ray binary pulsars emit hard power-law spectra (α_{ph} ∼1.5) with a rapid cut-off at energies above ∼20–30 keV. These two different spectral shapes are clearly related to the strength of the magnetic field of the neutron star, which guides the accretion flow in its vicinity. Because regular X-ray pulsations are not usually observed in LMXRBs (only a few exceptions exist, which then show the properties of X-ray pulsars), their neutron stars have "weak" magnetic fields (∼10^9 G), which means that the matter accretes onto a large portion of their surface. The observed soft X-ray emission is thought to come from a boundary layer of matter between the neutron-star surface and the inner accretion disk, and from the inner accretion disk itself. It shows the typical features of free-free emission of electrons in a hot plasma. Very recently hard power-law tails, not physically connected to the soft emission, with exponential cutoffs at 50–70 keV have been observed by the BATSE, OSSE, and SIGMA experiments from about 10 LMXRBs (e.g., Tavani and Barrett

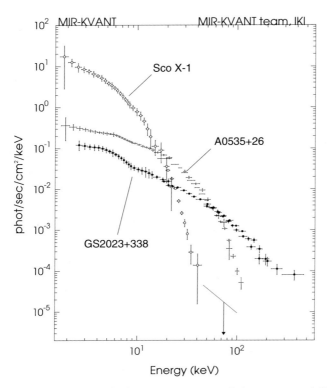

Fig. 7.8. Typical high-energy spectra of three types of Galactic XRBs with different compact objects. The LMXRB Sco X-1, the brightest persistent source in soft X-rays, contains a neutron star with a weak magnetic field, the X-ray pulsar A 0535+26 a neutron star with a strong field, and GS 2023+338 (most likely) a black hole. Only the black-hole system maintains its emission into the soft γ-ray range. (From Gilfanov et al. 1995, reproduced with kind permission of Kluwer Academic Publishers)

1997, and references therein). These tails occur typically in anticorrelation with the X-ray intensity, i.e., they are only observed during periods of low X-ray intensity. This emission is currently discussed in the framework of thermal Comptonization of soft photons in a hot plasma (e.g., an accretion disk corona) or in the framework of nonthermal models which consider particle acceleration by transient electric fields in the magnetosphere of the neutron star. In contrast to nonpulsing LMXRBs, neutron stars in XRB pulsars have strong magnetic fields ($\sim 10^{12}$ G) at their surface. It is generally believed that the strength of the neutron-star magnetic field is connected to its age, i.e., neutron stars in XRB pulsars are considered to be young objects, while those in LMXRBs are old objects. The strong magnetic field is able to channel the accreted matter near the neutron-star surface into one or several accretion columns. The matter then accretes only onto a small portion of the neutron star. The bottom of such an accretion column has a size of about $1\,\mathrm{km}^2$. The hard power-law type X-ray emission with its exponential cut-off is thought to be the consequence of shock acceleration of matter in such an accretion column.

Cataclysmic variables, binary systems with a white dwarf as compact object, are not known to emit at γ-ray energies. The gravitational energy release from mass accretion onto a white dwarf near its surface is not sufficient to yield plasma temperatures hot enough to generate continuum emission reaching up to γ-rays. If such a system were to be detected at γ-ray energies, this emission would rather be connected to nonthermal processes occurring, for example, in the magnetosphere of a magnetic white dwarf than to mass accretion onto its surface.

7.4 Peculiar Cases

7.4.1 Cygnus X-3

Cygnus X-3 (Cyg X-3) is one of the brightest X-ray sources in the sky and shows a roughly sinusoidal 4.8 h modulation of its X-ray flux (e.g., Parsignault et al. 1972). Due to the high stability of this periodicity, it is interpreted as the orbital period of a close binary system. The exact nature of the compact object, neutron star or black hole as well as the nature of the companion star is still unclear. Despite the fact that Cyg X-3 has been observed for about 30 years now, it remains an enigmatic source. After a series of intense radio bursts from the system in 1973 the source was observed worldwide throughout the complete electromagnetic spectrum from radio to high-energy γ-rays. These observations resulted in claims for its detection at γ-ray energies and, subsequently, Cyg X-3 received much attention. The source was reported to emit TeV and even PeV γ-rays which showed the characteristic 4.8 h periodicity. One group claimed to detect a 12.59 ms periodicity during some data stretches, which was interpreted to be the rotational period of a neutron star. However, other groups did not detect the source at all in these high-energy γ-rays, leaving an inconsistent picture. For more details on the TeV and PeV emission of Cyg X-3 see Sect. 7.5. The first γ-ray satellite, SAS-2, also reported detection at a significance level of 4.5σ in the energy band 35 MeV to 100 MeV in 1973 (Lamb et al. 1977). A flux value of the order of 10^{-5} photons $\mathrm{cm}^{-2}\,\mathrm{s}^{-1}$ was derived, showing evidence of a 4.8 h periodicity, which confirmed the association with Cyg X-3. Between 1976 and 1982 the COS-B satellite observed the source several times, covering roughly the same energy band as SAS-2. A positive signal was however never found (Hermsen et al. 1987), leaving an inconsistent picture also at these lower γ-rays. The question of whether the γ-ray emission of Cyg X-3 is time variable or whether the source was in fact never a γ-ray emitter made it an important target for CGRO.

OSSE, on board CGRO, observed the source several times in the hard X-ray band at energies above 50 keV. Cyg X-3 is always significantly detected up to energies of ~200 keV, showing a soft power-law spectrum with photon indices between 2.2 and 3.0, depending on source intensity (Matz et al. 1994). This soft spectrum together with the observed flux makes the source undetectable above ~250 keV for OSSE. The 4.8 h periodicity is always clearly detected, and no major hard X-ray flares have been reported so far. Generally, the hard X-ray emission observed by OSSE is the high-energy tail of the prominent and well-known emission in the soft and medium X-ray band. COMPTEL, at energies around 1 MeV and above, also observed the Cygnus region several times. Despite a major search in the COMPTEL data of the first 3.5 years of its mission, no hint at all of Cyg X-3 could be found (Collmar et al. 1995). Above 100 MeV, EGRET also observed the Cygnus region several times. An analysis of the first 3.5 years of EGRET Cygnus-region data revealed four significant γ-ray sources. One of them is positionally coincident with Cyg X-3. This source has a time-averaged flux of $(8.2\pm0.9)\times10^{-7}$ photons cm^{-2} s^{-1} (E > 100 MeV) and a spectral power-law shape with photon index of 1.9$\pm$0.1 (Mori et al. 1997). The probability for a chance coincidence is estimated to be of the order of 5%. However, no periodicity at 4.8 h has been found, which would have definitely confirmed the association of this γ-ray source with Cyg X-3. A search for the 12.59 ms period was also negative. If this EGRET γ-ray source is the counterpart of Cyg X-3, its observed emission is much too high to be the continuation of the spectrum as observed in hard X-rays by OSSE; instead it would have to be a different spectral component related to a different emission process. Therefore, despite several CGRO observations of Cyg X-3, the question of whether this enigmatic binary X-ray source is a γ-ray emitter still remains.

7.4.2 Centaurus X-3

Centaurus X-3 (Cen X-3) is a prominent binary X-ray pulsar. Roughly 30 years ago, it was in fact the first object for which coherent X-ray pulsations had been detected (Giacconi et al. 1971), which put Cen X-3 in a key position for our understanding of the class of accreting binary pulsars. The Cen X-3 system consists of a neutron star which rotates in ~4.8 s around its axis and which orbits an O-type supergiant in ~2.1 days. Controversial reports about possible detections and nondetections at TeV γ-rays by ground-based Cherenkov telescopes in the 1970s and 1980s generated a situation in which it was a matter of personal opinion whether this object is occasionally a source of high-energy γ-rays or not.

Recently, during a 12-day observation of the Centaurus sky region in 1994, the EGRET telescope found a temporary 5σ γ-ray source which was positionally coincident with Cen X-3 (Vestrand et al. 1997). Moreover, timing analysis showed clear evidence for a 4.8 s modulation of the γ-rays, which provided compelling evidence for its identification with Cen X-3. The flux was calculated to be of the order of 10^{-6} photons cm^{-2} s^{-1} at energies above 100 MeV, and a hard power-law spectrum between 70 MeV and 10 GeV with a photon index of 1.8±0.4 was measured. No significant modulation with respect to the 2.1-day binary orbit was found, although an excess of γ-ray events near orbital phase 0.05 was noted. Analyses of further EGRET observations of Cen X-2 covering roughly 5 years of data did not reveal any other significant detections. Only one other observation showed evidence for a γ-ray source positionally consistent with Cen X-3 at the 2.3σ level. The derived upper flux limits together with the significant detection show that the γ-ray emission is time variable on time scales less than a few months.

This observation, in fact, provided the first convincing evidence that neutron-star XRBs can also be the source of significant γ-ray emission on occasions. This in turn means, that, at least sporadically, particles are accelerated to GeV energies in such binary systems, because the observed γ-rays are not simply the high-energy tail of the X-ray emission. An extrapolation of the measured X-ray spectra falls far below the measured γ-ray fluxes. An estimation of the energetics shows that the particle acceleration scenario is possible, because less than $\sim$10% of the total accretion power is needed to generate the observed γ-ray fluxes. More details on the emission scenarios for such γ-rays are given in the next section.

7.5 TeV Observations of X-Ray Binaries

In the 1970s and 1980s XRBs were prominent objects in the VHE ($\sim$0.3 to 50 TeV) and UHE (50 TeV to 100 PeV) γ-ray bands of ground-based γ-ray astronomy. About 10 XRBs, binary pulsars (being the majority) as well as LMXRBs, and two cataclysmic variables (AE Aquarii and AM Hercules) had been reported to emit such high-energy γ-rays, which gave a tremendous "boost" to this field. For a review of these observations see, for example, Weekes (1992) and references therein. The most prominent representatives were the enigmatic object Cyg X-3 and the famous binary pulsar Hercules X-1 (Her X-1).

For Cyg X-3, a close binary system with an orbital period of 4.8 h for which the nature of the two stellar components is still unknown (for more details see Sect. 7.4.1), TeV emission at orbital phases of 0.1–0.2 and 0.6–0.8 has been reported by several Cherenkov telescopes. In addition, PeV emission has been claimed to have been observed several times by different air shower arrays clustering around orbital phase 0.6. In contrast to Cyg X-3, the physical nature of Her X-1 is well known: a 2.2 M$_\odot$ A9 star is orbited in about 1.7 days by a 1.4 M$_\odot$ neutron star which has a magnetic field of several 10^{12} G (Trümper et al. 1978) and which rotates in 1.24 s around its axis. Many episodes of TeV and PeV flaring have been reported which lasted between a few and roughly 100 minutes. The outbursts did not show a clear correlation with any binary-system parameters and occurred at all orbital phases.

In the 1990s the detector technology for these energies improved significantly. This is especially true for the Cherenkov telescopes, which measure in the VHE band around TeV energies. These improvements were – at least in part – stimulated by the detection of many γ-ray sources at MeV/GeV energies by the EGRET telescope aboard CGRO, which proved that there are many celestial sources radiating γ-rays at a detectable level. These new and much more sensitive instruments however, could not confirm any of the TeV/PeV XRB γ-ray sources claimed in previous decades. Either long-term intensity trends in different sources occur or false source detections, probably due to unknown background contributions, had been reported. Currently the situation is unclear and the question of whether XRBs are high-energy γ-ray emitters or not is still an open issue. Therefore these high-energy γ-ray source candidates are still monitored on a regular basis. Measurements provided hints, although not statistically convincing (2.5σ effects), for emission from Vela X-1, a binary pulsar, and AE Aquarii, a cataclysmic variable. However in 1997, a significant detection (6.4σ) of the X-ray pulsar Cen X-3 was reported for energies above 400 GeV and at orbital binary phases near 0.25 and 0.75 of the 2.1-day orbital period by the Durham Mark-6 experiment (Chadwick et al. 1998). In contrast to earlier measurements, where flaring episodes have been reported from this as well as from other sources, the detected flux level appeared to be stable over a period of 2 to 3 months. No evidence for any pulsar periodicity was found, implying an emission site either near the neutron star or somewhere out in the accretion disk, but not on the neutron star itself. This result indicates that the observed radiation must originate in an extended volume of the system. After all the nondetections in the early 1990s, this detection is in fact very promising for the whole field and may lead to an era where, apart from flares, high-energy γ-ray DC emission for these sources can also be studied. However I would like to note that this detection has not yet been confirmed by any VHE telescope, so this astronomical field "contains" currently (July 1999) only one source, which was detected by only one single instrument so far.

General scenarios for the generation of VHE and UHE γ-rays in these binary systems exist. They are thought to be the result of the decay of energetic π^0, which are produced by interactions of accelerated protons (or nuclei) with matter (pp reactions) or background radiation (pγ reactions). Because the decay modes and branching ratios of all reactions involved are well known, in principle one can compute the resulting γ-ray spectrum. The basic idea is that the protons (or nuclei) are accelerated somewhere in the vicinity of the neutron star – probably by forces originating in its magnetosphere – and on interacting with a target (e.g., the atmosphere of the companion star, matter of the accretion disk, or a thin photon field) produce π^0, which in turn decay into two γ-rays. For details of this emission process see Chaps. 2

and 9. To derive a detailed picture of the geometry and energy spectrum, for example, detailed measurements (e.g., γ-ray emission pattern with respect to the orbital phase) would be necessary. Many such detailed models had been developed for the abundance of early measurements. However, at the time of writing, only one convincing measurement is available, which has not yet been modelled in detail.

7.6 Summary and Conclusions

Prior to CGRO only a few measurements of XRBs between $\sim$100 keV and MeV/GeV γ-rays existed. Most of them resulted from short-term balloon flights of the order of several hours, which provided often inconclusive or even controversial results. Therefore our knowledge about the high-energy emission of XRBs was governed by single sources and individual reports, and hence was very limited. Recent observations of XRBs in this high-energy band, predominantly carried out by the more sensitive CGRO instruments, resulted in a significant improvement of our phenomenological knowledge and also in our understanding of the high-energy emission of these sources. Due to the many observations, long exposures, and the multifrequency capability provided by this satellite mission, a more clear, complete and consistent picture begins to emerge. Instead of the one or two sources previously known, several are measured now up to $\sim$500 keV and a few even up to 1 MeV and above.

It is clearly found that the systems which most likely host black holes as compact objects have harder spectra than the neutron-star systems and thereby maintain their emission throughout hard X-rays well into the soft γ-ray band, Cyg X-1 even significantly into the γ-ray band. In hard X-rays and soft γ-rays these black-hole systems show two distinct spectral states, which are connected to the spectral states in X-rays. In one state (sometimes called the "breaking state"), which is related to the X-ray hard state, the hard X-ray power-law spectrum ($\alpha_{ph} \sim 1.7$) cuts off in the hard X-ray band around $\sim$100 keV. This shape is generally explained by Comptonization of soft accretion-disk photons in a hot thermal plasma, although the exact emission geometry is still debated. Significant γ-ray emission in this hard X-ray state is only observed from Cyg X-1. The γ-ray spectrum then shows a power-law shape up to several MeV without any obvious cut-off at the high-energy end and is significantly above the extrapolation of standard thermal Comptonization models, which fit well at lower energies (Fig. 7.5). This suggests a different emission scenario, where either a second and higher-temperature plasma or nonthermal processes are involved. If the latter one is valid and Cyg X-1 can be considered as a prototype source, then a transition from a thermally dominated to a nonthermally dominated emission scenario would occur in XRBs around $\sim$1 MeV.

In the second state (sometimes called the "power-law state"), which is connected to the X-ray soft state, the hard X-ray spectrum maintains its soft power-law shape ($\alpha_{ph} \sim 2.5$) up to the visibility limit for current detectors, in several cases even up to 500 keV and above. This soft power-law shape is less well understood theoretically. It is still debated whether thermal and/or nonthermal processes are involved and, subsequently, no consensus about the emission geometries has been reached so far. Future accurate measurements of the spectral shape between $\sim$500 keV and several MeV will provide crucial information on this spectral state, in particular to answer the questions about the relationship between thermal and nonthermal processes.

Neutron-star XRBs are typically *not* γ-ray emitters, at least not from the point of view of current γ-ray detectors. Their spectra, resulting from the accretion process, cut off at a few tens of keV at most and therefore their emission becomes "invisible" for current detectors in the 100–200 keV range. However, with the detection of Cen X-3 by EGRET, first evidence has been found that these objects can be, at least sporadic, emitters of GeV γ-rays. Ground-based instruments for the detection of VHE and UHE γ-rays have continuously improved their sensitivities during the last decade. Unfortunately these improved instruments of the 1990s have not confirmed the many source claims from the 1970s and 1980s, providing doubts about those earlier reports. However in 1997 the first such source, the MeV/GeV source Cen X-3, was convincingly redetected. This result provides some confidence in these earlier reports and also supports the hope of a reopening of this high-energy field for XRBs in the near future. Such γ-ray measurements are especially important to study the sites and mechanisms of particle acceleration near the neutron star in the binary system and perhaps the emission and plasma geometries of these systems.

Further improvements in our knowledge of the γ-ray emission of XRBs can be expected in the forseeable future. The upcoming INTEGRAL (International Gamma-Ray Astrophysics Laboratory) mission will provide improved measurements from the hard X-ray to the γ-ray band, i.e., from $\sim$20 keV to $\sim$8 MeV. They will allow us to study the accretion process, especially in black-hole systems, in more detail. INTEGRAL will also provide improved sensitivity for γ-ray lines. The detection of a γ-ray line in such a source would be a very powerful diagnostic tool to study the emission processes at work. The GLAST (Gamma-Ray Large Area Telecope) mission, covering the energy range between $\sim$15 MeV and $\sim$300 GeV with significantly improved sensitivity and enlarged field of view with respect to EGRET, clearly offers the possibility of detecting further sources like Cen X-3. The prospects are very promising for the future in the VHE/UHE γ-ray domain. Because these ground-based instruments can be improved continuously, instruments with improved sensitivities, lower threshold in energy, and larger fields of view are already under construction (e.g., VERITAS (Very Energetic Radiation Imaging Telescope Array System) of the Whipple collaboration and

MAGIC (Major Atmospheric Gamma Imaging Cherenkov telescope) of the HEGRA [High-Energy Gamma-Ray Astronomy] collaboration). Several significant source detections and detailed measurements are expected in the GLAST and VHE γ-ray bands in the near future. If really achieved they will most likely provide new insights into the physics of particle acceleration, the conversion of the particle energy into radiation, and, if some orbital dependence of the emission or other details is detected, also the source geometries, i.e., the distribution of matter, in these systems.

Generally, XRBs provide natural laboratories for studying fundamental properties of matter in extreme physical conditions which are unattainable on earth by man-made machines. Such conditions are the strong magnetic fields in the vicinity of neutron stars, the strong gravitational fields in the vicinity of neutron stars and black holes, the large amount of mass transferred from a "normal" companion star to the compact object via the accretion process, the generation and acceleration of strong – in some cases even relativistic – jets, and the compact object itself, which – in the form of a neutron star or black hole – contains matter in states which cannot be achieved in our laboratories. Moreover, the detection of high-energy γ-rays from neutron-star binaries provided strong evidence that such fast rotating degenerate stars are powerful accelerators of particles up to energies well above those of terrestrial accelerators.

References

Böttcher, M., Liang, E.P., 1998, Astrophys. J. **506**, 281
Chen, W., Livio, M., Gehrels, N., 1993, Astrophys. J. **408**, L5
Chadwick, P.M., Dickinson, M.R., Dipper, N.A. et al., 1998, Astrophys. J. **503**, 391
Collmar, W., McConnell, M.L., Bennett, K. et al., 1995, Proc. of the 24th ICRC, Vol. 2, 170–173
Cordier, B., Paul, J., Ballet, J. et al., 1993, Astron. Astrophys. **275**, L1
Ebisawa, K., Titarchuk, L., Chakrabarti, S.K., 1996, PASJ **48**, 59
Esin, A.A., McClintock, J.E., Narayan, R., 1997, Astrophys. J. **489**, 865
Frank, J., King, A.R. & Raine, D.J., 1992, in *Accretion Power in Astrophysics*, 2nd edition, Cambridge University Press, Cambridge
Giacconi, R., Gursky, H., Paolini, F, Rossi, B., 1962, Physical Review Letters **9**, 439
Giacconi, R., Gursky, H., Kellogg, E. et al., 1971, Astrophys. J. **167**, L67
Gierlinsky, M., Zdziarski, A.A., Done, C. et al., 1997, MNRAS **288**, 958
Gilfanov, M., Churazov, E., Sunyaev, R. et al., 1995, in M.A. Alpar, U. Kiziloglu, J.van Paradijs (Eds), *The Lives of Neutron Stars*, Kluwer Academic Publishers, Dordrecht, p. 331
Goldwurm, A., Ballet, J., Cordier, B. et al., 1992, Astrophys. J. **389**, L79
Grove, J.E., Johnson, W.N., Kinzer, R.L. et al., 1994, AIP Conference Proc., **304**, 192
Grove, J.E., Johnson, W.N., Kroeger, R.A. et al., 1998a, Astrophys. J. **500**, 899
Grove, J.E., Strickman, M.S., Matz, S.M. et al., 1998b, Astrophys. J. **502**, L45

Hermsen, W., Bloemen, H., Jansen, F.A. et al., 1987, Astron. Astrophys. **175**, 141
Hameury, J.-M., King, A.R., Lasota, J.-P., 1986, Astron. Astrophys. **161**, 71
Herrero, A., Kudritzki, R.P., Gabler, R. et al., 1995, Astron. Astrophys. **297**, 556
Jung, G.V., Kurfess, D.J., Johnson, W.N. et al., 1995, Astron. Astrophys. **295**, L23
Kazanas, D., Hua, X.-M., Titarchuk, L., 1997, Astrophys. J. **480**, 735
Lamb, R.C., Fichtel, C.E., Hartman, R.C. et al., 1977, Astrophys. J. **212**, L63
Lasota, J.P., 1997, in *Accretion Phenomena and Related Outflows*, Wickramasinghe, D.T., Bicknell, G.V., Ferrario, L. (Eds), ASP Conf. Series Vol 121, p. 351
Liang, E.P., 1998, Physics Reports **302**, 67
Ling, J.C. et al., 1987, Astrophys. J. **321**, L117
Matz, S.M. et al., 1994, AIP Conf. Proc. **308**, 263
McConnell, M.L., Forrest, D.J., Owens, A. et al., 1989, Astrophys. J. **343**, 317
McConnell, M.L., Forrest, D.J., Ryan, J. et al., 1994, Astrophys. J. **424**, 933
McConnell, M.L., Bennett, K., Bloemen, H. et al., 1999, Proc. of 26th ICRC, Vol. 4, Kieda, D., Salamon, M., Dingus, B. (eds), Salt Lake City, p. 119
Miyamoto, S., Kitamoto, S., Mitsuda, K., Dotani, T., 1988, Nature **336**, 450
Mori, M., Bertsch, D.L., Dingus, B.L. et al., 1997, Astrophys. J. **476**, 842
Nowak, M.A., Vaughan, B.A., Wilms, J. et al., 1999, Astrophys. J. **510**, 874
Narayan, R., Yi, I., 1995, Astrophys. J. **452**, 710
Orosz, J.A., Jain, R.K., Bailyn, C.D. et al., 1998, Astrophys. J. **499**, 375
Parsignault, D.R., Gursky, H., Kellogg, E. M. et al., 1972, Nature Phys. Sci. **239**, 123
Poutanen, J., Coppi, P.S., 1998, Physica Scripta **T77**, L5
Sunyaev, R., Churazov, E., Gilfanov, M. et al., 1992, Astrophys. J. **389**, L75
Tanaka, Y., 1989, in *Proc. 23rd ESLAB Symp. on Two Topics in X-Ray Astronomy*, White, N.E., Hunt, J.J., Battrick, B. (Eds), ESA SP-296, Noordwijk, p. 3
Tanaka, Y., Lewin W.H.G., 1995, in *X-Ray Binaries*, Lewin, W.H.G., J. van Paradijs, E.P.J. van den Heuvel (Eds), Cambridge University Press, Cambridge, England, p. 126
Tavani, M., Barrett, D., 1997, in *Proc. of the Fourth Compton Symposium*, AIP Conference Proceedings 410; eds. Dermer, C., Strickman, M., Kurfess, J., p. 75
Trümper, J., Pietsch, W., Reppin, C. et al., 1978, Astrophys. J. **219**, L105
van Dijk, R., Bennett, K., Collmar, W. et al., 1995, Astron. Astrophys. **296**, L33–36
van der Klis, M., 1995, in *X-Ray Binaries*, Lewin, W.H.G., J. van Paradijs, E.P.J. van den Heuvel (Eds), Cambridge University Press, Cambridge, p. 252
van Paradijs, J., 1998, in *The many Faces of Neutron Stars*, Buccheri, R., Alpar, M.A. (Eds), Kluwer Academic Press, Dordrecht, p. 279
Weekes, 1992, Space Science Reviews **9**, 315
Webster, B.L., Murdin, P., 1972, Nature **235**, 37
Vestrand, W.T., Srekumar, P., Mori, M., 1997, Astrophys. J. **483**, L49
Zhang, S.N., Mirabel, I.F., Harmon, B.A. et al. 1997, in *Proc. of the Fourth Compton Symposium*, AIP Conference Proceedings 410; eds. Dermer, C., Strickman, M., Kurfess, J., p. 141

8 Continuum Gamma Ray Emission from Supernova Remnants

Anatoli Iyudin and Gottfried Kanbach

8.1 Introduction

Supernova remnants are the dramatic objects produced by the violent explosions of massive stars at the end of their life. This explosion, called a supernova, is one of the most energetic events in the universe, and causes a single star to briefly outshine the entire Galaxy in which it is located. Supernovae in our own Galaxy have produced spectacular light shows in the night (and often in the daytime) sky, but the last supernova in the Milky Way that was observed from Earth was in 1604. This is why the supernova discovered in the Large Magellanic Cloud (a nearby Galaxy at a distance of ~50 kpc) in 1987 was so exciting to astronomers: it was the only supernova visible with the naked eye in the last 400 years!

For more than 35 years, it has been generally accepted that there are two different ways to make a supernova: either from the gravitational collapse of the dense core of a massive star at the end of its evolution, or by explosie thermonuclear burning of a less massive but equally dense white dwarf. By observing spectra of supernovae (SNe) near their maximum brightness it has been established that some supernovae never show hydrogen emission lines and consequently these were named Type I SNe. Apart from the spectrally different signature, Type I SNe were found not to favor any particular type of galaxies, or spiral arms in our own Milky Way. At the same time, the corecollapse type SNe were found to have a preference to explode in star-forming regions, for example, in the spiral arms of our Galaxy. Later investigations have led to the establishment of a much larger diversity of supernova types. A description of the presently known supernova types was given by Harkness and Wheeler (1990).

Astronomers believe that supernovae occur in our Galaxy roughly two to three times every hundred years on average. Why are such infrequent events so interesting and important? The enormous amount of energy released in a supernova explosion has major effects on the interstellar medium (ISM; the gas and dust between the stars). The explosion itself involves the core of the star, which is primarily composed of nickel (for the massive star) or carbon and oxygen (for a white dwarf with a mass near $1.4\,M_\odot$) by the time it explodes. When the star is born, it is made of ~75% hydrogen and ~25% helium by mass. The nuclear fusion that occurs in the interior of the star combines hydrogen nuclei (protons) to form helium nuclei, releasing the energy that fuels the star during most of its life. Once the hydrogen in the core is exhausted, the helium is fused to form carbon, nitrogen, and oxygen, releasing more energy. For massive stars this process continues, with the inner core of the star being converted to heavier and heavier nuclei, until finally the star is composed of an Fe/Ni core surrounded by shells of Si/S, Ne/Mg, CNO, helium, and hydrogen. The structure is like an onion (see also Chap. 10). Once the inner core is converted to Fe/Ni, no more energy is available from the fusion process and the inner core collapses catastrophically to form a neutron star or a black hole.

For the case of a white dwarf which has accreted a critical mass of $\sim 1.4\,M_\odot$ from a binary companion, a thermonuclear runaway is believed to be inevitable, and the star explodes very much like a thermonuclear bomb. A runaway ignition will occur when carbon burning begins to provide an excess of nuclear energy over that which can be carried away by neutrinos. In both scenarios, i.e. for the massive star and for the white dwarf, the resulting explosion blows out the outer layers of the star into space with a velocity of up to $10000-25000\,\mathrm{km\,s^{-1}}$.

Such an enormous explosion produces primarily three effects. In the first place, it has a strong influence on the ISM. It blows a hole in the ISM that rapidly expands until it reaches up to several hundred light years in diameter. The interior of this bubble in the ISM is extremely hot (typically several million K). As a result, the ISM is strongly modified by these supernova explosions, which are believed to have a strong effect on the distribution of the gas and dust in our Galaxy.

In the second place, the shock waves from these explosions are important for the acceleration of cosmic rays and may also be the agent that trigger the collapse of existing interstellar clouds to form new stars, thus closing the cycle of stellar evolution. Eventually, the remnants cool and form interstellar clouds from which new stars can form.

Finally, supernova remnants are also important for distributing various elements throughout interstellar space. The Big Bang produced practically only hydrogen and helium; yet we know that the mass of our planet Earth is mainly composed of other, heavier elements. These other elements were produced inside stars and during supernova explosions, and were dispersed into interstellar space as supernova remnants (SNRs). These SNRs have been discovered as spatially extended sources mostly in radio surveys. A catalogue of the known Galactic radio SNRs can be found at www.nrao.cam.ac.uk/surveys/snrs/. They are distinguished from HII regions (clouds of ionized hydrogen, usually associated with luminous hot stars) by the shape of their radio spectra. The radio spectrum of SNRs generally has a negative spectral index, indicating a nonthermal origin for the radio emission, and the emission itself is highly polarized. Many SNRs carry the names assigned after the radio discovery. Cas A and Tau A (the Crab Nebula) are the brightest radio sources in the Cassiopeia and Taurus constellations, and Vela XYZ is compound of three radio-bright regions. Catalogues of such nonthermal extended radio sources now contain about 220 objects in our Galaxy and about 60 objects in the Magellanic Clouds. Most of these sources appear to be approximately circular in shape and are limb brightened as if we are looking at emission from a large hollow sphere which is transparent to its own radiation.

This is the shape expected for the expanding debris from a spherically symmetric supernova explosion into a medium of uniform and constant matter density. It is the simplest picture if the progenitor star was embedded in a uniform medium. The shell of ejected material expands rapidly and sweeps up the surrounding matter. A low-density region is left in the interior and behind the expanding shell. The expanding material in the shell itself soon becomes of low enough density for the shell to be transparent or "thin." During this phase the mass of the swept-up material is negligible compared to the mass of the ejecta, and the expansion proceeds at a uniform velocity. This "free" expansion is the first phase in the life of an SNR. The total ejected mass might be 1 $M_\odot$ and the density of the surrounding medium 0.3 atoms cm^{-3}. If so, this phase will last until the radius is $\sim$3 pc, when the swept-up mass becomes equal to that of the ejecta. If the initial velocity is 15000 km s^{-1}, the age of the remnant at this time will be $\sim$200 yr.

In the assumed uniform ISM the speed of sound is about 10 km s^{-1}. It is much less than the velocity of the stellar ejecta. A shock wave consequently forms at the leading edge of the ejecta. Atoms caught by the shock will be ionized and the temperature increases to 10^7–10^8 K. All the material is propelled outward, in the direction of the shock.

As time passes the expansion slows and the SNR enters the second phase, an adiabatic expansion. This is known as the Sedov phase or "blast-wave" phase. The mass of the swept-up material is now large compared to the original mass of the ejecta, but the energy radiated by the material in the shell is still small compared with its internal (kinetic) energy. Hence, the rate of expansion is determined only by the initial energy deposited by the explosion, E_0, and the density of the ISM, n. As it expands, the SNR sweeps up cold ISM and becomes cooler as its mass increases. More general discussions of the blast-wave evolution in a different context are given in the reviews by Chevalier (1981) and by Ostriker and McKee (1988). The simplest, so-called Sedov–Taylor solutions (Sedov 1946, 1959; Taylor 1950), are presented below.

The radius, R, of the blast wave is

$$R = 14 \cdot (E_0/n)^{\frac{1}{5}} \cdot t^{\frac{2}{5}} \quad \text{pc} , \tag{8.1}$$

where E_0, n, and t are in units of 10^{51} erg, cm^{-3}, and 10^4 yr, respectively. The shock temperature is given as

$$T_s = 1.0 \times 10^{10} \cdot (E_0/n) \cdot R^{-3} \quad \text{K} , \tag{8.2}$$

where R is in units of parsecs. The velocity of the shock wave is derived as

$$V_s = \left(\frac{E_0}{\pi n}\right)^{1/2} \times R^{-3/2} \quad \text{km s}^{-1}. \tag{8.3}$$

After eventual cooling of the material in and behind the shock (down to $T \sim 2 \times 10^5$ K), the SNR reaches phase 3 in its evolution, the "radiative" phase. During the duration of this phase ($\sim 10^5$ yr), most of the SNR internal (kinetic) energy is radiated away. The shell coasts through interstellar space becoming fainter and fainter until it is indistinguishable from the surrounding medium.

Generally, three basic types of SNRs are known:

8.1.1 Shell-type SNRs

As the shock wave from the supernova explosion plows through space, any interstellar material it encounters becomes heated and stirred up – thus producing a big shell of hot material in space. The *Tycho SNR* as shown in Fig. 8.1 is an example of such a shell-type remnant. It shows nicely the ring-like structure whose appearance is explained by the fact that when we look at the edge of the shell, there is more hot gas along the line of sight than when we are looking through the center of the remnant. Astronomers call this limb brightening. Shell-type remnants comprise more than 80% of all SNRs.

8.1.2 Crab-like SNRs

Figure 8.2 shows a picture of the Crab Nebula – the most intriguing and most spectacular supernova remnant. The image shows what is left of a star that "exploded" about 900 years ago – it was in fact seen as a "guest star" by Chinese Imperial astronomers in 1054 A.D. A guest star is the sudden appearance of a "new" star, and these new stars could even be seen in daylight! The Chinese records show that the progenitor of the Crab Nebula actually exploded on July 4, 1054, and was visible during the day for three weeks. Such events were viewed as very important by astrologers, since it marked a sudden and fairly dramatic change in the normally constant night sky.

SNRs that resemble the Crab Nebula (roughly spherical with a filled center) are called plerions. The appearance of a plerion (also as an interior substitute in a shell-like SNR) is thought to indicate the presence of a pulsar.

A pulsar is a very dense and compact star usually rapidly rotating and composed almost entirely of neutrons (a neutron star!) (see Chap. 6). This is what remains of the core of the original star – a 20 km diameter "atomic nucleus." Most of the bright "glow" that fills the center of the SNR is actually *sychrotron radiation* – produced by ultrarelativistic electrons travelling in the strong magnetic fields that "thread" the nebula. The short lifetime of these energetic electrons requires a steady resupply to maintain the synchrotron luminosity. Even before the discovery of pulsars the hypothesis that the nebular

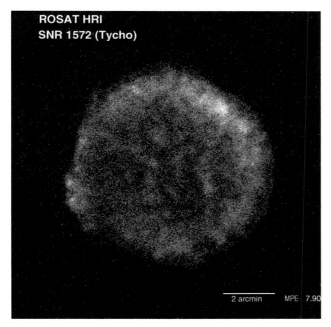

Fig. 8.1. ROSAT map of the Tycho SNR (SN1572) at 0.1–2.4 keV. (Reproduced with permission from B. Aschenbach, MPI, Garching)

electrons are produced by a central object, namely by a rotating, strongly magnetized compact star, was made (Pacini 1967, 1968). The picture on the right in Fig. 8.2 shows a Hubble Space Telescope image of the inner parts of the Crab. The pulsar itself is visible as the left of the pair of stars near the center of the frame. Surrounding the pulsar is a complex of sharp knots and wisp-like features. In the outer regions of the Crab Nebula many emission lines, including the red glow from hydrogen, are present.

8.1.3 Composite SNRs

These SNRs are a cross between the first two types. A composite SNR has the appearance of shell-like shock-heated hot gas with a small central synchrotron nebula. They appear either shell-like or Crab-like, depending on the part of the electromagnetic spectrum in which they are observed. Often it is believed that the shock wave is moving out making the shell, while the hot gas still fills the central part of the SNR. The Vela SNR serves as an example of a composite SNR, shown at X-ray energies 0.1–2.4 keV in Fig. 8.3. It contains a central source (pulsar) with a 1′ diameter X-ray synchrotron nebulae around it. It is interesting that the same SNR shows γ-ray emission at very high energies as a plerion at the Vela pulsar birth position, with a possible extended nebulae around it (see Fig. 8.6).

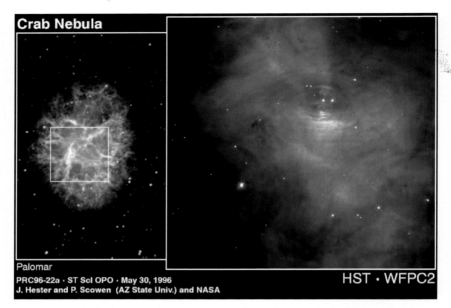

Fig. 8.2. Hubble Space Telescope image of the Crab Nebula (SN1054) taken with the Wide Field Planetary Camera 2 (see for more details Space Telescope Science Institute press release PRC96-22a); images are not north-oriented

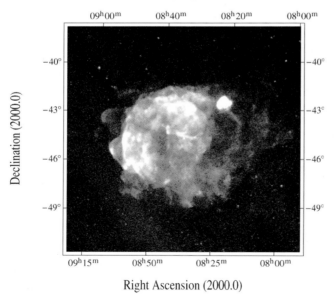

Fig. 8.3. ROSAT map of the Vela SNR at 0.1–2.4 keV. (Reproduced with permission from Aschenbach 1998)

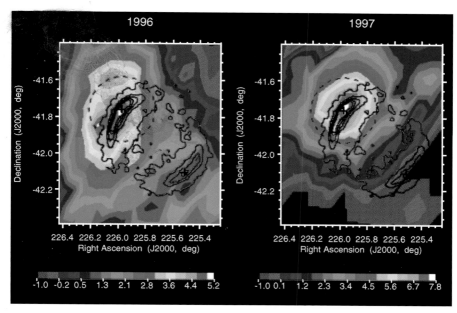

Fig. 8.5. CANGAROO images of SN1006 taken in 1996 and 1997. The cross marks the maximum flux point in the 2–10 keV band from the ASCA satellite. (Reproduced with permission from Tanimori et al. 1998a)

Apart from this shell-like remnant which does not show any central source and where TeV γ-ray emission spatially coincides with a bright spot in hard X-rays (Fig. 8.5), two more plerion-type SNRs have been detected by the same TeV telescope: Crab and Vela (Fig. 8.6). The Vela SNR is usually considered to be a composite-type remnant, but appears as a plerion-type SNR in very high-energy γ-rays. Caution may be advisable in the interpretation of this result, however, since the angular resolution of CIT CANGAROO is much larger than the 1' X-ray size of the Vela pulsar synchrotron nebula. The most interesting feature of these detections is that both plerions show steady (no pulsation) persistent emission at TeV energies (Tanimori et al., 1998b and Yoshikoshi et al., 1997, respectively).

From the spectrum of the nonthermal X-rays (assumed to be synchrotron emission), the detected TeV γ-rays are readily shown to be generated by inverse-Compton scattering of the same very energetic electrons on 2.7 K cosmic background photons. The calculated fluxes of TeV γ-ray emission based on these assumptions are consistent with the observed γ-ray fluxes (at $\sim$100 MeV) and the radio and X-ray emission (SN1006 – Yoshida and Yanagita 1997; IC443 – Sturner et al. 1997). However, these TeV results only confirm the acceleration of *electrons* in SNRs. What about the nucleonic component of the cosmic rays? It is decades since it was proposed that

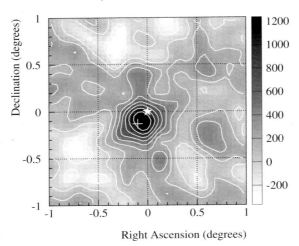

Fig. 8.6. CANGAROO off-set map of the Vela SNR nebula γ-ray excess. The "star" at the origin of map indicates the position of the Vela pulsar. (Reproduced with permission from Yoshikoshi et al. 1997)

Galactic cosmic rays are accelerated by supernova remnants (Ginzburg and Syrovatskii 1964), but the evidence remains inconclusive.

The four radio-bright remnants detected by EGRET (γ Cygni, IC 443, W44 and W28) are associated with molecular clouds, which provide the natural target material for the production of γ-rays via the spallation products of the proton–proton interactions, in line with the suggestion by Aharonian et al. (1994). The flux extrapolation to TeV energies from the typical EGRET SNR detection points in the energy range 100 MeV–10 GeV should result in detectable TeV γ-rays in general. However, Whipple and HEGRA (High-Energy Gamma-Ray Astronomy) observations have failed to reveal any positive TeV emission from the EGRET-detected remnants. The explanation of this nondetection at TeV energies was found in the separation of the electron-generated γ-rays from those generated in proton–gas interactions (by using multi-wavelength spectral fitting), with the contribution of the latter being at the level of 20% of the total γ-ray emission above 100 MeV (for example, W44; de Jager and Mastichiadis 1997), or by the great age of the SNR and accordingly large particle mean free path (∼SNR size) in the case of IC 443 (Sturner et al., 1997). It should also be noted that, for example, in γ Cyg the γ-ray source is point-like at GeV energies while the SNR extends over ∼1°. This could indicate a compact point source inside the remnant.

There are more potential candidates for supernova remnant identifications among the EGRET unidentified γ-ray sources. It has been estimated that the probability of chance coincidence of unidentified EGRET sources at low Galactic latitudes ($-10° \leq b \leq 10°$) with supernova remnants is less than $\sim 5 \times 10^{-4}$. The EGRET source catalogue includes ∼40 low-latitude unidentified sources for which no known counterparts exist in other spectral bands (Reimer et al., 1997). Some of the low-latitude unidentified EGRET sources could be old SNRs that are bright in γ-rays but have very uncertain radio or IR signatures due to the diffuse Galactic background and because of their spatial extent.

8.3 SNRs and Particle Acceleration

The dynamics of supernova remnants and associated particle acceleration physics is complex, especially due to the richness of the interactions between clumpy stellar, interstellar and circumstellar media. There is observational evidence that cosmic ray electrons are accelerated in SNRs, as was shown above, and that their acceleration mechanism is, in general, consistent with the so-called diffusive shock acceleration (DSA) theory (see also Chap. 5). However, many observational details, such as spectral index variations in time and space for specific remnants, remain to be explained by a fully nonlinear, self-consistent treatment.

Since the invention of the DSA process 20 years ago, there has been major progress in its predictive power in relation to particle acceleration in SNRs and to the origin of the ionic and electronic components of Galactic cosmic rays. The simplest, test-particle steady-state versions of the DSA theory make reasonable predictions of power-law momentum distributions of particles accelerated by the shock. Those power laws should lead to relativistic particle energy distributions approximately of the form $N(E) \sim E^{-2}$ for strong adiabatic shocks with a density-jump of $r \sim 4$ across the shock front. After correction for propagation through the interstellar medium, this shape may be consistent with the observed Galactic cosmic ray energy distribution. This spectrum is also similar to the implied mean energy distribution of relativistic electrons in shell supernova remnants, although there is a whole range of spectra actually seen in radio observations. Thus, SNRs have become almost universally accepted as the source of Galactic cosmic rays. There remain, however, important, unsolved issues in that relationship.

The theoretical studies have shown that 10–50% of the supernova explosion energy can be transferred to cosmic rays through DSA in the supernova blast shock. This is sufficient to account for the cosmic ray energy replenishment rate of 10^{42} erg s^{-1} required for diffusive escape from the Galaxy (Jones et al., 1998). In DSA theory charged particles gain energy by repeated pitch-angle scattering across the velocity jump in the shock (see illustration of the process given in Fig. 5.13). The scattering magnetohydrodynamic (MHD) waves can be excited by the high-energy particles themselves. Effective energy transfer by DSA from a shock to the cosmic rays requires a fully nonlinear treatment and there several complications arise, such as an enhanced density jump (r) exceeding the nominal factor of 4, and a smooth precursor due to "backreaction" from cosmic rays diffusing upstream. Nevertheless, quite encouragingly, the DSA theory gained strong support from comparisons that have been recently made between direct measurements of particle spectra in heliospheric shocks and predictions of nonlinear DSA theory using Monte-Carlo (Baring et al., 1995) and diffusion–convection methods (Kang and Jones 1997).

The remaining unsolved problem in basic DSA theory itself is an understanding of cosmic ray "injection" from the thermal plasma at shocks. Recently it was shown that there are some ways out of this problem, such as the "thermal leakage" of ions in shocks (Malkov and Völk 1995), which could be a natural part of collisionless shock formation (Ellison et al., 1996).

The bulk of Galactic cosmic rays detected on Earth should be accelerated in supernova remnants during their Sedov phase, though acceleration during the early free-expansion SNR phase could also be significant and should influence the supernova remnant dynamics. The cosmic rays are mostly released from within the supernova remnant after the radiative-cooling phase, when the shock speed eventually drops to the Alfvén speed of the interstellar medium. The full, nonlinear DSA problem is difficult to solve, especially applied to supernova remnants, where shocks are neither steady nor planar. Nevertheless, solutions with cosmic rays obeying Bohm diffusion exist, which assume the scattering length to be proportional to the particle gyroradii. The resulting source spectrum of cosmic rays escaping from supernova remnants in the warm or hot phases of the interstellar medium is proportional to $E^{-2.1}$ up to the so-called *knee* energy at 10^{14} eV (Berezhko et al. 1996). The observed spectrum on Earth in this energy range has a form proportional to $E^{-2.7}$. The commonly accepted interpretation of isotopic composition in Galactic cosmic rays as modified by propagation through and escape from the interstellar medium leads to an energy-dependent ISM column density $X(E) \sim E^{-0.6}$. These models reproduce both the observed Galactic cosmic ray spectrum and chemical abundances for ions up to 1 PeV. The maximum energy attainable by cosmic ray ions in a supernova remnant of age t_{SNR} under the assumption of constant shock speed is discussed in Berezhko et al. (1996) and by Baring et al. (1999), whose notation we follow in the formulae given below:

$$E_{\max}(t_{\text{SNR}}) \simeq E_{\text{trans}} \frac{t_{\text{SNR}}}{t_{\text{trans}}}; \quad t_{\text{SNR}} < t_{\text{trans}}, \tag{8.9}$$

where

$$E_{\text{trans}} \sim \frac{r-1}{r(1+gr)} \frac{Q}{\eta} \left(\frac{B_1}{3 \, \mu\text{G}}\right) \left(\frac{n_{p,1}}{1 \, \text{cm}^{-3}}\right)^{-1/3}$$
$$\times \left(\frac{E_{\text{SN}}}{10^{51} \, \text{erg}}\right)^{1/2} \left(\frac{M_{\text{ej}}}{M_\odot}\right)^{-1/6} \quad \text{TeV}, \tag{8.10}$$

is the ions maximum energy achieved at t_{trans}, e.g. at the time of SNR transition from the free-expansion phase to the Sedov phase. Here η is a constant, independent of ion species, energy and position relative to the shock, which relates the ion gyroradius $r_g = pc/(QeB)$ to the the ion scattering mean path (λ_i) by $\lambda_i = \eta r_g$. Q is the charge number of the ion, and g relates the upstream and downstream diffusion coefficients via the equation $k_2 = gk_1$. The transition time can be estimated from the expression

$$t_{\text{trans}} \simeq 90 \left(\frac{n_{p,1}}{1 \text{ cm}^{-3}}\right)^{-1/3} \left(\frac{E_{\text{SN}}}{10^{51} \text{ erg}}\right)^{1/2} \left(\frac{M_{\text{ej}}}{M_\odot}\right)^{5/6} \text{ yr}, \quad (8.11)$$

where $n_{p,1}$ is the unshocked proton-number density, E_{SN} is the supernova explosion energy in units of 10^{51} erg, and M_{ej} is the supernova ejected mass in solar mass units. Additionally, B_1 is the shock upstream magnetic field.

In order to estimate the maximum energy of the accelerated electrons, one has to take into account synchrotron and inverse-Compton losses of high-energy electrons. In contrast to inverse-Compton and synchrotron losses, Coulomb losses are important only in old supernova remnants. By assuming that the magnetic-field strength (B_{cbr}) has the same energy density as the 2.73 K background radiation ($B_{\text{cbr}} \approx 3 \times 10^{-6}$ G), one can write a simple formula describing both synchrotron and inverse-Compton losses (Baring et al. 1999):

$$\left(\frac{dE_e}{dt}\right)_{\text{tot}} \simeq -0.034 \left(\frac{\sqrt{B^2 + B_{\text{cbr}}^2}}{3 \text{ μG}}\right)^2 \left(\frac{E_e}{1 \text{ TeV}}\right)^2 \text{ eV s}^{-1}. \quad (8.12)$$

In oblique shocks it is not possible to obtain a precise measure of the loss rate without detailed knowledge of the shock geometry, but by taking an upper limit of the mean magnetic field $B = B_1$ one can evaluate the electron cut-off energy ($E_{\text{cut-off}}$) in the Sedov phase as

$$E_{\text{cut-off}} \sim 180 \left[\frac{r-1}{r(1+gr)} \frac{Q}{\eta}\right]^{1/2} \left(\frac{B_1}{3 \text{ μG}}\right)^{1/2} \left(\frac{\sqrt{B_1^2 + B_{\text{cbr}}^2}}{3 \text{ μG}}\right)^{-1}$$
$$\times \left(\frac{n_{p,1}}{1 \text{ cm}^{-3}}\right)^{-1/5} \left(\frac{E_{\text{SN}}}{10^{51} \text{ erg}}\right)^{1/5} \left(\frac{t_{\text{SNR}}}{10^3 \text{ yr}}\right) \text{ TeV}. \quad (8.13)$$

A similar cut-off energy was obtained by Sturner et al. (1997). One can clearly see from the above expression that the electron maximum energy will be essentially unaffected in the high-density interstellar medium throughout the Sedov phase, but will be truncated in the lower-density regions at all times.

For the Cas A SNR X-ray measurements at energies up to 120 keV (Allen et al. 1997; see also Fig. 8.7) have helped to establish a value of $E_{\text{max}} \geq 50$ TeV. One may use (8.9–11) to estimate E_{max} for very young remnants like SN 1987A. For $E_{\text{SN}} \sim 10^{51}$ erg, $M_{\text{ej}} \sim 1 M_\odot$, and $T_{\text{SNR}} = 10$ yr, $E_{\text{max}} \sim 7$ TeV. Protons of this energy produce pion-decay photons with energies a few

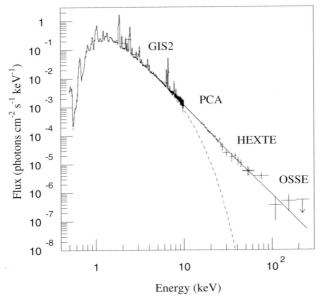

Fig. 8.7. Continuum emission spectrum of the Cas A SNR. (Reproduced with permission from Allen et al. 1997)

times lower than the energies of the interacting protons. Similarly, electron-bremsstrahlung photons have on average about one-third of the energy of the primary electrons, and inverse-Compton photons produced by 7 TeV electrons have energies less than 1 TeV. From the above simple estimates it follows that SN 1987A will not be a bright TeV source in the next decades (Baring et al., 1999).

8.4 Discussion and Outlook

In order to explain the observed Galactic cosmic rays, one usually invokes the idea of nuclear cosmic rays being accelerated in supernova remnants. This is an attractive and reasonable idea in terms of the energy budget of our Galaxy. Predictions of the DSA theory seem to explain most essential aspects of Galactic cosmic rays and appear convincingly self-consistent. There have recently been more efforts to predict the γ-ray emission from π° decay that results from proton–nucleon interactions. These predictions will face direct observational tests as γ-ray observations attain a better sensitivity.

Recent observations of supernova remnants in X-rays and γ-rays and their theoretical interpretations seem to indicate that there are selection effects; we mostly observe SNRs in dense environments. According to the standard DSA theory, the cosmic ray particles can be accelerated up to the *knee* energy at

SNRs only in the hot, tenuous interstellar medium. This implies that there are remnants in the low-density ISM which are not yet observed in X- and γ-rays, and these supernova remnants are the main acceleration sites for the bulk of the high-energy Galactic cosmic rays.

Magnetic fields and their structure in supernova remnants are very important for an understanding of the acceleration processes. Strongly turbulent magnetic fields in the vicinity of shocks are a key element for DSA. In addition, the brightness of the synchrotron emission is at least as dependent on the strength of the magnetic field as it is on the concentration of relativistic particles. Thus, to disentangle acceleration issues from magnetic-field structure issues, one must understand the field configuration. In most young SNRs the mean magnetic field direction has a clear radial orientation. However, old shell-type remnants display a variety of magnetic field orientations, sometimes including a mixture of radial and tangential field regions (Milne 1987). The field strength can be estimated by the rotation measures of background radio sources, from the Zeeman effect measurement at different wavelengths (see, for example, Schwarz et al. 1986) and from the X-ray and γ-ray spectra (Cowsik and Sarkar 1980). Generally, magnetic fields in SNRs seem to be in equipartition with the relativistic electron energy, within an order of magnitude (Jones et al. 1998 and references therein). However, since the relativistic electron component is energetically minor and its coupling to the local MHD behavior is unclear, the reasons to expect an equipartition between these two components are not obvious. Measurements of the supernova remnant broadband γ-ray spectrum may help to disentangle this problem (Cowsik and Sarkar 1980; Esposito et al., 1996). An example of such a comparison is shown in Fig. 8.8, as calculated by Ellison et al. (1999) for the Cas A SNR. It is clear from the model SNR spectra (Fig. 8.8) that only γ-ray flux measurements at 10^6–10^7 eV and at 10^{10}–10^{13} eV could give an answer to the question of the magnetic-field value in Cas A SNR and to settle the question of the relative importance of the IC or π°-decay components at TeV energies and/or of relativistic bremsstrahlung at MeV energies (see Atoyan et al. 2000; Ellison et al. 2000).

The Galactic cosmic ray and γ-ray spectra, however, are not the only constraints on the SNR acceleration model. The cosmic ray composition also plays an important role in the understanding of SNRs. Recently, Webber (1997), in his review of the relative composition of cosmic ray particles, emphasized that the composition at the source(s) is not simply ISM or supernova ejecta material. Differences with the ISM include a C/O ratio of unity, the large abundance of the isotope ^{22}Ne, the underabundance of N, and the low abundance of H and He in the cosmic rays. The correlation of the ratio of the cosmic ray source abundance to the interstellar medium abundance with the first ionization potential of the elements (e.g., Wefel 1991; Ferrando 1993) brings up the injection problem. Quasiparallel shocks have been observed to accelerate thermal particles in the heliosphere, and plasma simulations of

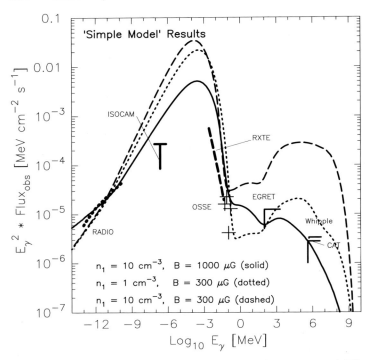

Fig. 8.8. Comparison of the "Simple Model" (Berezhko and Ellison 1999) calculations of the Cas A SNR continuum emission spectrum with the measurements in radio, IR, X-ray and γ-ray energy domains. (Reproduced with permission from Ellison et al. 1999)

quasiparallel shocks clearly show injection occurring (Ellison et al. 1990). Any ions which are pre-energized before encountering the shock will be enhanced in the acceleration process. However, the question remains as to how the apparent first ionization potential selection is introduced: Are the thermal ions already preselected, or is there a pre-energization which produces the selection, or are the composition biases intrinsic to the acceleration process itself (e.g. Ellison 1990)? These effects remain as major unanswered questions for the SNR model. It would appear that the material to be accelerated cannot be only supernova ejecta, which does have a large flow velocity and turbulent motion, but must also involve swept-up circumstellar or interstellar matter.

The recent progress in acquiring new experimental data on the elemental abundances of cosmic rays and of the anomalous cosmic ray component has helped to answer the question of the origin and of preferable acceleration. Meyer et al. (1997) introduced the idea that the cosmic ray elemental composition is better organized by volatility and mass-to-charge ratio than by the first ionization potential. At the same time Ellison et al. (1997) introduced a quantitative model of Galactic cosmic ray origin and acceleration based on the supernova blast-wave acceleration of a mixture of interstellar and/or circumstellar gas and dust. The gas and dust are accelerated simultaneously, but the acceleration efficiency is dependent on the fraction of ions locked into grains or in the gas phase.

The model by Ellison et al. (1997) is based on the nonlinear shock treatment and includes:

- the direct acceleration of interstellar ions;
- a simplified model for the direct acceleration of weakly charged grains to ~ 100 keV amu^{-1}, simultaneous with the acceleration of the gas ions;
- the energy losses of grains colliding with the ambient gas;
- the sputtering of grains; and
- the simultaneous acceleration of the sputtered ions to GeV and TeV energies.

See Fig. 8.9 for a sketch of the process suggested by Ellison et al. (1997). This model produces cosmic ray source abundance enhancements of the volatile, gas-phase elements that are an increasing function of mass, as well as a net, mass-independent enhancement of the refractory, grain elements over protons, consistent with cosmic ray observations. The cosmic ray ^{22}Ne excess is accounted for in this model in terms of acceleration of ^{22}Ne-enriched pre-supernova Wolf–Rayet stellar wind material surrounding the most massive supernovae. It is also shown that cosmic ray spectra are matched well by the model, at least below 10^{15} eV. High-resolution spectroscopic studies of SNRs

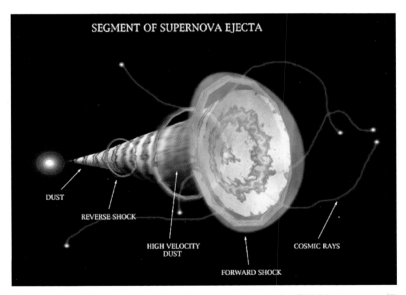

Fig. 8.9. Sketch of the dust grains acceleration by SN blast wave. (Reproduced with permission from NASA)

are needed to clarify and verify this new model. Additionally, studies of diffuse continuum γ-ray emission in our Galaxy provide important information on the properties of cosmic rays on the Galactic scale (see Chap. 9).

8.5 Conclusions

The prediction that SNRs are sites of cosmic ray acceleration will be directly confirmed only when γ-ray sources associated with SNRs are shown to be spatially extended and exhibit a constant flux. These could be a population of old ($\sim 10^5$ yr) SNRs in the vicinity of dense molecular clouds, whose γ-ray luminosity between ~ 10 MeV and ~ 100 GeV dominates their multi-wavelength spectra. Some of the low-latitude, unidentified EGRET and COMPTEL (Compton Telescope) sources could be such old SNRs that are bright in γ-rays but have a very uncertain radio or IR signature due to the diffuse Galactic background and because of their spatial extent. Generally, supernova remnants exhibit an enormous variety of morphologies and properties, and each one requires individual modelling. Therefore complementary X-ray and γ-ray observations of SNRs by telescopes more sensitive than those operational at present (e.g. future experiments such as XMM in X-rays and GLAST (Gamma-ray Large-Area Space Telescope), H.E.S.S. in γ-rays) are crucial in the ongoing effort to prove directly that SNRs are the sites of Galactic cosmic ray acceleration.

References

Aharonian, F.A., O'Drury, L.C., Völk, H.J., 1994, Astron. Astrophys. **285**, 645
Allen, G.E., Keohane, J.W., Gotthelf, E.V. et al., 1997, Astrophys. J. **487**, L97
Aschenbach, B., 1998, Nature **396**, 141
Aschenbach, B., Hahn, H.-M., Trümper, J., 1998, *The invisible sky: ROSAT at the age of X-ray astronomy* Springer-Verlag New York
Atoyan, A.M., Aharonian, F.A., Tuffs, R.J., Völk, H.J., 2000, Astron. Astrophys. **355**, 211
Baring, M.G., Ogilvie, K.W., Ellison, D.C., Forsyth, R., 1995, Adv. Space Res. **15**, 397
Baring, M.G., Ellison, D.C., Reynolds, S.P., Grenier, I.A., Goret, P., 1999, Astrophys. J. **513**, 311
Berezhko, E.G., Ksenofontov, L.T., Yelshin, V.K., 1996, Zh. Eksp. Teor. Fiz. [Sov. Phys. JETP] **109**, 3
Berezhko, E.G., Ellison, D.C., 1999, Astrophys. J. **526**, 385
Bethe, H.A., Heitler, W. 1934, Proc. R.Soc. London, Ser. A **146** 83
Blumenthal, G.R., Gould, R.J., 1970, Rev. Mod. Phys. **42**, 237
Chevalier, R.A., 1981, Fundam. Cosmic Phys. **7**, 1
Cowsik, R., Sarkar, S., 1980, Mon. Not. R. Astron. Soc. **191**, 855
de Jager, O.C., Mastichiadis, A., 1997, Astrophys. J. **482**, 874
Dorfi, E. A., 1991, Astron. Astrophys. **251**, 597

O'Drury, L.C., Aharonian, F.A., Völk, H.J., 1994, Astron. Astrophys. **287**, 959
Ellison, D.C., 1990, Proc. of 21st Int. Cosmic Ray Conf. Adelaide, Univ. of Adelaide, **11**, 133
Ellison, D.C., Möbius, E., Paschmann, G., 1990, Astrophys. J **352**, 376
Ellison, D.C., Baring, M. G., Jones, F., 1996, Astrophys. J. **473**, 1029
Ellison, D.C., O'Drury, L.C., Meyer, J.-P., 1997, Astrophys. J. **487**, 197
Ellison, D.C., Goret, P., Baring, M.G., Grenier, I.A., Lagage, P.-O., 1999, Proc. of 26 Int. Cosmic Ray Conf. Salt Lake City, Utah, **OG 2.2.09**
Ellison, D.C., Berezhko, E.G., Baring, M.G., 2000, Astrophys. J. **540**, 292
Esposito, J.A., Hunter, S. D., Kanbach, G., Sreekumar, P., 1996, Astrophys. J. **461**, 820
Ferrando, P., 1993, Proc. of 23th Int. Cosmic Ray Conf. (Calgary), Invited, Rapporter and Highlight Papers, World Scientific, Singapore, p. 279
Ginzburg, V.L., Syrovatskii, S.I., 1964, The Origin of Cosmic Rays, Pergamon Press, Oxford
Harkness, R.P., Wheeler, J.C., 1990, "Classification of Supernovae," in *Supernovae*, ed. by Petschek, A.G., Springer-Verlag, Heidelberg
Hartman, R.C. et al., 1999, Astrophys. J.SS **123**, 79
Haug, E., 1975, Z. Naturforsch. Teil A **30**, 1099
Jones, T.W., Rudnick, L., Jun, B.-I. et al., 1998, Publ. Astron. Soc. Pac. **110**, 125
Kang, H., Jones, T.W., 1997, Astrophys. J. **476**, 875
Malkov, M.A., Völk, H.J., 1995, Astron. Astrophys. **300**, 605
Meyer, J.-P., O'Drury, L.C., Ellison, D.C., 1997, Astrophys. J. **487**, 182
Milne, D.K., 1987, Aust. J. Phys. **40**, 771
Montmerle, T., 1979, Astrophys. J. **231**, 95
Ostriker, J.P., McKee, C.F., 1988, Rev. Mod. Phys **60**, 1
Pacini, F., 1967, Nature **216**, 567
Pacini, F., 1968, Nature **219**, 145
Reimer, O., Dingus, B.L., Nolan, P.L., 1997, Proc. of 25th Int. Cosmic Ray Conf. Durban, Vol. **3**, 97
Schwarz, U.J., Troland, T.H., Albinson, J.S. et al., 1986, Astrophys. J. **301**, 320
Sedov, L.I., 1946, Prikl. Mat. Mekh. **10**, No. 2, 241
Sedov, L.I., 1959, Similarity and Dimensional Methods in Mechanics Academic, New York
Shklovsky, I.S., 1953, Dok. Akad. Nauk USSR [Sov. Phys. Dokl.] **90**, 983
Sturner, S.J., Skibo, J.G., Dermer, C.D., Mattox, J.R., 1997, Astrophys. J. **490**, 619
Tanimori, T., Hayami, Y., Kamei, S. et al., 1998a, Astrophys. J. **497**, L25
Tanimori, T., Sakurazawa, K., Dazeley, S.A. et al., 1998b, Astrophys. J. **492**, L33
Taylor, G.I., 1950, Proc. R. Soc. London, Ser. A **201**, 175
Webber, W.R., 1997, Space Sci. Rev. **81**, 107
Wefel, J.P., 1991, in: *Cosmic Rays, Supernovae, the Interstellar Medium*, ed. by Shapiro, M.M., Silberberg, R., Wefel, J.P., Reidel, Dordrecht, p. 29
Yoshida, T., Yanagita, S., 1997, in Proc. Second INTEGRAL Workshop on The Transparent Universe, (ESA SP-382), p. 85
Yoshikoshi, T., Kifune, T., Dazeley, S.A. et al., 1997, Astrophys. J. **487**, L65

9 Diffuse Galactic Continuum Gamma-Rays
A Tracer of Cosmic-Rays

Andrew W. Strong and Igor V. Moskalenko

9.1 Introduction

Diffuse continuum γ-rays are produced by the interactions of high-energy nucleons and electrons (cosmic-rays, CR) with the gas and radiation in the interstellar medium. Therefore, the study of this emission provides a direct way to obtain information on the properties of CR on a Galactic scale. The main part of the continuum emission which we observe is believed to originate in CR processes but some part is certainly contributed by the combined emission of faint point sources which are undetectable as individual objects.

In this chapter we first review the physics of the most important emission mechanisms, then we describe the relevant properties of CR, the interstellar medium and Galactic structure. Next, we turn to details of the γ-ray observations, comparing them with models and deriving some basic conclusions. Implications for the molecular gas in the interstellar medium are then discussed. Gamma-ray properties of some local molecular clouds and of external galaxies conclude our review.

9.2 Theoretical Background

9.2.1 Basic Processes of Gamma-Ray Production

The important processes are

- decay of π^0 mesons produced in collisions of CR nucleons with interstellar gas particles;
- bremsstrahlung from relativistic electrons interacting with gas;
- and inverse Compton emission from relativistic electrons interacting with the interstellar optical, infrared and microwave radiation fields.

A detailed summary of all the radiation processes is given in Moskalenko and Strong (1998), Moskalenko and Strong (2000) and Strong et al. (2000). The characteristic spectral form and approximate relative contributions of these processes can be seen in Figs. 9.11 and 9.12.

Pion decay

CR nucleons, predominantly p and α, produce γ-rays via $N + N \to \pi^0 + X$ with subsequent decay $\pi^0 \to 2\gamma$. The production spectrum of secondary γ-rays (photon cm^{-3} s^{-1} MeV^{-1}) can be obtained if one knows the distribution of pions $F_\pi(\epsilon_\pi, \epsilon_p)$ from a collision of a proton of energy ϵ_p, and the distribution of γ-rays $F_\gamma(\epsilon_\gamma, \epsilon_\pi)$ from the decay of a pion of energy ϵ_π

$$\frac{\mathrm{d}f_\gamma}{\mathrm{d}\epsilon_\gamma} = n_H \int_{\epsilon_p^{\min}}^\infty \mathrm{d}\epsilon_p \, J_\mathrm{p}(\epsilon_\mathrm{p}) \, \langle \eta\sigma_\pi(\epsilon_\mathrm{p}) \rangle \int_{\epsilon_\pi^{\min}}^\infty \mathrm{d}\epsilon_\pi \, F_\gamma(\epsilon_\gamma, \epsilon_\pi) F_\pi(\epsilon_\pi, \epsilon_\mathrm{p}) \,, \quad (9.1)$$

where n_H is the atomic hydrogen number density, $J_p(\epsilon_p)$ is the proton flux, $\langle \eta\sigma_\pi(\epsilon_p) \rangle$ is the inclusive cross section of π^0 production, which grows logarithmically at high energies [$\langle \eta\sigma_\pi \rangle \approx 10, 30$, and 90 mbarn at $p_p = 3, 10$, and 100 GeV/c respectively (Dermer 1986b)], and $\epsilon_\pi^{\min}(\epsilon_\gamma) = \epsilon_\gamma + m_{\pi^0}/(4\epsilon_\gamma)$ is the minimum pion energy that contributes to the production of γ-rays with energy ϵ_γ.

Pion production in pp collisions is usually described using isobaric and scaling models (Dermer 1986a,b). The isobaric model has been shown to work well at low energies, while at high energies the relevant model is based on scaling arguments. In the isobaric model, it is assumed that the outgoing Δ isobar travels along the initial direction of the colliding protons in the center-of-mass system (CMS) and decays ($\Delta \to N + \pi$) isotropically. In the scaling model, the Lorentz invariant cross sections for charged and neutral pion production in pp collisions are inferred from experimental data at high energies.

For elements heavier than hydrogen, the inclusive cross section of π^0 production should be scaled well above the threshold with a factor $\sim (A_1^{3/8} + A_2^{3/8} - 1)^2$, where $A_{1,2}$ are the atomic numbers of the target and CR nuclei. The main contributor besides hydrogen is helium, with a ratio in the interstellar medium of He/H ~ 0.1 by number.

The distribution of γ-rays from π^0 decay is given by

$$F_\gamma(\epsilon_\gamma, \epsilon_\pi) = \frac{2}{m_\pi \gamma_\pi \beta_\pi}, \text{ and } \frac{m_\pi}{2}\gamma_\pi(1 - \beta_\pi) \leq \epsilon_\gamma \leq \frac{m_\pi}{2}\gamma_\pi(1 + \beta_\pi) \,, \quad (9.2)$$

where γ_π and β_π are the pion Lorentz factor and velocity, and the factor 2 accounts for 2 photons per decay.

The γ-ray spectrum (9.1) obtained after integration over the CR nucleon spectrum peaks at 70 MeV and is symmetric about its maximum on a logarithmic energy scale. The asymptotic high-energy spectrum is approximately the same as the nucleon spectrum, viz., a power law in energy with an index around -2.7.

Bremsstrahlung

Bremsstrahlung is a well-known QED process, the continuum photon emission from particles decelerating in the electric field of each other. In the interstellar medium consisting of the neutral gas (H and He) and the ionized medium, electron–nucleon bremsstrahlung is the most important, although electron–electron and nucleon–electron bremsstrahlung may play a role under certain conditions. A collection of appropriate formulae for the bremsstrahlung emissivity is given in Koch and Motz (1959).

The important parameter is the so-called screening factor, defined as $\delta = \epsilon_\gamma/(2\gamma_0\gamma)$, where ϵ_γ is the energy of the emitted photon, γ_0 and γ are the initial and final Lorentz factor of the electron in a collision. If $\delta \to 0$, the distance of the high-energy electron from the target atom is large compared to the atomic radius. In this case screening of the nucleus by the bound electrons is important. For low-energy electrons only the contribution of the nucleus is significant, while at high energies the atomic electrons can be treated as unbound targets in the same way as protons.

The contribution of bremsstrahlung to the Galactic diffuse emission is important in the energy range < 200 MeV. At energies $\epsilon_e \geq 2$ MeV the production cross section is given by (Koch and Motz 1959)

$$\frac{d\sigma}{d\epsilon_\gamma} = r_e^2 \alpha_f \frac{1}{\epsilon_\gamma}\left[\left(1 + \frac{\gamma^2}{\gamma_0^2}\right)\phi_1 - \frac{2}{3}\frac{\gamma}{\gamma_0}\phi_2\right], \tag{9.3}$$

where r_e is the classical electron radius and α_f is the fine structure constant. If the scattering system is an unshielded charge, the functions satisfy $\phi_1 = \phi_2 = \phi_u$,

$$\phi_u = 4Z^2\left[\ln\left(\frac{2\gamma_0\gamma}{\epsilon_\gamma}\right) - \frac{1}{2}\right], \tag{9.4}$$

where Z is the nucleus charge. For the case where the scattering system is a nucleus with bound electrons, the expressions for $\phi_{1,2}$ are more complicated and depend on the atomic form factor. Corresponding expressions for one- and two-electron atoms have been given by Gould (1969). For $\delta \gg \alpha_f Z^{1/3}$ the functions $\phi_{1,2}$ approach the unshielded value.

The production spectrum of electron bremsstrahlung for a distribution of electrons can be obtained by integrating over the spectrum of CR electrons and summing over the species of the ISM:

$$\frac{df_\gamma}{d\epsilon_\gamma} = \sum_{i=H,HI,HII,He\, etc.} c\, n_i \int d\gamma\, f_e(\gamma) \frac{d\sigma_i}{d\epsilon_\gamma}, \tag{9.5}$$

where c is the speed of light, n_i is the number density of the corresponding species, and $f_e(\gamma)$ is the spectrum of CR electrons.

Inverse-Compton Scattering

Inverse-Compton (IC) scattering is another QED process in which a high-energy electron interacts with low-energy background photons, transferring some part of its energy so that the scattered photons become more energetic. The Thomson cross section is a good approximation when the photon energy in the electron rest system (ERS) is small, $\epsilon_0 \gamma(1-\beta \cos\theta) \ll m_e c^2$, where ϵ_0 is the energy of the background photon and θ is the angle between the momenta of the electron and the photon. As soon as the energy of the electrons is high enough that the ERS energy of photons becomes comparable with the electron rest mass, the more general Klein–Nishina cross section should be used.

Simple calculation shows that the average energy of scattered photons is $\langle \epsilon_\gamma \rangle \sim \gamma^2 \epsilon_0$. This allows us to estimate the contribution of different background radiation fields to the resulting spectrum of γ-rays, taking into account that $\epsilon_0 \sim 10^{-4}$ eV for microwaves, $\sim 10^{-2}$ eV for infrared radiation and ~ 1 eV for optical light.

If the electron is energetic enough, the incoming photons can be considered as a unidirectional beam in the ERS since the angular distribution of the photons in that system is confined to angles $\sim 1/\gamma$ rad. In this case, for isotropic distributions of monoenergetic electrons and photons the spectrum of upscattered photons is given by (Jones 1968)

$$\frac{\mathrm{d}R(\gamma,\epsilon_0)}{\mathrm{d}\epsilon_\gamma} = \frac{2\pi r_e^2}{\epsilon_0 \gamma^2}\left[2q\ln q + (1+2q)(1-q) + \frac{1}{2}\frac{(4\epsilon_0\gamma q)^2}{(1+4\epsilon_0\gamma q)}(1-q)\right], \quad (9.6)$$

where ϵ_γ is the photon energy after scattering, $q = \epsilon_\gamma/[4\epsilon_0\gamma^2(1-\epsilon_0/\gamma)]$ and $1/4\gamma^2 < q \leq 1$.

The resulting spectrum for a distribution of electrons can be obtained after integration over the spectra of electrons and the background radiation

$$\frac{\mathrm{d}f_\gamma}{\mathrm{d}\epsilon_\gamma} = c\int \mathrm{d}\epsilon_0 \int \mathrm{d}\gamma\, f_\nu(\epsilon_0) f_e(\gamma)\frac{\mathrm{d}R(\gamma,\epsilon_0)}{\mathrm{d}\epsilon_\gamma}, \quad (9.7)$$

where $f_{\nu,e}$ are the spectra of the interstellar radiation field and the CR electrons.

If the distributions are anisotropic, the formulas become more complicated. The case when the background photons are emitted mostly by stars and dust in the Galactic plane, electrons are distributed in the Galactic halo and the observer is at an arbitrary point in the Galactic plane, is considered in Moskalenko and Strong (2000).

9.2.2 Cosmic-Rays

CR are observed with energies from MeV up to more than 10^{20} eV; those of relevance for the γ-ray continuum are electrons from MeV to TeV and protons and α particles from GeV to TeV energies. Direct measurements from balloons and satellites give the spectra in the local region of the Galaxy. For energies below 1 GeV the modulation by the heliosphere is large; thus, γ-ray and radio observations are our main source of information on the interstellar spectra.

The *primary* CR are those produced in sources such as supernova remnants and hence have the composition characteristic of the material accelerated by the sources. *Secondary* CR are produced in the interstellar medium by spallation of nucleons (e.g. $^{12}C \to {}^{11}B$) or other nuclear reactions such as charged pion production leading to secondary positrons and electrons. The γ-ray emission is dominated by interactions of primaries, with secondaries giving only a minor contribution.

The ambient spectrum of primaries is the result of a combination of the injection spectrum at the sources and the complex effects of interstellar propagation. It is now generally believed, on the basis of composition studies, that the primary CR represent accelerated material of the general interstellar medium. The particles reside for at least 10 million years in the Galaxy, spending most of the time in a region several kpc wide around the disk; the most direct evidence for this wide region is the observation of synchrotron radiation from CR electrons, which indicates a scale height much larger than that of the disk. More indirect determinations from radioactive CR nuclei, especially ^{10}Be, suggest a height of at least 4 kpc.

The dominant propagation mode is probably diffusion due to scattering on magnetic irregularities in the interstellar medium; however convective transport due to an outflowing Galactic wind is also likely to play a rôle. Energy losses have a major impact especially on electrons (via ionization, bremsstrahlung, inverse-Compton emission, and synchrotron emission) and are also important for nucleons below a few GeV (via Coulomb and ionization losses). In addition the diffusive scattering of CR on magnetohydrodynamic turbulence inevitably leads to a certain amount of redistribution of energy by second-order Fermi processes, usually referred to as 'diffusive reacceleration,' and this may have an important influence on the relation between injection and ambient spectra. The effect of supernova shocks pervading the Galaxy may also play a role in reaccelerating particles after injection.

Figure 9.1 shows the spectrum of protons as observed directly and as computed from a propagation model for various injection spectra as described in Sect. 9.4.3; at low energies the effect of solar modulation is evident. Above 1 GeV the spectrum is represented well by a power law in momentum with an index -2.75; at low energies it turns over due to energy losses. It should be emphasized that the direct observations refer only to the local region of the Galaxy, which is not necessarily representative or even typical of the rest of the Galaxy; hence the importance of γ-ray observations which sample much larger regions. The injection spectrum of nucleons is thought to be a power law in momentum with an index between -2.0 and -2.3; the spectrum is steepened at high energies by the effects of propagation, the most important of which is the decrease of residence time in the Galaxy with increasing energy.

Figure 9.2 shows the spectrum of electrons, again as observed directly and as computed from a model for various injection spectra as described

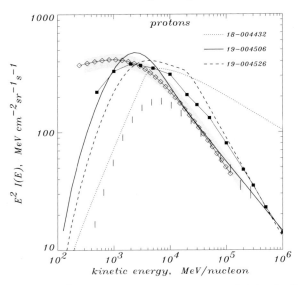

Fig. 9.1. Cosmic-ray proton spectra as obtained after propagation in the models described in the text, compared with experimental data and published estimates of the interstellar spectrum. *Solid line*: using power law injection spectrum; *dashed line*: power law with break in injection spectrum at 20 GeV; *dotted line*: hard nucleon spectrum. *Vertical bars*: IMAX (Isotope Matter-Antimatter Experiment) balloon experiment direct measured values (Menn et al. 2000). Evaluations of the interstellar spectrum: *shaded area*, based on IMAX data (Menn et al. 2000); *connected filled squares*, Webber and Potgieter (1989) and Webber (1998); *connected open diamonds*, based on LEAP (Low Energy Antiproton) and IMP-8 (Interplanetary Monitoring Platform 8) satellite data (Seo et al. 1991). (Reproduced with permission from Strong et al. 2000)

in Sect. 9.4.3. As for protons, solar modulation strongly affects the direct observations below 1 GeV. During propagation the effect of energy losses is dominant, causing the spectrum to steadily steepen from low to high energies. At low energies, ionization losses progressively reduce the flux, and at high energies IC and synchrotron emission cause a steepening by about 1 in the spectral index. Around a few GeV the ambient index is about –3.0, as deduced from radio synchrotron observations. At higher energies the direct observations give an index of –3.3, but this may not be representative of the Galaxy as a whole. The injection index of electrons is considered to lie between –1.7 and –2.1, and this is consistent with the theory of shock acceleration in supernova remnants, which are the favoured candidate sources. New direct evidence for high-energy electrons in supernova remnants comes from X-ray observations of SN 1006, which show a power-law spectrum indicating synchrotron radiation from very high-energy electrons in the magnetic field of the remnant.

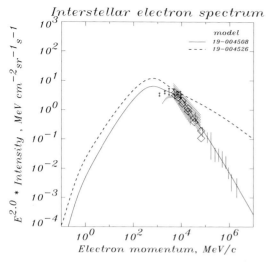

Fig. 9.2. Cosmic-ray electron spectra as obtained after propagation in the models described in the text, compared with direct measurements. *Solid line*: 'conventional' model; *dashed line*: hard electron injection spectrum. Data: Taira et al. (1993) (*vertical lines*), Golden et al. (1984, 1994) (*shaded areas*), Ferrando et al. (1996) (*small diamonds*), Barwick et al. (1998) (*large diamonds*). Adapted from Strong et al. (2000)

9.2.3 Galactic Structure

The Galaxy is a barred spiral with a radius of about 20 kpc. From the point of view of γ-ray diffuse emission the important components are the gas and the interstellar radiation. These are also relevant for the energy losses of CR. The gas content is dominated by atomic and molecular hydrogen, which are present in approximately equal quantities ($10^9\ M_\odot$) in the inner Galaxy, but with very different radial distributions (Fig. 9.3); the atomic gas (HI) extends out to 30 kpc, with a rather uniform density and a scale height of about 200 pc, while the molecular hydrogren (H_2) is concentrated within $R < 10$ kpc, with a peak around 4 kpc and a smaller scale height, about 70 pc. The HI gas has typical densities of 1 atom cm^{-3}. It is mapped directly in its 21 cm radio line, which gives both distance (from the Doppler-shifted velocity and Galactic rotation models) and density information. The H_2 hydrogen is concentrated mainly in dense clouds of typical density of 10^4 atom cm^{-3}. It cannot be detected directly on large scales but the abundant molecule ^{12}CO is a good 'tracer,' since it forms in the dense clouds where the H_2 resides. The 2.6 mm ^{12}CO line has been mapped over the entire Galactic plane, and as for HI the velocity measurement gives spatial information. The derivation of H_2 density from the CO data is problematic; normally a linear relation is assumed and the conversion factor derived from independent estimates

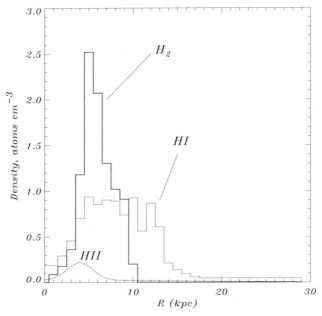

Fig. 9.3. Radial distribution of gas density in the plane of the Galaxy. The atomic (HI), molecular (H$_2$) and ionized (HII) components are shown. (Reproduced with permission from Strong and Moskalenko 1998)

of the mass of gas, including the assumption of virial equilibrium and γ-ray analyses. The γ-ray method has the advantage of sampling large regions of the Galaxy and requiring only the assumption that CR freely penetrate molecular clouds; further details are given in Sect. 9.4.5. In addition to hydrogen, the interstellar gas contains heavier elements, dominated by helium, with a ratio of ~10% by number relative to hydrogen. Helium is therefore an important contributor to the gas-related γ-ray emission.

Ionized hydrogen (HII) is present at lower densities but with much larger vertical extent. The 'warm ionized medium' has densities $\sim 10^{-3}$ atom cm^{-3} and a scale height of 1 kpc. This gas makes a small contribution to the γ-ray emission, but is nevertheless of interest because it produces a much broader latitude distribution than the neutral gas.

The interstellar radiation field (ISRF) is essential for electron propagation (energy losses) and γ-ray production by IC emission. It is made up of contributions from starlight, emission from dust, and the cosmic microwave background. Estimation of the spectral and spatial distribution of the ISRF relies on models of the distribution of stars, absorption, dust emission spectra and emissivities and is therefore in itself a complex subject. New data from infrared surveys by the IRAS (Infrared Astronomy Satellite) and COBE (Cosmic Background Explorer) satellites have greatly improved our knowledge of

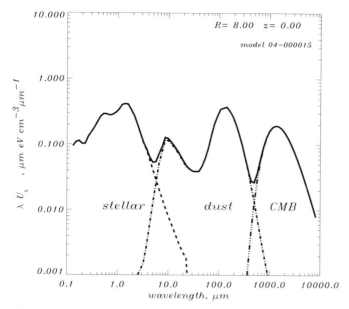

Fig. 9.4. Estimated spectrum of the interstellar radiation field at $R = 8$ kpc. Contributions from stars ('stellar'), dust emission ('dust') and the cosmic microwave background ('CMB') are shown. The calculation is described in Strong et al. (2000)

both the stellar distribution and the dust emission. Figure 9.4 shows a recent estimate of the spectrum at $R = 8$ kpc, near the solar position. Stellar emission dominates from $0.1\,\mu$m to $10\,\mu$m, and emission from very small dust grains contributes from $10\,\mu$m to $30\,\mu$m. Emission from dust at $T \sim 20$ K dominates from $20\,\mu$m to $300\,\mu$m. The 2.7 K microwave background is the main radiation field above $1000\,\mu$m. The ISRF has a vertical extent of several kpc, since the Galaxy acts as a disk-like source of radius ~ 10 kpc. The radial distribution of the stellar component is also centrally peaked, since the stellar density increases exponentially inwards with a scale-length of ~ 2.5 kpc until the bar is reached. The dust component is related to that of the neutral gas (HI + H_2) and is therefore distributed more uniformly in radius than the stellar component.

9.3 Gamma-Ray Observations

Observational techniques and missions are covered in Chaps. 1 and 3.

Today most of our information for energies above 1 MeV is based on the full-sky surveys by the COMPTEL (Compton Telescope) and EGRET (Energetic Gamma-Ray Experiment Telescope) instruments on NASA's Compton Gamma-Ray Observatory (CGRO). Figures 9.5 and 9.6 show representative

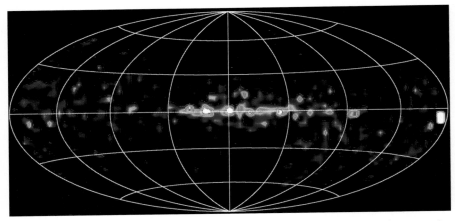

Fig. 9.5. COMPTEL all-sky map in continuum γ-ray emission for energies 10–30 MeV. The coordinates are Galactic centered at $l = 0°$, $b = 0°$, with $45°$ and $30°$ grid spacing in l and b respectively. (Reproduced with permission from Strong et al. 1999)

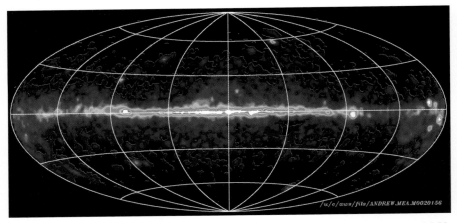

Fig. 9.6. EGRET all-sky map in continuum γ-ray emission for energies >100 MeV. Coordinates as in Fig. 9.5. (A.W. Strong, unpublished data)

whole-sky maps from these instruments, at low energies (10–30 MeV) and high energies (>100 MeV). The Galactic plane stands out as the dominant source of emission; most of this is believed to be indeed of diffuse interstellar origin, except perhaps at low energies. Discrete sources are also visible in these maps, as described in detail in Chap. 4. While the maps give an overview of the general distribution of the emission, quantitative evaluation is made using the full energy and angular information contained in the original data. COMPTEL data analysis techniques for diffuse continuum emission are described in Bloemen et al. (1999) and Strong et al. (1999).

Below 1 MeV the OSSE (Oriented Scintillation-Spectrometer Experiment) instrument on CGRO provides spatial and spectral information for the inner Galaxy (Kinzer et al. 1999), which connects to the X-ray regime measured by instruments such as ROSAT (Röntgen Satellite), RXTE (Rossi X-ray Timing Explorer) and the GINGA satellite.

9.4 Interpretation

There are two approaches to the analysis of diffuse continuum emission. One method attempts to fit the data to a combination of components which are as far as possible model independent, using HI and CO surveys as tracers of the gas and treating the emissivity as a function of energy and galactocentric radius as free parameters to be determined. Interpretation in terms of emission processes can then be made on the basis of these emissivities. The non-gas-related component (e.g. IC emission, sources) must however still be modelled, with the scaling factor being determined in the fitting. This approach has the advantage of allowing direct fits to the data rather than skymaps derived from the data. It is especially useful in determining the CR nucleon distribution and the CO-to-H_2 relation. At low energies where IC emission and unresolved sources are probably important, this method is not always so useful since it is based on the assumption that the truly diffuse emission dominates. However model-fitting techniques can be used to advantage for COMPTEL to determine spectra and skymaps (Bloemen et al. 1999). The other approach is direct modelling of the distribution and spectrum of CR in the Galaxy based on theoretical considerations and observational CR data. The models are compared with skymaps in the form of longitude and latitude profiles; this has the advantage of allowing the models to be judged directly. Physical models which attempt to model all the relevant components have been developed (Hunter et al. 1997; Strong et al. 2000). We will first use one of these models to illustrate the salient features of the interpretation. This model has a halo height (in the direction perpendicular to the Galactic plane) of 4 kpc and the modified nucleon and hard electron spectra described in Sect. 9.4.3, which take into account direct measurement of CR and radio synchrotron data, as well as the γ-ray constraints. The CR propagation model is adjusted to agree with the observed CR B/C and $^{10}Be/^9Be$ ratios (Strong and Moskalenko 1998). The distribution of sources is chosen to be compatible with the γ-ray data, since a priori distributions such as supernova remnants are not entirely satisfactory (see discussion in Sect. 9.4.4). This model is not supposed to be unique but provides a reasonable fit to the data over a wide energy range, with deviations at low and high energies which are discussed in Sect. 9.4.3. Since cylindrical symmetry is assumed for the CR distribution, the longitude profiles cannot be expected to be reproduced exactly if there are large-scale deviations from symmetry in either the CR sources or propagation. The gas distribution is however based directly on HI and CO surveys, and so the Galactic structure is correctly included in this aspect of the model.

9.4.1 Longitude and Latitude Distributions

Figures 9.7–9.10 show sample longitude and latitude distributions for 10–30 MeV and 70–100 MeV from COMPTEL and EGRET, together with the model predictions. In the 70–150 MeV range the contributions from bremsstrahlung, π^0 decay and IC emission are roughly equal, and the model reproduces the EGRET data reasonably well, including the extension of the Galactic emission to high latitudes. At lower energies IC emission becomes more important and eventually dominates, but the model accounts for at most 50% of the COMPTEL intensities; the remainder may be attributable to unresolved point sources, as described later. However the general trend is reproduced, which is the principal goal of this approach. At higher energies π^0 decay dominates and the distribution follows closely that of the gas

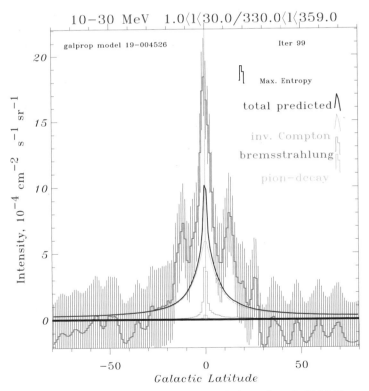

Fig. 9.7. Latitude profile of the inner Galaxy for 10–30 MeV γ-rays, $|l| < 30°$, based on COMPTEL data (Strong et al. 1999), compared with model predictions

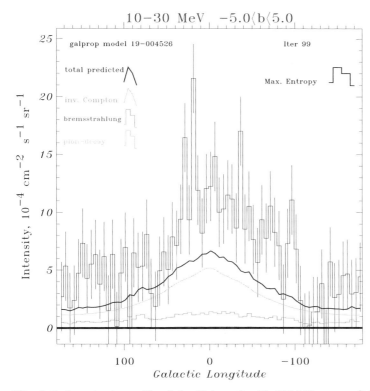

Fig. 9.8. Longitude profile of the Galaxy for 10–30 MeV γ-rays, $|b| < 5°$, based on COMPTEL data (Strong et al. 1999), compared with model predictions

(Hunter et al. 1997); however the intensities above 1 GeV are higher than predicted by π^0 decay, an effect which will be discussed in Sect. 9.4.3.

9.4.2 Spectra of the Inner Galaxy

Spectra of selected regions of the sky are complementary to the intensity profiles. Figure 9.11 shows the spectrum of the region of the inner Galaxy ($330° < l < 30°$, $-5° < b < 5°$) based on EGRET, COMPTEL and OSSE data, compared with a 'conventional' model. This model has an electron injection spectrum with an index of -1.6 below 10 GeV and -2.6 above 10 GeV, which is required to fit the local directly measured electron spectrum after propagation and for consistency with radio synchrotron data. The model reproduces the observed intensities within a factor of three over three orders of magnitude in energy, 10 MeV to 10 GeV, which indicates that the general framework is correct. At low energies, IC emission dominates, at intermediate energies bremsstrahlung becomes more important, and at high energies π^0 decay produces a large fraction of the emission. The deviations from the

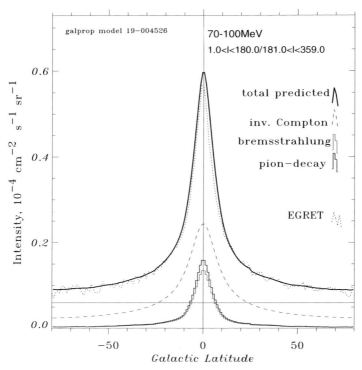

Fig. 9.9. Latitude profile of the Galaxy for 70–100 MeV γ-rays averaged over all longitudes, from source-subtracted EGRET data, compared with model predictions (Reproduced with permission from Strong et al. 2000)

model are however significant and allow some general conclusions to be drawn. First, above 1 GeV the π^0-decay emission falls short of that observed by a factor of ∼2. This is a consequence of the observed shape of the CR nucleon spectrum; possible variations in the normalization (e.g. from CR gradients in the Galaxy) do not improve the fit. Possible ways of resolving this problem are discussed in Sect. 9.4.3. At low energies, the predicted spectrum is compatible with the observations only down to about 30 MeV, below which an additional component seems to be required. Figure 9.11 shows the spectrum extended down to 50 keV using OSSE data; the OSSE observations (Kinzer et al. 1999) show a steeply increasing spectrum, which is confirmed by other hard X-ray observations (GINGA, RXTE). If truly diffuse this emission must be bremsstrahlung, but this would require a very steep low-energy electron injection spectrum (to overcome ionization losses). It is more likely that this component is related to some kind of source population; supernova remnants containing freshly injected CR electrons are possible candidates.

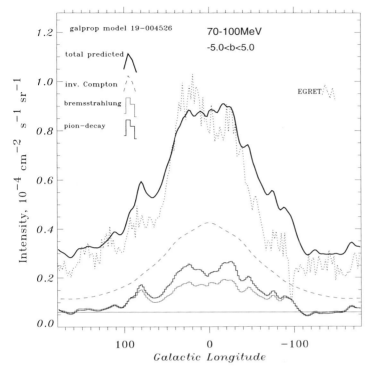

Fig. 9.10. Longitude profile of the Galaxy for 70–100 MeV γ-rays, |b| < 5°, from source-subtracted EGRET data, compared with model predictions. (Reproduced with permission from Strong et al. 2000)

9.4.3 Implications for Cosmic Rays

Nucleon spectrum

It is normally assumed that the nucleon (p, α) spectrum observed in the heliosphere is characteristic of the local interstellar medium (after allowance for solar modulation), and hence can be used directly to compute the π^0 decay emission. The high-energy nucleon spectrum is well represented by a power law, $E^{-2.75}$. We have seen that this leads to a γ-ray deficit above 1 GeV; a possible solution is to allow a harder nucleon spectrum (e.g. $E^{-2.4}$), implying that the heliospheric value is not typical of the average Galactic spectrum. It is then possible to fit the observed γ-ray spectrum. However this would upset the foundations for many aspects of CR physics such as the secondary/primary ratios, secondary positrons, etc., which are known to be consistent with the idea that the measured spectra are typical of the Galactic environment. Independent proof of such a fundamental effect would therefore be necessary. A possible test is the production of secondary antiprotons and

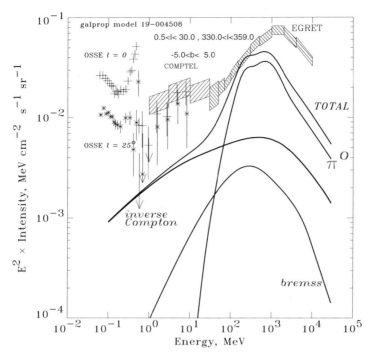

Fig. 9.11. γ-ray spectrum of the inner Galaxy in the 'conventional' model, compared with data. (Reproduced with permission from Strong et al. 2000)

positrons by the required large nucleon fluxes, and in fact this appears already to allow a disproof of this explanation (Moskalenko et al. 1998; Strong et al. 2000).

Electron spectrum

An electron injection spectrum of the approximate form E^{-2} is consistent with supernova shock acceleration, but produces an ambient spectrum (after propagation and energy losses) which is much harder above 10 GeV than that observed locally. A break in the injection spectrum at about 10 GeV to $E^{-2.4}$ is required for agreement with the direct observations up to 1 TeV. However it has been pointed out (Pohl and Esposito 1998) that electrons with energies above 100 GeV undergo such rapid energy losses (time scales $< 10^5$ yr) that large spatial variations are expected; close to a CR source the flux is larger and the spectrum is flatter. Thus we do not really expect the direct measurements to be typical of the local average on kpc scales. Dropping the constraint from the direct observations allows us to continue the hard injection spectrum to TeV energies, and the IC emission from such high-energy electrons can then help to explain the excess diffuse γ-ray emission above 1 GeV. Such

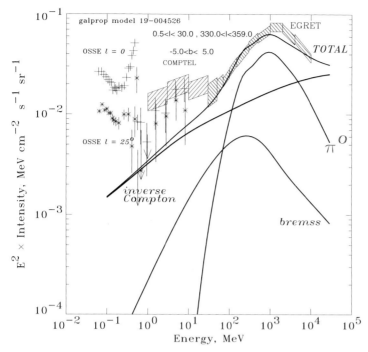

Fig. 9.12. γ-ray spectrum of the inner Galaxy for a model with a hard CR electron injection spectrum, compared with data. (Reproduced with permission from Strong et al. 2000)

a spectrum is also consistent with radio synchrotron data. The model shown in Fig. 9.12 has an electron injection spectrum of the form $E^{-1.8}$ (see Fig. 9.2) and also adopts a slightly modified nucleon spectrum (see Fig. 9.1), which improves the fit to γ-rays while still being consistent with direct nucleon measurements, allowing for some moderate variation with position in the Galaxy.

9.4.4 Multi-component Fitting Approach

Here we describe a method which is an alternative to the explicit physical modelling discussed up to now and which allows a more model-independent analysis of the γ-ray data, as described by Strong and Mattox (1996). The distribution and spectrum of the γ-ray emissivity and the calibration of the CO surveys in terms of molecular hydrogen density are directly determined by this method. The technique depends on the fact that HI and CO surveys contain velocity information which enable the gas to be assigned to a distance from the Galactic Center; assuming the γ-ray emissivity is only dependent on radius, then we have the basis for the gas-related emission.

Other components (IC emission, γ-ray sources and isotropic emission) are also included. The survey data is described in Strong et al. (1988); the kinematic information is used to assign the gas to six Galactic annuli bounded by $R = 2, 4, 8, 10, 12, 15$ and $30\,\mathrm{kpc}$.

The intensity is modelled by the sum of components:

$$I_\gamma(l,b,E) = \sum_i \frac{q_i(E)}{4\pi}[N_{\mathrm{HI},i} + 2X_\gamma(E)W_{\mathrm{CO},i}]$$
$$+ f_{\mathrm{IC}}(E)I_{\mathrm{IC}} + I_B(E), \tag{9.8}$$

where $I_\gamma(l,b,E)$ is the γ-ray intensity (cm^{-2} sr^{-1} s^{-1}), $q_i(E)$ is the γ-ray emissivity per H atom in the ith ring, $N_{\mathrm{HI},i}$ is the atomic hydrogen gas column density in the ith ring, $W_{\mathrm{CO},i}$ is the integrated CO temperature in the ith ring and $I_\mathrm{B}(E)$ is the isotropic background term. We assume that the molecular hydrogen column density is related to the integrated CO temperature by $X_{\mathrm{CO}} = N_{\mathrm{H}_2}/W_{\mathrm{CO}}$ molecules cm^{-2} (K km s^{-1})$^{-1}$, but since the relation of the 'γ-ray' conversion factor to the true factor may be subject to other effects (e.g. CR production or exclusion from molecular clouds) the symbol X_γ is used for this parameter. (Note that X_γ can be energy-dependent while X_{CO} by definition cannot.) I_{IC} is a model-based IC intensity, which is scaled by the free parameter f_{IC}. The free parameters of the model for each energy range

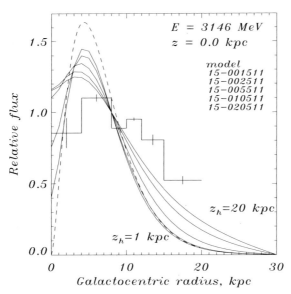

Fig. 9.13. Radial distribution of γ-ray emissivity of the Galaxy for $E_\gamma > 100\,\mathrm{MeV}$ (Strong and Mattox 1996) compared to the distribution predicted for various halo sizes (*solid lines*) for a source distribution like that of supernova remnants (*dashed line*). (Reproduced with permission from Strong and Moskalenko 1998)

are the $6q_i(E)$, $X_\gamma(E)$, $f_{IC}(E)$, $I_B(E)$. The method has been applied to EGRET data covering 30 MeV–10 GeV. The radial distribution of emissivity for $E > 100$ MeV is shown in Fig. 9.13. The most important conclusion is the presence of a variation with R, which illustrates the Galactic origin of CR (otherwise we would not expect a variation), and the fact that the gradient is rather small. For comparison the distribution of supernova remnants is also plotted; although this distribution is uncertain, the difference between this and the emissivity gradient is striking, and it can also be shown that propagation in a large halo hardly improves the agreement. For this reason it seems necessary to conclude that either (a) supernova remnants are not the main sources of CR, or (b) the distribution of supernova remnants is much flatter than that derived from radio catalogues, due to selection effects, or (c) CR propagation in the Galaxy is not isotropic but is more rapid in the radial direction. Since from many other lines of evidence supernova remnants are indeed likely to be the CR sources, case (b) has been favoured and case (c) has yet to be investigated.

9.4.5 Implications for ISM: H_2/CO Calibration

An important outcome of the method described above and other similar approaches is the empirical derivation of X_γ. Figure 9.14 shows the values as a function of energy from the study of EGRET data (Strong and Mattox 1996). The lack of a strong energy dependence gives confidence that the measured values are a good estimate of the 'true' conversion factor and that effects such as energy-dependent CR penetration into molecular clouds are not important. Some γ-ray values are listed in Table 9.1; sample estimates based on non-γ-ray methods are listed in Table 9.2. For the large-scale values the agreement between the estimates is satisfactory. In general X_{CO} and X_γ

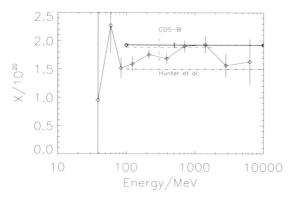

Fig. 9.14. Gamma-ray estimates of X_γ as function of energy from EGRET data, compared with values from Hunter et al. (1997) (EGRET) and Strong et al. (1988) (COS-B). (Reproduced with permission from Strong and Mattox 1996)

Table 9.1. Some published γ-ray estimates of X_γ.
[Units: 10^{20} molecules cm^{-2} (K km s^{-1})$^{-1}$]

Reference	Region	Energy (GeV)	X_γ	Comments
Strong et al. (1988)	Galaxy	0.15–5	1.9 ± 0.3	COS-B; gradient model
Hunter et al. (1997)	Galaxy	0.10–10	1.56 ± 0.05	EGRET; coupling model
Strong and Mattox (1996)	Galaxy	0.10–10	1.9 ± 0.2	EGRET; gradient model
Hunter et al. (1994)	Ophiuchus	0.10–10	1.1 ± 0.2	EGRET
Digel et al. (1999)	Orion	0.10–10	1.35 ± 0.15	EGRET
Digel et al. (1996)	Perseus arm	0.10–10	2.5 ± 0.9	EGRET; outer Galaxy
Digel et al. (1996)	Cepheus, Polaris	0.10–10	0.92 ± 0.14	EGRET; local

Table 9.2. Some published non-γ-ray estimates of X_{CO}.
[units: 10^{20} molecules cm^{-2} (K km s^{-1})$^{-1}$]

Reference	Method	X_{CO}
Lee et al. (1990)	Virial, LTE	1.6–2.8 (inner Galaxy)
Sodroski (1991)	Virial	1.8–2.6 (inner Galaxy)
Sodroski (1991)	Virial	4.7–6.7 (outer Galaxy: $R \approx 11$ kpc)
Digel et al. (1990)	Virial	4–14 (outer Galaxy: $R \approx 13$ kpc)
Magnani and Onello (1995)	CH molecule	2.0 (average for 12 translucent clouds)
Magnani and Onello (1995)	CH molecule	3.5 (average for 13 dark clouds)
Sodroski et al. (1995)	FIR (COBE)	0.2–0.7 Galactic Center $R < 400$ pc

are consistent, suggesting again that, on average, CR can penetrate molecular clouds without significant attenuation. However from the non-γ-ray methods there is evidence for an increase in X_{CO} with Galactocentric distance, and for a significantly smaller value in some local clouds. Evidence for a higher X_γ in the outer Galaxy has been found in the Perseus arm (Digel et al. 1996), and the radial dependence of X_γ should be investigated in future studies. Using X_{CO} it is possible to estimate the total mass of molecular hydrogen in the Galaxy to be $\sim 10^9 M_\odot$, about equal to the mass of atomic hydrogen in the inner Galaxy (although the atomic and molecular hydrogen distributions are very different, as described in Sect. 9.2.3).

9.4.6 Local Clouds

Some nearby molecular clouds lie at latitudes outside the intense Galactic plane and hence can be detected as separate extended sources. Two, the Orion and ρ-Oph clouds, were detected by COS-B and have been studied in more detail using EGRET. The Orion clouds lie at a distance of about 500 pc and have a mass of $\sim 10^5$ M$_\odot$. According to Digel et al. (1999) the γ-ray emission is approximately as expected from the assumption of normal CR intensities, and $X_\gamma = 1.35 \pm 0.15 \times 10^{20}$ molecules cm^{-2} (K km s^{-1})$^{-1}$, slightly less than the large-scale value for the Galaxy. A similar result was found for the nearer but less massive ρ-Oph cloud (Hunter et al. 1994).

9.4.7 Contribution of Unresolved Sources to 'Diffuse' Emission

The rather close correlation between gas column densities and γ-rays at least above 100 MeV combined with the general agreement between expected and observed intensities leads to the conclusion that most of the high-energy γ-ray emission from the Galactic plane is indeed diffuse in origin. However even when the known catalogued sources are removed, a residual component from undetected/unresolved point sources must exist, and it is an open question just how large this contribution to the 'diffuse' intensity can be. The main contributors are expected to be pulsars since these dominate the Galactic point sources. Estimates based on the 12 known γ-ray pulsars indicate that they may contribute 6–10% above 1 GeV and 3% at lower energies (Pohl and Esposito 1998). However emission from hard-spectrum pulsars of the Geminga type can hardly explain the excess high-energy emission discussed in Sect. 9.4.3. The contribution from supernova remnants is likely to be even less, since only two (γ Cyg and IC443) have been identified in the EGRET data. It has not been excluded that other, unexpected populations of fainter sources exist which contribute more to the plane emission. Below 10 MeV the source contribution may be dominant and may well be supernova remnants, as discussed in Sect. 9.4.3.

9.4.8 High Latitudes

The determination of the extragalactic, isotropic γ-ray spectrum is dependent upon knowledge of the Galactic emission at high latitudes. The latter is rather uncertain and model dependent. Figure 9.15 shows the high-latitude intensity spectrum from COMPTEL and EGRET, and for comparison we show the expected Galactic emission for a halo size of 4 kpc and the hard electron spectrum model. The IC emission makes a quite significant contribution, which would be larger for larger halos. If the excess above 1 GeV in the Galactic spectrum is indeed due to a hard electron spectrum as discussed above, then the contribution to the high-latitude emission will be very important (Pohl and Esposito 1998; Moskalenko and Strong 2000). The derivation

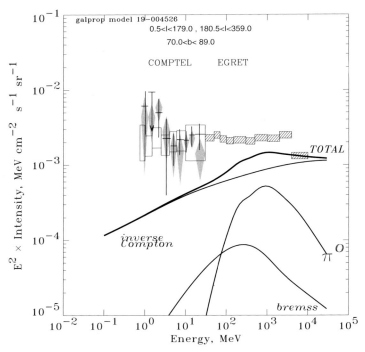

Fig. 9.15. Gamma-ray spectrum at high latitudes: model with 4 kpc halo. For data references see Strong et al. (2000). (Reproduced with permission from Strong et al. 2000)

of the extragalactic spectrum has generally used models predicting a smaller Galactic foreground (Sreekumar et al. 1998). For more details of the extragalactic spectrum see Chap. 14.

9.5 Global Gamma-Ray Properties of Our Galaxy

So far some 70 extragalactic sources have been observed with the EGRET telescope (Hartman et al. 1999), and 10 of these have been seen with COMPTEL (Schönfelder et al. 2000). Most of these sources are blazars. Such data usually serve as a basis for estimates of the extragalactic γ-ray background radiation. However, the number of normal galaxies far exceeds that of active galaxies, so it is interesting to calculate the total diffuse continuum emission spectrum of our Galaxy as an example. Using cosmological evolution scenarios, this spectrum can be used as the basis for estimates of the contribution from normal galaxies to the extragalactic background. The luminosity spectrum of the diffuse emission from the Galaxy is shown in Fig. 9.16, based on a model with a 4 kpc halo. The total γ-ray diffuse luminosity of the Galaxy is

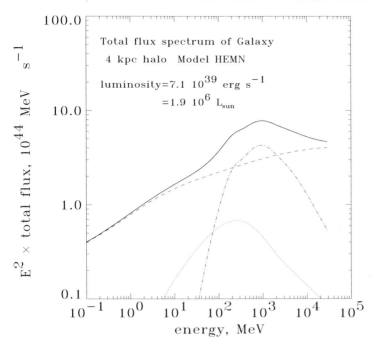

Fig. 9.16. Gamma-ray luminosity spectrum of diffuse emission from the whole Galaxy (4 kpc halo). The total is shown as a *solid line*. Separate components: IC (*dashed line*), bremsstrahlung (*dotted line*), and π^0 decay (*dot-dashed line*). (Reproduced with permission from Strong et al. 2000)

$L_G = 7.1 \times 10^{39}$ erg s^{-1} above 1 MeV and 5.4×10^{39} erg s^{-1} above 100 MeV. These values are higher than previous estimates [e.g. Bloemen et al. (1984): $(1.6–3.2) \times 10^{39}$ erg s^{-1} above 100 MeV] due to the large halo and the fact that this model incorporates the EGRET GeV excess.

9.6 Extragalactic Objects

9.6.1 The Large and Small Magellanic Clouds

The Large and Small Magellanic Clouds (LMC and SMC) are the only 'normal' galaxies apart from our own for which the sensitivity of current γ-ray instruments gives some hope of detection. The detection of the LMC (Sreekumar et al. 1992) and the upper limit on the SMC (Sreekumar et al. 1993) based on EGRET data are therefore of great importance and relevant to the interpretation of the Galactic diffuse emission. Sreekumar et al. (1992) found that the flux from the LMC appears quite consistent with a CR intensity similar to that in our Galaxy. The non-detection of the SMC provides the long-awaited 'proof' that CR cannot be of universal (extragalactic) origin;

the upper limit (>100 MeV) is only 20% of the expected flux based on the Galactic emissivity. The meaning of these results has been discussed in the context of a possible energy equipartition between CR and magnetic fields. Sreekumar et al. (1992, 1993) concluded that equipartition is possible in the LMC but not in the SMC. Chi and Wolfendale (1993) claimed that the CR density in the LMC is much lower that in our Galaxy and hence that very large magnetic fields are required to give the observed synchrotron emission; in this case the situation is far from equipartition in the LMC. These conclusions have been contested by Pohl (1993), who concluded that, when the energy density of CR electrons is taken into account, equipartition cannot be excluded in either system as long as the average e/p ratio is higher than in the Solar neighborhood.

The LMC will be resolved well by the new mission GLAST (Gamma-Ray Large-Area Telescope; see Chap. 3). The CR distribution can then be studied in detail by analyzing the γ-ray data together with 21 cm HI and 2.6 mm CO surveys of this galaxy. GLAST data should also reveal the degree of enhancement of CR density in the vicinity of the massive star-forming region 30 Dor and the associated superbubble LMC2. Detection or non-detection of the SMC by GLAST will be useful to verify conclusions about the Galactic origin of CR.

9.6.2 Other Galaxies

The prospects for observing other spiral galaxies with existing instruments are poor; Pohl (1994) estimates the flux (>100 MeV) from M31 and M82 as less than 2×10^{-8} cm^{-2} s^{-1}, well below the sensitivity of EGRET (see Chap. 3). The EGRET 2σ upper limit for the >100 MeV γ-ray flux from M31 is 1.6×10^{-8} cm^{-2} s^{-1} (Blom et al. 1999), which is much less than the flux of the Milky Way at the distance of M31. The CR densities in M31 are hence lower than in our Galaxy, and it has less ongoing massive star formation. At comparable flux levels, GLAST will be able to resolve the diffuse γ-ray emission along the major axis of M31 (Digel et al. 2000), to provide information about the relationship between CR, star formation rate, and interstellar gas on a large scale. Spectral measurements may allow assessment of the various contributions to the interstellar emission. The puzzling GeV excess in our Galaxy described in Sect. 9.4 will also be detected if present in M31. From the γ-ray spectra of the Galaxy and M31 the contribution of normal galaxies to the extragalactic γ-ray background can begin to be assessed (see Chap. 14).

References

Barwick, S.W. et al., 1998, Astrophys. J., **498**, 779
Bloemen, H., Blitz, L., Hermsen, W., 1984, Astrophys. J., **279**, 136

Bloemen, H. et al., 1999, Astrophys. Lett. Comm. **39**, 205/[673]
Blom, J.J., Paglione, T.A.D., Carramiñana, A., 1999, Astrophys. J., **516**, 744
Chi, X., and Wolfendale, A.W., 1993, Nature, **362**, 610
Dermer, C.D., 1986a Astron. Astrophys., **157**, 223
Dermer, C.D., 1986b, Astrophys. J., **307**, 47
Digel, S.W., Bally, J., Thaddeus, P., 1990, Astrophys. J., **357**, L29
Digel, S.W. et al., 1996, Astrophys. J., **463**, 609
Digel, S.W. et al., 1999, Astrophys. J., **520**, 196
Digel, S.W. et al., 2000, in: *Acceleration and Transport of Energetic Particles in the Heliosphere: ACE-2000 Symposium*, ed. by Mewaldt, R.A. et al. AIP Conf. Proc., New York **528**, 449
Ferrando, P. et al., 1996, Astron. Astrophys., **316**, 528
Golden, R.L. et al., 1984, Astrophys. J., **287**, 622
Golden, R.L. et al., 1994, Astrophys. J., **436**, 769
Gould, R.J., Phys. Rev., 1969, **185**, 72
Hartman, R.C., et al., 1999 Astrophys. J.S, **123**, 79
Hunter, S.D. et al., 1994, Astrophys. J. **436**, 216
Hunter, S.D. et al., 1997, Astrophys. J. **481**, 205
Kinzer, R.L., Purcell, W.R., Kurfess, J.D., 1999, Astrophys. J. **515**, 215
Jones, F.C., 1968, Phys. Rev. **167** 1159–1169
Koch, H.W., and Motz J.W., 1959, Rev. Mod. Phys. **31**, 920
Lee, Y., Snell, R.L., Dickman R.L., 1990, Astrophys. J., **355**, 536
Magnani, L., Onello, J.S., 1995, Astrophys. J., **443**, 169
Menn, W. et al., 2000, Astrophys. J., **533**, 281
Moskalenko, I.V., Strong, A.W., 1998, Astrophys. J., **493**, 694
Moskalenko, I.V., Strong, A.W., 2000, Astrophys. J., **528**, 357
Moskalenko, I.V., Strong, A.W., Reimer, O., 1998, Astron. Astrophys., **338**, L75
Pohl, M., 1993, Astron. Astrophys., **279**, L17
Pohl, M., 1994, Astron. Astrophys., **287**, 453
Pohl, M., Esposito, J.A., 1998, Astrophys. J., **507**, 327
Schönfelder, V. et al., 2000, Astron. Astrophys. Supp., **143**, 239
Seo, E.S. et al., 1991, Astrophys. J., **378**, 763
Sodroski, T.J., 1991, Astrophys. J., **366**, 95
Sodroski, T.J. et al., 1995, Astrophys. J., **452**, 262
Sreekumar, P. et al., 1992, Astrophys. J., **400**, L67
Sreekumar, P. et al., 1993, Phys. Rev. Lett, **70**, 127
Sreekumar, P. et al., 1998, Astrophys. J. **494**, 523
Strong, A.W., Mattox, J.R., 1996, Astron. Astrophys., **308**, L21
Strong, A.W., Moskalenko, I.V., 1998 Astrophys. J., **509**, 212
Strong, A.W. et al., 1988, Astron. Astrophys., **207**, 1
Strong, A.W. et al., 1999, Astrophys. Lett. Comm. **39**, 209/[667]
Strong, A.W., Moskalenko, I.V., Reimer, O., 2000, Astrophys. J., **537**, 763
Taira, T. et al., 1993, in Proc. 23rd Int. Cosmic Ray Conference (Calgary), 2, 128
Webber, W.R., 1998, Astrophys. J., **506**, 329
Webber, W.R., Potgieter, M.S., 1989, Astrophys. J., **344**, 779

10 Nucleosynthesis

Roland Diehl

10.1 Introduction

Radioactive decay of unstable nuclei is one of the main γ-ray source processes. The radioactive nuclei are a by-product of nuclear reactions in energetic environments; the weak interaction is in most cases responsible for their transformation into daughter isotopes, with a characteristic 'decay time' τ or 'half life' $T_{1/2} = \tau/\ln(2)$. The daughter nucleus is generally created in an excited state; its transition to the ground state often includes electromagnetic transitions, hence emission of line γ-rays with characteristic energy values (see Chap. 11). In radioactive decays of the 'β decay' type, positrons (e^+) are produced, which annihilate upon encounter with their antiparticles, the electrons, to produce characteristic annihilation γ-rays.

Gamma-ray lines thus are causally connected to the nuclear processes of element formation, and their observation makes possible the study of physical conditions in nucleosynthesis sites. The measured γ-ray line intensity translates into the abundance of a specific isotope; hence it constrains the parameters of nuclear reaction networks most directly. In comparison, abundance measurements from other astronomical line observations, atomic or molecular transitions observed in the X-ray to radio regime, generally involve additional assumptions or models of line excitation; therefore the inferred abundances depend on the applicability of these often complex line excitation models.

The 'energetic environments' for nucleosynthesis reactions may be stellar interiors, explosive events such as novae and supernovae, but also high-energy collisions of cosmic ray nuclei and the early universe shortly after the Big Bang.

The initial elemental composition in the universe was established by primordial nucleosynthesis shortly after the Big Bang, producing all the hydrogen and deuterium of the universe (the total hydrogen mass fraction is denoted by X with $X_0 \sim 76$ %), some of the present-day ^{3}He, the major part of ^{4}He (total primordial He mass fraction $Y_0 \sim 24$ %), some ^{7}Li, ^{6}Li, and negligible traces of heavier isotopes ($Z_0 \sim 10^{-5}$ %). Nucleosynthesis inside stars is believed to be the origin of the bulk of elements heavier than He, called 'metals' ('Z') by astronomers. Nuclear reactions inside stars also destroy deuterium and lithium isotopes ('astration'), reducing them significantly below their initial abundances. 'Standard abundances' (Fig. 10.1) have

Standard Abundances

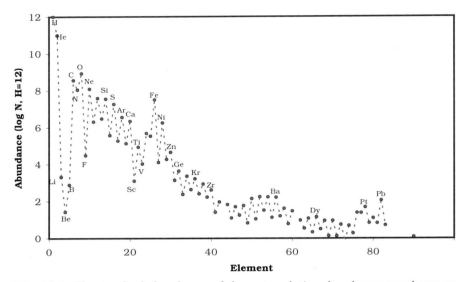

Fig. 10.1. The standard abundances of elements: relative abundances are shown on a logarithmic scale, normalized to hydrogen ($= 10^{12}$). The large dynamic range, but also the regular patterns, are all to be explained by theories of nucleosynthesis. The abundances of specific places, such as the Earth's crust, the interstellar medium in different parts of the universe, or in cosmic rays, deviate in characteristic ways from these 'standard' abundances, which are mostly based on solar-atmosphere and meteoritic measurements

been assessed, mainly from stellar atmosphere absorption lines and meteoritic analyses (Anders and Grevesse 1989). This abundance pattern is believed to be representative for large parts of the evolved universe. A metal fraction of ~2–4% has been contributed by stars during the evolution of the universe. It is the subject of nucleosynthesis and chemical-evolution studies to understand how the primordial elements fit into our understanding of the early universe, and how in detail the enrichment of the full variety of elements in galactic and stellar evolution phases occurred. The observations of radioactivity γ-rays from isotopes with relatively long lifetimes, such that they escape their mostly dense productions sites, allow us to probe these enrichment processes for the recent history of the universe.[1]

Table 10.1 lists the radioactive-decay chains which turn out to be suitable for γ-ray studies of nucleosynthesis. Isotopes with lifetimes ≥hours have a chance to escape a rapidly evolving nucleosynthesis site before decay. On the other hand, when the lifetime reaches millions of years, the sparse amounts

---

[1] 'Recent' here is determined by the radioactive decay time of the observed isotope, which determines the effective 'exposure' of a gamma-ray line measurement.

Table 10.1. The isotopes relevant for γ-ray line astronomy. ('S' = stars, 'SN' = supernovae, 'N' = novae, 'CR' = cosmic rays, 'Ps' = Positronium)

Isotope	Lifetime	Decay chain	Gamma-rays	Source
^{7}Be	77 days	^{7}Be→^{7}Li*	478 KeV	N
^{56}Ni	111 days	^{56}Ni→^{56}Co*→^{56}Fe*	847, 1238 KeV	SN
^{57}Ni	390 days	^{57}Ni→^{57}Co*→^{57}Fe*	122 KeV	SN
^{22}Na	3.8 yrs	^{22}Na→^{22}Ne* + e^+	1275, 511	N
^{44}Ti	89 yrs	^{44}Ti→^{44}Sc* →^{44}Ca*	1156, 68, 78 KeV	SN
^{26}Al	1.04×10^6 yrs	^{26}Al→^{26}Mg* + e^+	1809, 511 KeV	S,SN,N
^{60}Fe	2.0×10^6 yrs	^{60}Fe→^{60}Co*	59, 1173, 1332 KeV	SN
e^+	$\sim 10^5$ yrs	$e^+e^- \to^{(Ps)} \to \gamma\gamma(\gamma)$	511 KeV	SN,N,CR

of this radioactive trace material from a nucleosynthesis site is insufficient to generate the required γ-ray luminosity to overcome the sensitivity threshold of γ-ray telescopes (see Chap. 3).

The intensity in a specific γ-ray line allows us to derive the present amount of radioactive nuclei of type X (number of nuclei) through

$$n_X = I_\gamma \cdot n_\gamma \cdot 4\pi d^2 \cdot \tau_X , \qquad (10.1)$$

with I_γ as measured line flux, n_γ the number of line photons emitted per decay with decay time τ_X of a nucleus of type X, and d the source distance. The originally produced number of radioactive nuclei X then is

$$n_{X0} = n_X \cdot e^{t/\tau_X} . \qquad (10.2)$$

For a source located at a distance of d pc, a measurement of γ-rays from species X with n_γ γ-ray line photons of the measured line energy created per decay, and a species X having a molecular weight of m_X and a radioactive lifetime of τ_X years, a line flux of I_γ (photons cm^{-2} s^{-1}) converts into a nucleosynthesis production yield of[2]

$$M_{X0} = 3.12 \times 10^{-12} \cdot I_\gamma \cdot \frac{1}{d^2} \cdot \frac{n_\gamma \tau_X}{m_X} \cdot e^{t/\tau_X} (M_\odot) . \qquad (10.3)$$

---

[2] The constant 3.12×10^{-12} arises from conversion of units, including Avogadro's constant ($N_A = 6.022 \times 10^{23}$ atoms mole^{-1}), the parsec astronomical distance unit (3.085×10^{18} cm), and the mass of the sun ($M_\odot = 1.99 \times 10^{33}$ g). As an example, a 1.156 MeV ^{44}Ti source at the Galactic Center with $I_\gamma = 10^{-5}$ photons cm^{-2} s^{-1} would correspond to a present ^{44}Ti amount of 9×10^{-6} $M_\odot$ at such a source, consistent with a typical expected Supernova Type II yield of 8.4×10^{-5} $M_\odot$ if the supernova had occurred $\simeq 200$ yr ago. Present-day instruments have sensitivities ranging down to a few 10^{-6} photons cm^{-2} s^{-1}. This corresponds to a capability of detecting typically an individual supernova in ^{56}Ni out to 10 Mpc, the near side of the Virgo cluster of galaxies, or 6×10^{-3} $M_\odot$ of ^{26}Al (corresponding to ~ 150 supernovae) at the distance of the Galactic Center.

In this chapter, I describe the individual sites of nucleosynthesis, and I discuss how γ-ray measurements relate to the physics in specific sources and, more indirectly from accumulated radioactivity in extended regions, to the evolution of stellar ensembles. For more in-depth study, I refer the reader to textbooks and recent review articles on nuclear astrophysics (Rohlfs and Rodney 1988; Arnould and Takahashi 1999), on nucleosynthesis in general (Clayton 1968; Arnett 1996; Pagel 1997; Wallerstein et al. 1997), and on γ-ray line astrophysics with radioactivities (Diehl and Timmes 1998). The history of the field of nucleosynthesis is reflected in its foundation paper 'B^{2}FH' (Burbidge et al. 1957), complemented by other pioneering work (Suess and Urey 1956; Cameron 1957; Clayton et al. 1969; Ramaty and Lingenfelter 1977; Arnett et al. 1977) and reviews (Trimble 1975, 1991).

10.2 Nucleosynthesis Processes

The energy regime of nuclear reactions may be estimated from the requirement of close encounters of nuclei to within the short range[3] of nuclear forces: Coulomb repulsion between protons places the reaction threshold in the regime of MeV energies. The binding energy of nucleons in the atomic nucleus is of the order 8 MeV nucleon^{-1}, typical nuclear energy-level intervals are in the 100 keV to 1 MeV range. Therefore typical line radiation energies of nucleosynthesis sites are of the order of MeV, about 3–5 orders of magnitude above the regime of atomic transitions. Atomic transitions are the basis of spectroscopy of stellar-envelope and interstellar gas, which allows (mostly elemental, as opposed to isotopic) abundances at those sites to be derived.

Nuclear reactions rearrange the configurations of nucleons inside an atomic nucleus via two physical processes:

- Nuclei collide and approach within the range of strong interactions, and
- Weak-force transitions convert neutrons and protons into each other (β decay).

The weak transitions are (almost) independent of density and temperature, while for collision-induced reactions, collision energy and frequency determine the rate of nuclear reactions. The phase space of daughter states in both cases determines the reaction rate.

In laboratory nuclear reactions, projectile energies are chosen conventionally such that the Coulomb barrier is clearly overcome. In astrophysical environments, on the other hand, the interacting particle population has a broad distribution in energies. In this case, the total reaction rate is a convolution of the reaction cross section (as a function of projectile energy) with the energy distribution of the projectiles. This convolution results in the 'Gamow peak' of the reaction rate versus energy (Gamow 1928). The reaction cross

---

[3] The range of nuclear forces is $\simeq$ fm = 10^{-13} cm.

section for astrophysical purposes must be known precisely at and around this relevant energy. Typically the Gamow peak energy (in keV) is[4]

$$E_0 = (15.65 \cdot Z_1 \cdot Z_2 \cdot \mu kT)^{\frac{2}{3}}, \qquad (10.4)$$

with $\mu = \sqrt{\frac{m_1 \cdot m_2}{(m_1 + m_2)}}$ as reduced mass.

'Nucleosynthesis' is the net effect of a complex interplay of many nuclear reactions, whereby material of some initial composition evolves to a different one. The boundary conditions (collision frequency and energy) for the nuclear reactions usually vary during the process. The type of environment has been used to define characteristic 'processes' (Burbidge et al. 1957) which are typical for specific source types, and allow approximations to solve the nucleosynthesis reaction network (see Table 10.2). In the general case, a large set of coupled nonlinear differential equations must be solved (Arnett 1996): For species i, the specific abundance is described as a balance between production and destruction,

$$\frac{\mathrm{d}Y_i}{\mathrm{d}t} = Y_k Y_l \rho N_\mathrm{A} \langle \sigma v \rangle_{kl,i} - Y_i Y_l \rho N_\mathrm{A} \langle \sigma v \rangle_{il,j} + Y_j \lambda_{j,i} - Y_i \lambda_{i,l} + \ldots, \qquad (10.5)$$

with terms for all reaction channels that lead to production or destruction of species i. Here Y_i denotes the relative abundance of species i at their center-of-mass velocities v, N_A the Avogadro number, ρ the mass density, $\sigma_{i,j}$ the reaction cross section between species i and j, and $\lambda_{i,j}$ the decay constant from species i into j. Reaction cross sections $\sigma_{j,i}$ are often steep functions of temperature. The general and time-dependent solution of this network of equations can be very complex. In many cases, however, identification of the most relevant reactions and isotopes reduce the network to a manageable size, thus defining local 'cycles' of reactions, which are only weakly or in a simple way coupled to the rest of the nuclear network (see Table 10.2).[5] Alternatively, thermodynamic treatment of steady-state situations and equilibria provides useful simplifications. Composition changes as part of the process, energy is transferred between its different forms of kinetic and internal energy, and therefore the thermodynamic equations include the chemical potentials:

$$\mathrm{d}U = T\mathrm{d}S + p\mathrm{d}V - \sum \mu_i \mathrm{d}Y_i. \qquad (10.6)$$

The idealized complete equilibrium between all species and fields is called thermodynamic equilibrium. Here matter would adopt its most stable form, Fe group isotopes as most tightly bound nuclei; the isotopic composition

---

[4] As an example, for the $^3\mathrm{He}(\alpha,\gamma)^7\mathrm{Be}$ reaction at 30 million K the Gamow peak energy is 36 keV.

[5] Examples are the CNO cycle in core hydrogen burning, or the Na–Mg–Al cycle (Fig. 10.2).

Table 10.2. The nuclear-reaction process categories relevant for nucleosynthesis in different astrophysical environments. (SN: supernovae; N: novae; CR: cosmic rays; NSE: nuclear statistical equilibrium; Ps: Positronium formation)

Process	Description	Site
NSE	All except weak interactions in thermal equilibrium	SNIa, (SNII)
Quasiequilibria	Equilibrium valid only in localized regimes of nuclear chart	Si-burning, hot H burning, etc.
Freeze-out	Equilibrium breaks down as region expands, only a few reactions remain	BBN, SNII/Ib hot bubble
Nonequilibrium	Reaction rate networks to be solved explicitly (no thermodynamic treatment)	Stellar nucleosynthesis
r/s process	Neutron captures and β decays only	SNII/Ib, pulsating stars
p process	Neutron extraction or p capture (also 'γ process')	Novae, SNII
ν process	ν triggered additional spallation	SNII/Ib hot bubble, BBN

then depends on the relative numbers of protons and neutrons available at a particular temperature.[6]

Neutrinos are produced in weak decays (e.g., $n \to p + e + \overline{\nu_e}$), and carry away energy from the reaction site. Therefore complete thermodynamic equilibrium is usually not attained. nuclear statistical equilibrium' (NSE) may be achieved, with all exept weak interactions being in equilibrium. In this NSE state, all strong nuclear reactions $[A(p,x)B, A(n,x)B, A(\alpha,x)B$, etc., but also $A(\gamma,x)B]$ are balanced by their inverse reactions, such that the overall composition is dynamically stable. Because of the different temperature dependencies of the various reaction cross sections, the resulting composition is a function of temperature.[7] NSE burning is realized in the most dense and

---

[6] The neutron excess is a characteristic parameter for nucleosynthesis product composition. It may also be measured through the relative abundance of electrons Y_e, as p+e and n are in weak-interaction equilibrium, assuming charge neutrality. For equal numbers of neutrons and protons, $Y_e = 0.5$.

[7] It is advantageous to characterize nucleosynthesis environments through the normalized 'entropy per baryon' rather than by 'temperature' (Meyer 1993). Then the relative importance of photodisintegration reactions better describe the characteristics of these usually radiation-dominated environments.

violent nucleosynthesis sites, in thermonuclear supernovae, probably also in inner parts of core collapse supernovae. Radioactive ^{56}Ni is produced in large amounts, when nuclear burning settles to most tightly bound species. Other explosive regimes evolve faster than these equilibration time scales, so that incomplete burning leaves a characteristic abundance pattern.

'Quasiequilibria' may serve as useful approximation to NSE: when changes of abundances Y_i are small compared to the actual rates of production or destruction within a local group of nuclei, equilibrium treatment within this group is allowed. Although NSE is not obtained globally, this significantly reduces the number of reactions that have to be followed explicitly in a reaction network (Arnett 1996; Hix and Thielemann 1999). Such a description applies in the r, s, and α processes (Meyer 1994).

Freeze-out from equilibrium is another useful approximation to characterize the composition from a nucleosynthesis site. Here, an initial equilibrium (or quasiequilibrium) situation experiences rapid dilution, so that nuclear reactions cannot keep up due to the decreasing collision frequency. This results in characteristic modifications of initial abundance patterns, which nevertheless derive from the equilibrium situation in a straightforward manner. Examples are primordial nucleosynthesis in the early universe, but also the 'α-rich freeze-out' from a supernova: at high entropy per baryon, photodisintegration reactions result in an abundance pattern where α nuclei are the most abundant species. The dilution of such an environment due to thermal expansion results in a relative lack of seed nuclei other than α particles, specifically those with odd numbers of nucleons; the most frequent collisions of α particles preferentially build up nuclei composed of α multiples, producing, for example, copious amounts of radioactive ^{44}Ti.

When 'cycles' of nuclear reactions dominate, isotopic abundances are critically determined from the individual nuclear reaction cross sections. Direct, or even indirect, experimental measurements are often difficult, in particular for charged-particle reactions, due to Coulomb repulsion and the dominating tunnel effect (Rohlfs and Rodney 1988). A prominent example is the ^{12}C$(\alpha,\gamma)^{16}$O reaction, a key reaction to generate the seeds for all heavier elements; it controls red giant evolution as well as the chemistry and dust formation in stellar envelopes, which critically depend on the ^{12}C/^{16}O ratio. This reaction's cross section results from details of two subthreshold resonances, and is still a main topic of research; its current value has been inferred from chemical evolution studies, more than from nuclear reaction experiments (Wallerstein et al. 1997). Similarly, measurements of abundances for other intermediate-mass isotopes can help to calibrate nuclear reaction cycles, such as γ-ray line observations of ^{7}Be, ^{22}Na, and ^{26}Al (here in particular the Mg–Al cycle illustrated in Fig. 10.2). For some of these isotopes, production in hot hydrogen burning, hence proton-rich environments, through the rp process can be significant (Schatz et al. 1997).

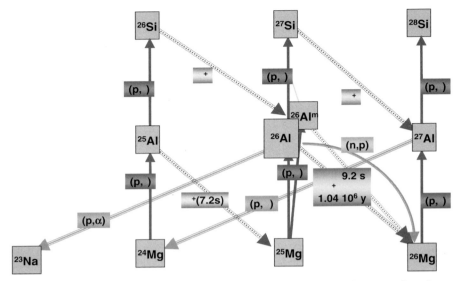

Fig. 10.2. The production of intermediate-mass nuclei occurs in networks of nuclear reactions, involving strong [e.g. (p,γ)] and weak reactions; (p,α) reactions feed back to lighter nuclei and close the cycles, if the (p,γ) break-out reactions are comparatively slow

Neutron capture reactions are of special importance in cosmic nucleosynthesis, because of the absence of Coulomb repulsion in this reaction type. Essentially all elements heavier than iron are produced through successive neutron capture, hence from the seeds of Fe-group elements. In the case of the 'r process,' neutron capture reactions are much more rapid than competing β decays. Isotopes are produced far out on the neutron-rich side of the valley of stability near the neutron drip line[8] through an intense (10^{20} neutrons cm^{-3}, $\tau \leq$ seconds) neutron flooding. They relax into the characteristic final r process abundance pattern through β decays (see Fig. 10.3). In the other extreme, 'slow' neutron capture drives the 's process' into its characteristic abundance pattern (Käppeler et al. 1989). Here the neutron collision frequencies are much lower (10^8 cm^{-3}), so that between neutron captures all isotopes have time to β decay to their stable daughter products. Therefore the s process populates isotopes along the valley of stability, on the neutron-rich side. The source process is unique for isotopes which are either 'shielded' towards the neutron-rich side by a stable isotope ('s only'), or separated by one or more unstable isotopes from the valley of stability ('r only').

---

[8] Here the neutron separation energy S_n reaches small values, often reflecting a configuration where all bound neutron states in a nucleus for a given proton number are filled.

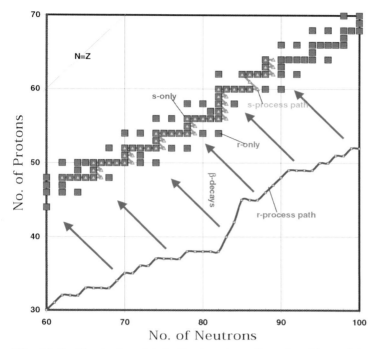

Fig. 10.3. Chart of nuclei, indicating the valley of stable nuclei, and the path of r- and s process nucleosynthesis processes.

For production of isotopes on the proton-rich side of the valley of stability, a 'p process' is invoked. This process is also termed a 'γ process', because (γ, n) reactions yield the same reaction path; it is unclear, which of these reactions produce the 'p isotopes.'

Unfortunately γ-ray-emitting radioactive r, s, or p isotopes are too rare to allow study of the r, s, or p processes through γ-ray line astronomy.

In the inner regions of core collapse supernovae, nuclear reaction sites are exposed to the most luminous neutrino source in the universe, the proto-neutron star. Interactions of neutrinos with nuclei result in spallation, thus releasing additional lighter nuclei, protons and α particles (Qian and Woosley 1996) and smoothing the abundance distribution. This enhancement of seed nuclei for nuclear reactions (ν process) may increase the production of specific intermediate-mass isotopes such as ^{26}Al by up to 50% (Woosley et al. 1990).

In models of isotopic yields from astrophysical sites, two approaches are common:

- The network is restricted to the nuclear reactions with significant contributions to the total energy budget. This approach is common in stellar-evolution models with a consistent treatment of hydrodynamic and nuclear physics.

- The nuclear processing within a complex network of isotopes is decoupled from the hydrodynamic-evolution and applied as post-processing. This approach is common in fast-evolving sites such as novae and supernovae, where the hydrodynamic treatment time scale is thus adapted to the problem, while the detailed composition is evaluated in a secondary step.

Because of these modeling steps, the relation of the measurement of a radioisotope γ-ray line to one of the above physical processes is often indirect.

10.3 Sites of Nucleosynthesis

10.3.1 Hydrostatic Nuclear Burning

In dense astrophysical nucleosynthesis sites, temperatures range over 3–4 orders of magnitude from 10^6 to 5×10^9 K, densities span even 12 orders of magnitude from 100 nuclei cm^{-3} (hydrogen burning) to 10^{14} cm^{-3} (Si burning in massive stars). The time scales for the different stages of stellar burning are quite different, not only because of the steep temperature dependence of most cross sections, but more so because of the different cooling mechanisms: In the hydrogen- and helium-burning stages, the bulk of the nuclear energy is generated in the form of γ-rays and thus is converted into local heating. This, in turn, results in thermal expansion of the burning site, such that the energy generation rate is reduced. Eventually, radiation transport and thermal expansion will have established hydrostatic equilibrium for the burning stage. This may last billions of years for stars as massive as the sun, while, for example, stars 25 times as massive would experience stable hydrogen burning only for a million years. From carbon burning onward, a large part of the nuclear-burning energy emerges in the form of neutrinos and hence escapes from the burning region. Therefore these burning stages are much shorter, exhausting their fuel locally at an almost maximum rate. For example, carbon burning for a 25 M$_\odot$ star lasts only 600 yr, and silicon burning just a day.

Even the seemingly stable hydrostatic burning inside stars will evolve significantly after the main-sequence phase of core hydrogen burning. A steady-state description is rarely adequate to model the nucleosynthesis, even less in intermittent nuclear burning in shell-burning stages of massive stars: nuclear-burning time scales become shorter than the hydrostatic-adjustment time scales, and pulsing instabilities are a common characteristic of these stars in their 'giant' stages (Iben and Renzini 1983).

Stars ascend to the Asymptotic Giant Branch (AGB) after core ^{4}He burning has been exhausted (Lattanzio and Boothroyd 1997). The stellar envelope is enriched with burning products by 'dredge-ups'. In AGB stars, both hydrogen and helium eventually burn concurrently in shells. For low-mass ($\leq$4 M$_\odot$) AGB stars, a flash of helium burning in the inner part will cause expansion of the outer part, and thus extinguish the shell hydrogen burning further out; only after a settling time will re-started hydrogen shell burning have produced new fuel for this inner-shell helium burning to resume. Thermal-pulsing cycles are the result of this basically unstable situation. Convective energy and material transport ('third dredge-up') characterizes this phase, and the physical environment for nuclear reactions will vary significantly both in time and space. For more massive ABG stars, the bottom of the convective envelope may reach down to the hydrogen-burning shell, so that fresh fuel is supplied to hydrogen burning, turning carbon-rich envelopes resulting from earlier burning stages into oxygen-rich envelopes and producing ^{7}Li and other intermediate-mass isotopes in this 'hot-bottom burning.' Typical envelope turnaround times are ~ 0.5 yr. The outer envelope of AGB stars is weakly bound; strong stellar winds expel substantial parts of the stellar envelope into interstellar space. Therefore the interstellar medium element abundances will be sensitive to how efficiently products from nuclear reactions deeper inside the star can be mixed into the envelope, where the wind is formed. Additionally, the cool outer envelopes favor formation of dust; therefore the laboratory study of presolar dust grains embedded in meteoritic material has been a rich field to analyze and constrain AGB nucleosynthesis (Bernatowicz and Walker 1997). Due to their relatively long stellar lifetime, in AGB stars even inefficient production processes can become important for the cosmic abundance budget. There is sufficient time for reactions operating in the far tails of the Gamow peak to produce intermediate-mass elements such as Ne, Mg and Al, or for slow neutron capture on prestellar heavy-element seeds to produce rare-earth elements in the s process. The detection of atomic lines from technetium in the atmosphere of an AGB star indeed provided the first unambiguous proof of nucleosynthesis[9] in stellar interiors (Merill 1952).

The γ-ray radioisotope ^{26}Al may be produced in hydrogen shell burning at temperatures $> 5 \times 10^7$K. Additionally, in the hot-bottom burning scenario the Mg(p,γ) reaction can produce ^{26}Al very efficiently (even from the more abundant ^{24}Mg isotope), mixing freshly produced ^{26}Al quickly into the wind. Therefore, AGB stars of intermediate mass are candidate ^{26}Al sources. This is true even without hot-bottom burning, due to the third dredge-up and thermal pulses; yet here some ^{26}Al also is expected to be mixed downward closer to the He burning shell, where n-capture reactions quickly destroy it. The large uncertainties of convective processes and intermittent shell burning translate into significant uncertainty in the interstellar nucleosynthesis yields from AGB stars (Forestini and Charbonnel 1997).

In more massive stars, stellar evolution is initially similar to that described above, with convective-core hydrogen burning. The more massive cores result in higher temperatures, however, so that for stars of $\geq 15 M_\odot$ proton capture reactions can produce fresh isotopes such as ^{26}Al from suitable seed nuclei. Thus, as a by-product of hydrogen burning, ^{26}Al is enriched in the

---

[9] Tc has only unstable isotopes, with radioactive decay times $\leq 10^6$ yr.

H-burning core and predominantly left behind in the nonconvective outer core as convection recedes during progressive H-burning. The enormous radiation pressure in these stars during the shell burning phase drives a strong wind of 10^{-5}–10^{-4} $M_\odot$ yr^{-1}, quickly removing the inert outer envelope and uncovering the hydrogen-burning layer on time scales of order 10^6 yr. Therefore, for such massive stars, this 'Wolf Rayet' (WR) phase helps to extract ^{26}Al from the earlier H-burning phase into the interstellar medium, before it can decay or be destroyed in He-burning, as is the case for stars with $M \leq 20$–40 $M_\odot$, the lower limit for a star to become WR (Meynet et al. 1997). There are interesting deviations from this first-order model for substantial ^{26}Al production by WR stars: stellar rotation may enhance the mixing processes, thus enhancing the supply of seed nuclei for proton capture nucleosynthesis. The high fraction of binary systems (more than 50% of massive stars are members of binary systems) may modify nucleosynthesis of the system: the companion's gravitational pull produces tidal effects, which affect convective mixing of envelope material and hence late evolution; the mass transfer to the secondary enriches its envelope with seed material for more efficient shell burning, with the products then ejected in its supernova.

10.3.2 Explosive Nucleosynthesis

In explosive events such as novae (Gehrz et al. 1998) and supernovae (Burrows 2000) large-scale hydrodynamical adjustment of the burning site due to the local release of nuclear energy cannot occur, in contrast to the hydrostatic nuclear burning described above. Much higher burning temperatures develop in these sites and result in substantial nuclear processing of material, in spite of the short duration of explosive nuclear burning. Thus, explosive events are prime sites of nucleosynthesis. Additionally, supernova explosions release large amounts of processed material instantaneously, while hydrostatic-burning products have to be mixed into stellar winds to be released into the interstellar medium during the lifetime of a star. Therefore much of the fresh material produced in explosive events ends up as observable chemical enrichment, while much of hydrostatically produced fresh elements are buried in white dwarfs.[10]

Core-Collapse Supernovae

The final stage of stellar evolution will, for stars more massive than about 8–10 $M_\odot$, result in a core of nuclei in their most stable configuration (iron-group elements), thus leaving no nuclear source of energy to counteract gravity. At this stage such a star, or more precisely its iron core, will undergo gravitational collapse (see Fig. 10.4 and Burrows 2000), compressing the nuclei until a new physical limit is reached, where the entire core is just one giant

---

[10] This is not true for the inner regions of a core-collapse supernova.

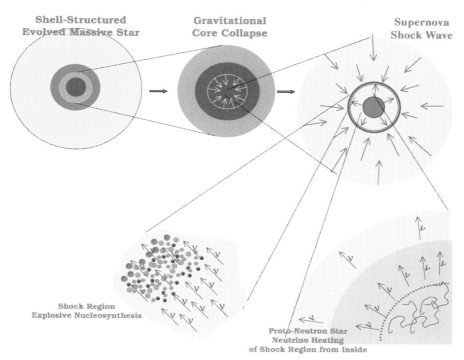

Fig. 10.4. Core-collapse supernovae result from the terminal gravitational collapse of massive star cores. The collapsing matter rebounds when reaching nuclear-matter density; intense neutrino heating from the forming neutron star helps the shock to expand against infalling matter and eject the mantle and envelope. Explosive nucleosynthesis occurs in the shock region

atomic nucleus, i.e. nuclear matter. At this moment, the gravitational collapse is stopped abruptly, resulting in a 'core bounce' shock wave travelling outward. Simultaneously, the protons in the core yield to the high electron pressure and undergo inverse β decay to become neutrons, which can be even more densely packed, thus reacting to gravitational pressure and decreasing electron thermal pressure. This 'neutronization' results in copious emission of neutrinos, radiating most of the gravitational energy ($\sim 10^{53}$ erg) into the infalling material. At these densities, neutrinos have a significant probability for interacting with nuclei of the surrounding matter within this inner core of ~ 300 km radius and thus support the struggle of the core-bounce shock wave against the matter falling in from the mantle, eventually reversing the gravitational collapse into an explosion. During this event, nucleosynthesis will occur in two very different environments: close to the proto-neutron star, infalling material is flooded with neutrons and neutrinos, spallation and neutron captures occur at very high rates; further out, an intense shock wave heats up the shells of material (which may have been burning hydrostatically already; see Fig. 10.4), modifying the nuclear reaction environment substantially out through the oxygen shell, for a transition period ('explosive nucleosynthesis'). The shock wave energy is eventually sufficient to eject all except the very inner parts close to the forming neutron star, thus enriching the surrounding medium with products of explosive burning and of prior hydrostatic nucleosynthesis.

The genesis of the central compact remnant in core-collapse supernovae includes several interesting physical problems, some of which may be constrained by radioactivity observations. Formation of a neutron star appears likely for main-sequence masses below ~ 20 $M_\odot$, while a black hole probably forms for main-sequence masses above this regime. The energy of the explosion, the placement of the mass cut,[11] and how much mass falls back onto the remnant shortly after the explosion all modify this neutron-star–black-hole bifurcation point. Each of these processes also affects the mass of nucleosynthetic products ejected from the inner regions of the supernova. Note that these supernova parameters are not yet understood and are empirically adjusted in different ways in current models of core-collapse supernova nucleosynthesis (Woosley and Weaver 1995; Thielemann et al. 1996).

The chief explosive-nucleosynthesis products are nuclei of the iron group, from the inner region where NSE conditions are obtained, and intermediate-mass elements such as oxygen and neon, from burning in the supernova shock wave. Enrichment of α elements is likely, from freeze-out due to the rapid expansion of the main inner burning region. Typically, 0.1 $M_\odot$ of radioactive ^{56}Ni is produced, its rapid decay ($\tau \sim$ days) powers the bright optical display of the supernova. The mass profile of these inner nucleosynthesis products is shown in Fig. 10.5.

In supernovae from massive stars, stable ^{44}Ca is produced chiefly, almost exclusively, as radioactive ^{44}Ti in the α-rich freeze-out from the inner part of the supernova. No other ^{44}Ca production process is compatible with the large observed ^{48}Ca/^{46}Ca ratio. Production in core-collapse events after wind loss of the envelope (the 'Type Ib' events)[12] should be more uniform because stellar-evolution models converge to a common presupernova mass in the narrow range 2.3–3.6 $M_\odot$. All of the models, whose ejecta all have $\sim 10^{51}$ erg of kinetic energy (at infinity), predict ^{44}Ti yields[13] between 1 and 15 $\times$ $10^{-5} M_\odot$. Production of more than 10^{-4} $M_\odot$ of radioactive ^{44}Ti is difficult to achieve in such spherically symmetric models adjusted to observables, but ejection

---

[11] The mass cut is the separation line between matter that will end up onto the compact remnant by collapse or post-supernova accretion and matter that will be ejected into interstellar space. For neutron-star remnants, it lies in the regime 1.3–1.9 $M_\odot$.

[12] If the envelope of a core-collapse supernova still contains hydrogen, a 'Type II' event results, while absence of a hydrogen envelope identifies 'Type Ib/c' events.

[13] Typical values are $\sim 3\times 10^{-5}$ $M_\odot$ for the Type II models, or twice that value for the Type Ib models.

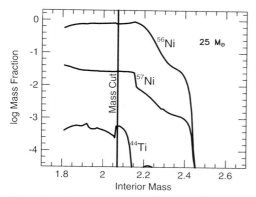

Fig. 10.5. Mass profiles of ^{44}Ti and ^{56}Ni for a 25 M$_\odot$ core-collapse supernova model. The *vertical line* shows the mass cut, separating ejecta from material that ends up on the compact remnant star. (Reproduced with permission from Timmes et al., 1996)

of much less seems easily possible. The interstellar ^{44}Ti yield is especially sensitive to how much mass falls back onto the remnant (see Fig. 10.5).

Supernova remnant observations suggest a nearly constant kinetic energy of the ejecta, which implies that the explosion energy in more massive events must be steadily increased in order to overcome the increased binding energy of the mantle. However, even in present 35 and 40 M$_\odot$ models nearly all the produced ^{44}Ti falls back onto the compact remnant. Unless the explosion mechanism, for unknown reasons, provides a much larger characteristic energy in more massive stars, it appears likely that stars larger than about 30 M$_\odot$ will have dramatically reduced ^{44}Ti yields and leave massive remnants ($M \geq 10\,\mathrm{M}_\odot$), which become black holes. Rotation, however, may modify this picture significantly by breaking the spherical symmetry of the explosion. Regions of material with larger entropy could develop behind an asymmetric shock front, and cause higher ^{44}Ti production, perhaps by as much as an order of magnitude. Plausibly, jets enriched in ^{44}Ti may be induced (e.g. in the polar regions of a rotating supernova), and still remain in agreement with above energy arguments. Birth velocities up to as high as $1000\,\mathrm{km\,s}^{-1}$ are observed for radio pulsars, and often taken as evidence for some small asymmetry in the explosion of core-collapse supernova ('kicks'), although magnetic field or ν wind asymmetry provide alternative explanations. Rotation and magnetic fields are a common characteristic of massive stars, and each could induce spatial asymmetries of nuclear-burning conditions. Consistent modelling of such asymmetry has not yet been achieved. Note that in particular nickel isotope ratios can be measured and provide a tight constraint on the entropy/neutron ratio in the inner nucleosynthesis region.

Thermonuclear Supernovae

A very different kind of supernova results from a terminal evolutionary phase of compact white dwarfs in binary systems (Fig. 10.6 and Nomoto et al. 1997, Thielemann et al. 1986). Accretion of further material from the secondary onto the white dwarf eventually ends up in conditions which ignite carbon inside the white dwarf, which in such a degenerate environment results in a runaway explosion with total disruption of the star. Generally no hydrogen envelope is present; therefore these supernovae are called 'Type Ia.'

We distinguish several scenarios from either the progenitor evolution or the trigger of carbon ignition. Progenitors may be relatively massive white dwarfs with a nondegenerate companion star in its high-mass loss phase ('single-degenerate'). In this case, the accreted hydrogen burns steadily, increasing the white dwarf density gradually above the carbon ignition density ('Chandrasekhar models'). Alternatively, the accretion onto a 0.6–0.9 $M_\odot$ carbon–oxygen white dwarf gradually accumulates a shell of helium, which

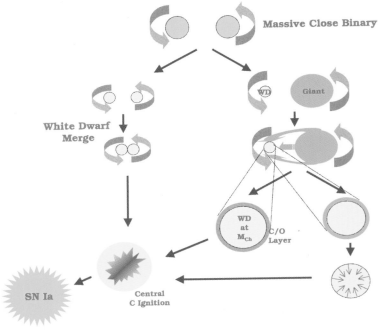

Fig. 10.6. Thermonuclear supernovae result from the explosion of white dwarfs after accretion of matter from a companion induces them to ignite. Central carbon burning in degenerate matter leads to the supernova. Ignition can occur through accretion beyond the Chandrasekhar stability limit or through a shock wave from flash burning of an accreted helium layer. Merging of white dwarfs may be an alternate model for thermonuclear supernovae

eventually becomes massive (0.15–0.20 $M_\odot$) and hot enough to ignite; the compression shock wave from this helium flash then may ignite carbon at the center of the white dwarf ('sub-Chandrasekhar' or 'helium cap' models). In both these models, accretion occurs at relatively high rates (10^{-8}–10^{-6} $M_\odot$ yr^{-1}), resulting in steady hydrogen burning (rather than accumulation as in the case of novae). 'Supersoft' X-ray emission ($\leq$ keV) may result from these progenitors, from nuclear burning heating the white-dwarf surface to unusually high temperatures; this could help us to recognize candidates for future thermonuclear supernovae. Note that the state of the companion determines how the mass transfer occurs. An AGB star companion with its strong wind might engulf the white dwarf, remeniscent of symbiotic stars; alternatively, Roche-lobe mass transfer might occur, in a cataclysmic system. Also, hydrogen burning could be avoided if the companion is a helium star already deprived of its H envelope in earlier binary evolution. A third class of models has two white dwarfs in a close-orbit binary system ('double-degenerate' or 'merger' models). Gravitational radiation incurs loss of orbital energy and eventually leads to merging of the two white dwarfs. The resulting object generally exceeds the Chandrasekhar stability limit and ignites carbon, as above. Rotation and strong tidal effects add special characteristics to this scenario, as does the specific composition of the igniting material.

In all three cases, the compactness of the fuel and the high thermal conductivity of the white-dwarf matter result in a runaway-type evolution, where the nuclear energy generation rate is too high for any adjustment processes: nuclear burning proceeds up to the iron peak, and the enormous energy density disrupts the entire star in a gigantic explosion. The nucleosynthesis conditions in this type of supernova are very different than for core collapse: nuclear burning occurs simultaneously with neutronization of matter, and the temperatures and densities most likely induce NSE. A large amount of radioactive ^{56}Ni (0.5 to 1 $M_\odot$) may be produced. Details of the supernova flame evolution determine the evolution of the light curve and the velocity profile of the ashes; a critical parameter is the duration of the subsonic deflagration phase relative to the supersonic detonation phase of the burning front. The abundances of neutron-rich isotopes of the iron group may provide sensitive diagnostics of this phase. Physical models of this phase of the supernova have not been obvious, although flame propagation in combustion engines shows promising similarities in details (Niemeyer 1999). In empirical full-scale models, parameters are tuned to the observed appearance of the supernovae; the empirical 'W7' model still serves as the best description (Iwamoto et al. 1999). Nevertheless, the evaluated total bolometric light for this class of supernovae has been found to be identical to within a few percent and forms the basis of the use of Type Ia supernovae as standard candles in cosmological studies (Branch 1998). The absence of an outer envelope implies that both the overall radioactivity produced and the nucleosynthesis abundance pattern are dominated by the white-dwarf configuration, which is hence expected to vary less than the core-collapse supernovae, which are the final stage for stars of masses between 8 and ~ 80 $M_\odot$.

Supernova Gamma-Rays

From their very different progenitors, core-collapse supernovae are expected to be found wherever massive stars form, with a relatively short time lag after initial star formation due to the short (million years) evolutionary time scale of such stars. On the other hand, thermonuclear supernovae occur only after intermediate-mass stars have terminated their stellar evolution ($\sim 10^8$ yr and more), and the remaining white dwarf within a binary system accumulates a critical surface layer from its companion.

How do γ-ray-emitting radioisotopes relate to these two supernova types? The large amounts of ^{56}Ni and ^{57}Ni radioactivity (~ 0.5 $M_\odot$ for thermonuclear, about 1/10 of that for core-collapse supernovae) provide the prime γ-ray diagnostics of the explosion morphology: when the supernova envelope gradually becomes transparent to γ-rays, decay γ-rays show up as a line feature above the Compton-scattered continuum γ-ray emission. Different relative γ-ray line intensities, as they evolve with time, encode the structure of the envelope and its mixing with fresh nucleosynthesis products. Gamma-ray line shapes measure the nucleosynthesis ejecta velocities, complementing the velocity information of atomic lines from the envelope [here ^{44}Ti adds late light curve information to the early light curve data from Ni isotopes (Chan and Lingenfelter 1991)]. The unique information from γ-ray measurements is a calibration of the total radioactivity produced in the event and powering the light curve after the initial flash in all spectral bands. The Ni isotopes again address the early light curve, where abundant optical/UV/IR light curves need such calibration; the ^{44}Ti information is unique for the late supernova/early supernova-remnant phase which is difficult to constrain otherwise. Additionally, the yields in the specific isotopes which are directly visible through γ-rays set constraints on the nucleosynthesis conditions and inner explosion dynamics (this adds ^{26}Al and ^{60}Fe to the list of relevant γ-ray probes of supernovae).

Novae

A classical nova is now understood to result from explosive burning of a shell of hydrogen which has been accreted onto a white dwarf from a companion star. This evolutionary path is reminiscent of the mass transfer path to a thermonuclear supernova (see Fig. 10.6). The rate of accretion must be tuned within a range $\leq 10^{-9}$ $M_\odot$ yr^{-1}, such that the gravitational energy release of the accretion flow remains below the critical heating towards hydrogen nuclear burning on the white-dwarf surface. On the other hand, accretion must be sufficiently rapid to build up an amount of hydrogen which can trigger the nuclear burning that starts a nova, within the time scale of galactic evolution. Once the energy generation rate of nuclear burning rises above the level that can be radiated away by the white-dwarf surface, a runaway starts, raising the temperature rapidly to 10–100 million K and burning the entire hydrogen envelope within a few minutes (Gehrz et al. 1998). The violent convection triggered during this evolution dredges up some matter even from deeper layers of the white dwarf, mixing carbon, oxygen, and in some cases even heavier elements into the burning region. This results in nucleosynthesis under proton-rich, hot burning conditions, which leave behind a characteristic abundance pattern (José and Hernanz 1997).

The details of nova nucleosynthesis are very hard to model; the short time scale of the event plus the criticality of convection for mixing fuels and determination of density and temperature of the burning region pose major challenges (Hernanz et al. 1997). Besides, the progenitor evolution is not understood, but determines the mass and composition of the underlying white dwarf (Kolb and Politano 1997). Nevertheless, nuclear networks have been run under the conditions resulting from hydrodynamic evolution of novae and have shown that intermediate-mass elements up to sulfur and silicon may readily be generated, consistent with the observed enrichments of such elements in the atmosphere of novae. Radioactivities important for γ-ray astronomy comprise a variety of β-decaying species with relative short lifetimes, such as ^{19}F, ^{15}N, and ^{13}C. These are produced on the proton-rich side of the nuclear valley of stability and hence decay through positron emission. This incurs intense nova γ-ray emission early on from positron annihilation, with a γ-ray energy of 511 keV. This 'annihilation flash' is of very short duration, so it can be observed only for the first day of the nova even before the nova emerges in optical light; therefore any detection would be fortuitous (Gomez-Gomar et al. 1998). On the other hand, substantial production of relatively short-lived ^{7}Be is expected to result in a γ-ray line at 478 keV, visible for months due to the 77 day decay time of this isotope. None of these radioactivities has been observed yet. Nevertheless, classical novae represent a site of nucleosynthesis with unique characteristics, and from chemical evolution studies we expect novae to be the main sources of the intermediate-mass elements neon, sulfur, and magnesium. Novae also are known to be significant dust producers and hence may deposit parts of their freshly produced elements onto dust grains, where they could be locked up and escape their discovery in the interstellar medium through their atomic radiation characteristics. There is however the prospect of more long-lived radioactivity generation from novae, which could be detected as diffuse γ-ray line emission for years, or millions of years, after the nova. Novae which occur on very massive white dwarfs could be substantial producers of ^{22}Na, which with its decay time of 3.8 yr is an excellent candidate to be seen after the nova has settled.

10.3.3 Other Nucleosynthesis Sites

Other sites of nucleosynthesis, beyond stellar interiors and explosive events, will be described briefly below for completeness; they are less relevant to γ-ray studies of nucleosynthesis (Rohlfs and Rodney 1988).

The Sun

Nucleosynthesis reactions inside the Sun provide the energy source that determines its structure and luminosity, yet in a rather complex way. It remains difficult to convert external observations of solar parameters (e.g. those obtained from helioseismology) into constraints on the Sun's core nuclear physics environment. However, the neutrinos produced as by-products of the hydrogen-burning reactions readily escape from the solar core. Major experiments have been performed or set up to record these, through chlorine, light-water Cherenkov, and gallium detectors (Kirsten 1999). It is the deficit in several energy channels of ν's which provides the most sensitive test of nuclear processing in the solar interior. The large overlying gas column prevents diagnostics of this inner region through γ-ray observations. Nevertheless, the surface and coronal regions of the Sun are a prominent source of γ-rays from nuclear excitations and continuum processes (see Chap. 5).

Cosmic-Rays in Interstellar Space

The study of Galactic cosmic-rays revealed an isotopic composition with lithium, beryllium, and boron elements greatly enriched with respect to standard abundances (see Fig. 10.1). Combined with the depression of these very elements in standard abundances, this is evidence that interstellar spallation occurs and contributes significantly to nucleosynthesis in the universe (Ramaty et al. 1999). Interstellar spallation nucleosynthesis for Li, Be, and B competes with $\alpha + \alpha$ reactions (for Li) and the neutrino process inside core-collapse supernovae (for ^{7}Li and ^{11}B), both being able to generate primary Li and B. For beryllium the spallation chain appears to be the predominant source, although the observational proof from low-metallicity stars remains difficult. Astration destroys some of these fragile elements.

Spallation reactions themselves are well understood; the interaction energies are well within the range of laboratory measurements, in the range above tens of MeV per nucleon. The relevance of this process in comparison to other nucleosynthesis channels is difficult to estimate, however, as the most relevant low-energy cosmic-rays cannot be observed otherwise; hence their flux and energy spectra remain uncertain (see Chap. 11).

Further proof of spallation nucleosynthesis in the interstellar medium exists from the direct measurements of the high-energy cosmic-ray composition with the Ulysses (Simpson and Connell 1998) and Advanced Composition Explorer (ACE) (Stone et al. 1998) instruments. Shortlived radioactive isotopes have been identified, such as ^{54}Mn, ^{10}Be, and ^{26}Al.

In order to estimate the global relevance of interstellar spallation nucleosynthesis for an isotope's abundance, a balance of the initial abundances and the source and destruction processes, a chemical evolution model including cosmic-ray propagation, is required.

More specifically, one may estimate the total nucleosynthetic yield for a nearby region of active massive star formation, such as the Orion region. It was found, however, that even in this favourable case the yield of a relatively abundant hence γ-ray-bright isotope such as ^{26}Al still remains below instrumental sensitivity limits.

In the solar system, enrichment of meteoritic samples in ^{26}Mg had been a striking puzzle for a long time (MacPherson et al. 1995). Some of the models to explain this enrichment invoke injection from a nearby nucleosynthetic event just before the formation of the solar nebula. Any γ-ray-emitting radioactivity would have decayed since then, however. Other models invoke local production through cosmic-ray interactions (Shu et al. 1987).

The Big Bang and the Early Universe

The early universe's evolution is commonly described by expansion from a hot Big Bang singularity. Initial global thermal equilibrium involves all possible states of matter; the high temperature and density maintains a thermal distribution throughout; and strong, weak, and electromagnetic reactions equilibrate all system components more rapidly than the global system evolves. This initial fireball expands rapidly. Nuclei begin to form when the temperature drops below the regime of nuclear binding energies ($\simeq 8\,\mathrm{MeV}$). At this time, the density of baryons has fallen to 10 g cm^{-3}, however, and rapid expansion soon constrains the nuclear reaction rate due to a lack of collisions. Hence the period of primordial nucleosynthesis is bounded from two sides, and it encompasses universe ages from 0.01 to 200 s.

Big Bang nucleosynthesis manifests itself in the primordial abundances of four species: the abundances of ^{4}He, D, ^{3}He, and ^{7}Li can all be expressed as a function of the universe's photon-to-baryon ratio (Schramm and Turner 1998). The relation of these abundances to the key cosmological parameters implies that our present understanding of the early universe is intimately coupled with our understanding of nucleosynthesis.

There are no γ-ray-emitting radioactive species left over from this early nucleosynthesis. The primordial composition, or at least an approximation of it, is conserved in the first generation of stars and in the gas that we can probe at very large redshifts (z $\simeq$ 4) with quasar absorption lines.

10.4 Gamma-Ray Lessons on Supernovae

10.4.1 Individual Nucleosynthesis Sources

Thermonuclear Supernovae

A goal for γ-ray astronomy for the last three decades has been the detection of characteristic lines from the decay of radioactive ^{56}Ni and its daughter ^{56}Co, produced in supernovae. Type Ia events are favored over the other supernova classes because they produce an order of magnitude more ^{56}Ni than the other types ($\sim 0.6\,M_\odot$), and they expand rapidly enough to allow the γ-rays to escape before all the fresh radioactivity has decayed. Even so, detection of these events has been difficult to achieve.

The only Type Ia supernova which has possibly been seen in γ-rays is SN 1991T in the galaxy NGC 4527. This galaxy is $\sim 17\,\mathrm{Mpc}$ away (determined from Cepheid variables) and in the direction of the Virgo cluster. The supernova was unusually bright at maximum ($0.7\,M_\odot$ of ^{56}Ni) and the light curve evolution was unusually slow. The classification as a peculiar Type Ia event is based on the absence of the silicon lines so typical in early Type Ia spectra. In addition, the Fe III lines were unusually strong at early epochs. Since no other iron group lines were observed at this time, this iron was probably not a fresh nucleosynthetic product. The spectra became more typical of Type Ia events at later epochs, when the expanding debris became more transparent. Detection of high velocity ($\sim 13\,000\,\mathrm{km\,s^{-1}}$) iron and nickel in the outer layers of SN 1991T favors models where the subsonic flame front propagates to larger distances from the white-dwarf core before making the transition to a detonation. These types of delayed-detonation models are also consistent with the velocity profile of most of the other ejecta (silicon, calcium) seen in SN 1991T. Detection of the 812 keV γ-ray line from the decay of ^{56}Ni in the early light curve would have been direct evidence for a delayed detonation, as the line cannot be seen when the ^{56}Ni is embedded deeper in other categories of Type Ia models, which have been discussed for SN 1991T as well. These models may be favorable for ejecting a larger than average ^{56}Ni mass, and seek to explain some of the other early light curve peculiarities as arising from interactions of the supernova debris with the thick disk of material which surrounded the merger.

The tentative detection by the COMPTEL Compton telescope (Morris et al. 1995) of the ^{56}Co decay γ-rays (Fig. 10.7), at 3–4σ significance only suggests that this isotope was present in the outer envelope and supports extensive-mixing scenarios. The COMPTEL flux value converts into an overly large ^{56}Ni mass, between $1.3\,M_\odot$ (for a distance of 13 Mpc) and $2.3\,M_\odot$, the value for the 17 Mpc favoured currently. This would require that almost all of the Chandrasekhar-mass white dwarf must have turned into radioactive ^{56}Ni. Upper limits from OSSE (Oriented Scintillation-Spectrometer Experiment on the Compton Obseervatory) may indicate that the COMPTEL ^{56}Co line flux value is too high.

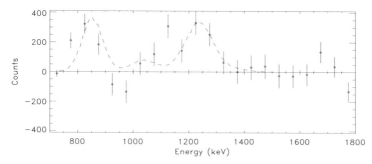

Fig. 10.7. SN 1991T spectrum as measured by COMPTEL, after subtraction of a background model. Reproduced with permission from Morris et al. 1995

SN 1998bu at only 8–11 Mpc distance was observed by the CGRO instruments for several months. Apparently it occurred deeper within its host galaxy, M96, as inferred from apparently local reddening. Appearing rather 'normal' from its light curve and spectra, the ^{56}Ni γ-ray lines seem dimmer than expectations, however [neither OSSE (priv. communication) nor COMPTEL (Georgii et al. 2000) have seen any of the ^{56}Ni decay chain γ-ray lines].

Rapid pointing of γ-ray telescopes at early times after the supernova is essential for proof/disproof of the helium cap models. INTEGRAL's spectrometer with its superior sensitivity would be able to add line shape (hence velocity) information for further diagnostics; note however that for broad lines the sensitivity of high-resolution instruments degrades. More detections of Type Ia supernovae in ^{56}Ni are required to clarify how typical SN 1991T was.

Type Ia supernovae of a rare sub-type could be important sources of ^{44}Ti. Predicted ^{44}Ti yields for sub-Chandrasekhar models (Tutukov et al. 1992) are 200–3000 times the relative solar ^{44}Ca proportion. Depending on how frequently these sub-Chandrasekhar mass white dwarfs explode, these large production factors open the possibility that these types of thermonuclear supernovae might be the principal source of ^{44}Ca, rather than the typical core-collapse event. The ^{44}Ti observation from Cas A, generally believed to be a core-collapse supernova, presents a counterexample of a ^{44}Ti source, however.

Gravitational-Collapse Supernovae

Core-collapse supernovae of Type Ib synthesize 5 to 10 times less ^{56}Ni than a typical Type Ia supernova, but expand almost as rapidly. Thus their early γ-ray signal should be intermediate between Type II and Type Ia. Any Ni or Co γ-ray line detection of a Type II supernova outside the Local Group is very improbable with present instruments.

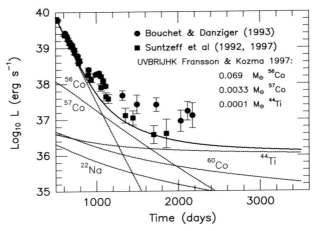

Fig. 10.8. The light curve of SN 1987A. The optical/UV emission is found to be consistent with energy input from the radioactivities seen in γ-rays, as indicated by the reference lines for ^{56}Co and ^{57}Co; ^{44}Ti energy input is also shown. (Reproduced with permission from Diehl and Timmes 1998)

The best studied supernova of all is the Type II supernova SN 1987A, (see Fig. 10.8) mainly because of its proximity, but also because instrumentation in IR to γ-ray regimes had matured and was ready for such observations (Phillips et al. 2000). Detection of ^{56}Co and ^{57}Co lines from SN 1987A by many experiments provided the first extragalactic γ-ray line signal from radioactive isotopes (Gehrels et al. 1988). The early appearance of ^{56}Co γ radiation presented evidence for enhanced mixing of supernova products within the envelope.

Later, OSSE reported detection of ^{57}Co radiation from SN 1987A, with a measured flux of $\sim 10^{-4}$ photons cm^{-2} s^{-1} between 50 and 136 keV (Kurfess et al. 1992). For models with low ^{57}Co optical depths, the observed γ-ray flux suggested that the ^{57}Ni/^{56}Ni ratio produced by the supernova was about 1.5–2.0 times the solar system ratio of ^{57}Fe/^{56}Fe. Estimates of SN 1987A's bolometric luminosity at optical and infrared wavelengths beyond day 2000 show that the cooling time of the remnant is longer than previously thought (≤ 1 magnitude per 1000 days). Fits of cooling models to these light curves yield 0.069 M$_\odot$ of ^{56}Co and 0.0033 M$_\odot$ of ^{57}Co, and hence a ^{57}Ni/^{56}Ni ratio in reasonable agreement with the γ-ray measurements.

If we exclude any energy input from a pulsar, an accreting compact object, or circumstellar interaction, SN 1987A is now in a phase where the dominant energy source should be from the decay of ^{44}Ti (see Fig. 10.8). Its thermal luminosity should derive mostly from e^+ kinetic energy. From the late-time bolometric light curves 10^{-4} M$_\odot$ of ^{44}Ti is suggested. The type of energy input may be difficult to resolve uniquely, however, if the atomic processes that convert energy from radioactivity into optical/infrared radiation are no longer in steady state.

For a distance of 50 kpc to SN 1987A and a ^{44}Ti e-folding lifetime of 89 yr, 1×10^{-4} M$_\odot$ of ^{44}Ti would produce a γ-ray line flux of 2×10^{-6} photons cm^{-2} s^{-1}. This line flux is too small for the instruments aboard CGRO, possibly achievable for INTEGRAL, ESA's International Gamma-Ray Astrophysics Laboratory to be launched in 2002. Spherically symmetric models of SN 1987A suggest ^{44}Ti ejection at very low velocities ($\leq$1000 km s^{-1}), and hence a very narrow γ-ray line. On the other hand, broad infrared lines of nickel, early appearance of X-rays, and smoothness of the bolometric light curve argue for mixing of ^{56}Ni out to velocities between 2000 and 4000 km s^{-1}. With ^{44}Ti and ^{56}Ni originating from similar processes, it will be interesting to measure the line shapes and centroid energy, in order to obtain valuable information about the explosion mechanism and multi-dimensional mixing.

The Cas A supernova remnant in our Galaxy is relatively close ($\simeq$ 3.4 kpc), young (explosion in 1668–1680), and accessible (physical diameter $\sim$4 pc, corresponding to an angular extent of $\sim$4 arc min), making it one of the prime sites for studying the spatial structure and kinematics of a supernova remnant as it ploughs into the interstellar medium. Cas A exhibits a rich variety of phenomena: fast-moving knots have been diagnosed as ejecta clumps from the supernova explosion, and jet-like structures also support spherical nonsymmetry. The discovery of the central compact object with the Chandra X-ray telescope, plus the signs of an intense pre-supernova wind in quasistellar flocculi on the outside, support the classification as a Type Ib event from a massive progenitor. The COMPTEL discovery of 1.157 MeV γ-rays from the $\sim$ 300 yr old Cas A supernova remnant (Fig. 10.9) (Iyudin et al. 1994) was a scientific surprise. Supernova models had indicated that $\sim 3\times10^{-5}$ M$_\odot$ of ^{44}Ti would be ejected (Timmes et al. 1996), which translates into a γ-ray intensity generally below instrument sensitivities. COMPTEL's detection of Cas A at $\sim$3 ($\pm$1)$\times10^{-5}$ photons cm^{-2} s^{-1} in the 1.157 MeV line, implies $(1-2)\times10^{-4}$ M$_\odot$ of ^{44}Ti. Conversion of the measured flux into mass limits must account for the uncertainties in the ^{44}Ti lifetime, the distance to the event, and the precise time of the explosion as well as a possible underlying γ-ray continuum from electrons accelerated in the remnant. Ionization-delayed radioactive decay has been proposed as a potential observational bias, but is probably unimportant.[14]

The abundance of ^{44}Ti and ^{56}Ni as a function of mass inside a 25 M$_\odot$ star, shown in Fig. 10.5 for a standard model, suggests that if ^{44}Ti is ejected ejection of ^{56}Ni is needed too and produces a bright supernova. If interstellar absorption did not attenuate the optical light curve, the Cas A supernova would have had a peak apparent magnitude of $m_V = -4$, easily recognizable

---

[14] Ionization by the supernova reverse shock requires major parts of ^{44}Ti ejecta to remain in rather dense clumps (Mochizuki et al. 1999).

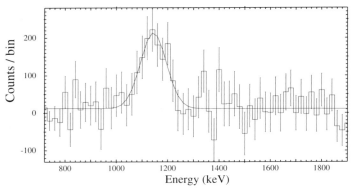

Fig. 10.9. The COMPTEL discovery of Cas A in the γ-ray line of ^{44}Ti demonstrates that nucleosynthesis products from the inner region of core-collapse supernovae may be ejected into the interstellar medium. (Reproduced with permission from Diehl and Timmes 1998)

in the sky. Cas A was not widely reported as such; some 10 magnitudes of visual extinction is required to make the γ-ray ^{44}Ti measurements consistent with these (absent, except for one hint) historical records.

There may indeed have been such a large visual extinction to Cas A at the time of the explosion (Hartmann et al. 1997): if Cas A was embedded in a dusty region, or experienced significant mass loss which condensed into dust grains before the explosion, the extinction could have been exceptionally large but not observed. Measurements of the X-ray scattering halo around Cas A with ROSAT (Röntgen Satellite) and ASCA (Advanced Satellite for Cosmology and Astrophysics) suggest an unusually large reddening correction. The X-ray scattering halo is unusually low for the derived N_H values ($N_H = 1.8 \times 10^{22}$ cm^{-2}), while this N_H is twice as large as what $A_V = 5$ usually implies. Extra material seems to be distributed close to Cas A, possibly it is the dusty shell of material ejected prior to the explosion as a Type Ib supernova. The supernova shock wave could have destroyed much of the dust as it propagated through the debris and the material surrounding the Cas A supernova. This scenario would be consistent overall, with the lack of optical detection, excess neutral hydrogen column density, dust-free and metal-rich debris, and ejection[15] of $\sim 10^{-4}$ M$_\odot$ of ^{44}Ti.

Notice that $\sim 10^{-4}$ M$_\odot$ of ^{44}Ti is inferred to have been ejected both in SN 1987A (a Type II event) and in Cas A (probably a Type Ib event). This

---

[15] ^{44}Ti yields may be enhanced relative to ^{56}Ni by explosion asymmetries or yet-unknown details of fall-back and mass accretion onto the compact remnant. The decay chain of ^{44}Ti involves electron capture and hence may be inhibited by highly ionized nucleosynthesis ejecta in the early remnant phase (Mochizuki et al. 1999).

agreement may be fortuitous; however, it may suggest that the inner core-collapse explosion/accretion mechanism is well regulated.

The COMPTEL search for ^{44}Ti sources in the Galaxy did not discover as many sources as would be suggested by a Galactic core-collapse supernova rate of about one event in even 30 yr (Dupraz et al. 1997; The et al. 2000). This suggests that ^{44}Ti-producing supernovae are exceptional events. Cas A appears to be one. ROSAT X-ray measurements had revealed another supernova remnant, RX J0852.0-4622 of diameter 2°, superimposed on the 7° diameter Vela supernova remnant (Aschenbach 1998); COMPTEL may have discovered ^{44}Ti γ-rays from this source (Schönfelder et al. 2000), which would place it very nearby at a distance of 200 pc only (Iyudin et al. 1998). This case illustrates the new potential of ^{44}Ti γ-ray observations to search for supernova remnants in otherwise difficult if not totally inaccessible regions.[16] Indeed, for very nearby sources ($d \leq 500$ pc) their ^{26}Al emission should be detectable individually, and thus a comparison of ^{44}Ti and ^{26}Al yields alone allows age and distance determination for such objects. Future observation/instruments may make this a realistic proposition.

From the initial COMPTEL discovery of ^{26}Al emission from the Vela region there was some hope to detect ^{26}Al from one single source, the nearby Vela supernova remnant. Refined analysis showed, however, that the main 1.809 MeV feature in this region is significantly offset from the Vela supernova remnant (see Fig. 10.12 left). The offset may be due to a contribution from the RX J0852.0-4622 supernova remnant (Aschenbach et al. 1999) and/or from the known populations of OB associations and active star-forming regions in this direction at larger distances (Diehl et al. 1999). A direct calibration of core-collapse supernova ^{26}Al nucleosynthesis on the Vela SNR, however, does not apply, and model predictions are consistent with the flux range seen by COMPTEL (see Fig. 10.10). Note that the distance to this SNR has been revised from the traditional 500 pc to $\sim$250 pc.

Novae

The origin and evolution of an accreting white dwarf which becomes a classical nova are essentially unknown, and yet have important consequences for any γ-ray signal which may originate from novae (Gomez-Gomar et al. 1998). For example, the composition of the nuclear burning region is often assumed to be a 50–50 mixture of accreted material and dredged-up white-dwarf material. Typically the accreted material is taken to have a solar composition, and the white dwarf material is assumed to be an oxygen–neon–magnesium mixture in mass proportions of 0.3:0.5:0.2; however, oxygen–neon–magnesium ratios of 0.5:0.3:0.05 have been suggested from evolutionary models as more appropriate (Ritossa et al. 1996), although the detailed abundances could

---

[16] Either confused with other structures or occulted in cases of supernovae occurring inside dense cloud regions.

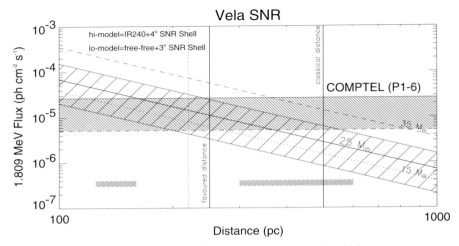

Fig. 10.10. The residual 1.809 MeV signal attributed to the Vela supernova remnant appears consistent with core-collapse supernova models, in particular if a distance of ∼250 pc is assumed

vary substantially with initial stellar mass. Since the yield of ^{26}Al and ^{22}Na from novae is sensitive to the initial ^{25}Mg and ^{20}Ne abundances, this latter white-dwarf composition eliminates most of the necessary seed material from which radioactive isotopes may be synthesized. Uncertainty in binary-star evolution and the binary mass distribution function correspond to uncertainty in the fraction of classical novae that originate from $\geq 8\,M_\odot$ main-sequence stars.

All nova models predict ejected masses that are by an order of magnitude too small when compared to observations. As the white dwarf becomes more massive, less material is accreted before the fuel ignites and the total mass of matter lifted to escape velocity, ^{22}Na and ^{26}Al in particular, is smaller. If the inconsistency between model and observed ejected masses is resolved by resorting to a less massive ($M \leq 1.1\,M_\odot$) white dwarf, then the ^{26}Al yields are expected to increase due to the lower burning temperatures. However, should a more violent explosion be required, and be achieved by increased mixing of core material, then the higher burning temperatures are expected to decrease the ^{26}Al mass ejected. While these examples show that our understanding of the physics is rather incomplete, the thermonuclear runaway model for classical novae has been a success of the first order. A nova event within 1 kpc would provide several important diagnostics (radioactive and otherwise) for refining the mixing and energetics aspects of the model.

Observationally, about 1/3 of the nearby novae which could be analyzed in detail have been found to show large enhancements in the abundance of neon, among other intermediate-mass elements. This has been taken as evidence that these novae occurred on white dwarfs with substantial enrichment in elements heavier than oxygen (thus attributed either to more massive progenitors or to a self-enrichment through successive and frequent surface nucleosynthesis events). From neon seed nuclei, ^{20}Ne(p,γ) and successive proton capture reactions can occur in the hot hydrogen-burning environment, to produce substantial amounts of ^{22}Na. Note however that the high temperatures required for efficient ^{22}Na production makes only the upper end of the white-dwarf mass spectrum (≥ 1 M$_\odot$) relevant for ^{22}Na production. For these massive white dwarfs, the amount of accreted material needed to trigger the nova is less, and it is still unclear how the observed large ejected masses from neon-rich novae should be understood. Therefore upper limits on the ^{22}Na yields per novae are interesting, even in the absence of any detection: the flux limit in the 1.275 MeV line of $\sim 3 \times 10^{-5}$ photons cm^{-2} s^{-1} corresponds to $\sim 4 \times 10^{-8}$ M$_\odot$ of ^{22}Na, if all individual limits are combined (Iyudin et al. 1995). Current models for more massive white-dwarf novae still predict values about an order of magnitude lower, though [e.g., 1.6×10^{-9} M$_\odot$ for a 1.25 M$_\odot$ white dwarf (José and Hernanz 1997)].

For the progenitor range considerably below 1 M$_\odot$, production of ^{26}Al should be enhanced, because burning occurs at lower temperatures and is less violent, so photodestruction of the freshly produced ^{26}Al is avoided. Some models predict enrichments of the ^{26}Al isotope with respect to stable ^{27}Al of up to 10^{-3}, which, combined with observed ejected masses of a few 10^{-4} M$_\odot$, could accumulate to a significant part of the observed diffuse glow of ^{26}Al emission from the Galaxy (see below). However this naive argument already outlines the difficulty in such estimates: no self-consistent evolutionary model has been carried through to predict the amount of Galactic ^{26}Al from such novae. Rather, specific models of nucleosynthesis have been calculated, often selecting the most favourable conditions. The fractional enrichments obtained have then been adjusted to the observed ejected masses, which are generally about one order of magnitude higher than those produced in standard models. The reason for this discrepancy is not understood. Moreover, such nucleosynthesis production for the modelled progenitor then is multiplied by an uncertain fraction of novae occurring on relatively massive white dwarfs (the mean white dwarf mass lies in the vicinity of 0.4–0.5 M$_\odot$). Only about a dozen novae have been observed with adequate spectroscopic data to determine atmospheric abundances; the values for the latter are uncertain due to the atmospheric excitation model that underlies interpretation of spectroscopic data in terms of elemental abundances. These dozen novae all are 'disk novae' at distances within a few kpc, hence not representing the Galactic Bulge. Any Galactic nova population estimate therefore must make assumptions about nova frequencies per class in the disk and bulge, mostly taken from observations of nearby galaxies such as M31 and the Large Magellanic Cloud. Obviously, there are substantial open issues. Gamma-ray line observations would be a paramount tool to help clarify these. Nova Velorum 1999 at a distance of 0.8–2 kpc could be another opportunity; it was being observed by CGRO instruments at the time this article was being written. Diagnostics in the 511 keV annihilation line and the 478 keV line from ^{7}Be await detection (Harris et al. 1991).

Stellar Interiors: Hydrostatic Nuclear Burning

At present, sensitivity levels of γ-ray telescopes identify only one massive star as a candidate source: the γ^2Velorum system in the Vela region at a distance of 260 pc. Indeed, this O-star/WR star binary system has been studied in great detail, but ^{26}Al emission at the level predicted by models is not observed (Oberlack et al. 1999) (see Fig. 10.11). Modification of the ^{26}Al ejected from WR11 due to the O star companion and an age of WR11 beyond its maximum ^{26}Al ejection phase are possible explanations of the discrepancy.

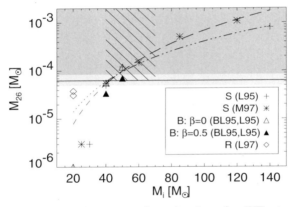

Fig. 10.11. The expected production of a WR star exceeds the sensitivity limit of COMPTEL's 1.8 MeV survey: the special conditions of the γ^2Velorum binary system and its current phase must be invoked to explain the absence of an 1.809 MeV signal from a WR star at this distance

10.4.2 Integrated Nucleosynthesis

^{26}Al in the Galaxy

^{26}Al with its radioactive decay time of 1.04×10^6 yr accumulates in the interstellar medium from many individual sources of nucleosynthesis. This results in a diffuse glow of active regions of nucleosynthesis in the Galaxy in the 1.809 MeV γ-ray line, imaged for the first time with the COMPTEL telescope (Diehl et al. 1995) (Fig. 4.1). Since the pioneering discovery in 1982 by HEAO-C, the third High-Energy Astronomy Observatory Satellite of NASA, many experiments have contributed to aspects of ^{26}Al astronomy, as summarized in a recent review (Prantzos and Diehl 1996). Instrumental capabilities differ substantially, but the integrated flux measured from the general direction of the inner Galaxy, integrated over latitude and the inner radian in longitude, has been used to roughly compare results. All measurements are consistent with values of 4×10^{-4} photons cm^{-2}s^{-1}. Note that the determination method varies between instruments, and in particular the flux values for the nonimaging instruments depend on the assumed spatial distribution: nonimaging instruments essentially assume the same (or equivalent) smooth spatial distribution narrowly following the plane of the Galaxy as had been derived from COS-B measurements of Galactic continuum γ-rays in the $\geq$100 MeV regime.

The distribution of 1.809 MeV emission seems significantly different from the continuum, however (see Fig. 4.1 in Chap. 4). Images derived from COMPTEL measurements show spatial structure in the emission. The ridge of the Galactic Plane dominates; however, there is asymmetry in the emission profile along the disk, with several prominent regions of emission, such as in the Vela region and Cygnus regions. All estimates of the absolute ^{26}Al mass in the Galaxy rest on assumptions about the spatial distribution of the sources, as the 1.809 MeV measurements themselves do not carry distance information. The COMPTEL team fitted a wide range of models for candidate source spatial distributions to their high-quality imaging data (over 30σ significance) (Knödlseder et al. 1999b). When localized regions of emission beyond the inner Galaxy are excluded, then all axisymmetric model fits yield a Galactic ^{26}Al mass of $\sim$2 M$_\odot$. The extent of spiral-arm emission can be estimated if a composite model of disk emission plus emission along spiral arms is adopted and compared to the disk-only model. Spiral structure appears significant, contributing between 1.1 M$_\odot$ and all of the ^{26}Al.

If massive stars are the candidate sources, they should follow the molecular gas distribution, and thus would be traced by CO survey data. Although generally compatible with the ^{26}Al map, other tracers were found to provide a better fit. One of these tracers is free electrons from ionizing effects related to nucleosynthesis sources. One may estimate the free-electron content of the interstellar medium from radio measurements of free–free emission, which can be obtained after subtraction of synchrotron emission. Alternatively, a semi-analytical model of spiral-arm structure based on HII region data, refined by free-electron measurements from pulsar signal dispersions, has been used. This tracer shows ridges similar to the ^{26}Al map at longitudes $\pm35°$, along with a prominent feature in Carina ($l = 280°$), and it is proportional to the 1.809 MeV map in all significant detail over the entire plane of the Galaxy. In particular, a calculation can reproduce the expected massive star population and the supernova rate from both maps consistently, if WR stars from high-metallicity regimes in the inner Galaxy provide the bulk of ^{26}Al (Knödlseder 1999). Other good candidate tracers were identified in warm dust maps, such as the long-wavelength maps from DIRBE (Diffuse Infrared Background Experiment) aboard the Cosmic Background Explorer (COBE), or the far-infrared cooling lines at e.g. 158 μm from the ionized interstellar medium. All these arguments confirm that tracers which measure the energy input into the interstellar medium from massive stars appear to represent an approximate representation of the ^{26}Al source distribution.

The evidence above may be taken to constrain ^{26}Al contributions from classical novae, where a smooth distribution of the emission with a pronounced peak in the central bulge region would be expected. The upper limit for such contributions is probably 1 M$_\odot$ of ^{26}Al. On the other hand, Ne-rich novae in our Galaxy may occur more frequently in the disk, and hence follow the Galactic distribution of massive stars more closely than the overall white-dwarf distribution. In this case, differentiating nova sources from massive star sources will rely on the consistency of the calculated yields with other lines of evidence (such as other radioactive isotopes), supplemented by 1.809 MeV line shape arguments.

More local ^{26}Al contributions may play a significant role: the slightly lower ^{26}Al flux value from COMPTEL is mainly based on Galactic-plane emission, while the large field-of-view instruments (100–160°) of GRIS (Gamma-Ray Imaging Spectrometer) and SMM (Solar-Maximum Mission) mainly sampled the sky along the plane of the ecliptic with relatively more exposure of the high-latitude sky; those instruments may include large-scale flux of low surface brightness that COMPTEL's image failed to capture. ^{26}Al emission from the solar vicinity had been predicted long ago, but was discarded when COMPTEL's image showed dominant emission along the plane of the Galaxy. Local contributions to the overall emission are supported by the existence of two nearby pulsars at ∼100 pc distances (Geminga and R CrA), the nearby Gould Belt structure apparent in UV through young and massive B stars, and other massive-star activity signposts such as Loop I, which is associated with the nearby Sco–Cen association; all these suggest that the ∼500 pc environment of the sun may well have experienced a higher-than-average star formation and supernova activity during the past ∼50 million years. In view of the underlying continuum emission, and also very different instrumental techniques, each with substantial systematic uncertainties, the 1.809 MeV line flux measurements must be consolidated and ensured to be comparable before such speculations are further pursued.

Imaging of diffuse sources of MeV γ-rays is far from straightforward due to the high instrumental backgrounds and the complex γ-ray detection methods. Consistency checks between different techniques have shown that some of the spikyness of the apparent emission in the COMPTEL result could be an artifact of analysis techniques. Nevertheless, significant emission from the Cygnus, Carina, and Vela regions appears consolidated (Knödlseder et al. 1999b).

About 80% of the prominent 1.809 MeV emission associated with the Cygnus region (see Fig. 10.12 right) can be understood in terms of the expected ^{26}Al signal from known sources (Del Rio et al. 1996). One may be

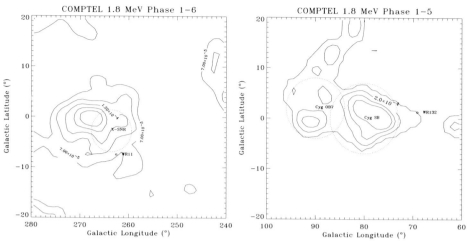

Fig. 10.12. *Left*: The Vela region image in the 1.8 MeV line from ^{26}Al shows no prominent emission from the nearest known candidate sources, the Vela supernova remnant and γ^2 Velorum. The peak of the emission rather coincides with several associations on the near side of the Vela molecular ridge, reflecting massive-star activity within these. *Right*: The Cygnus region 1.8 MeV image appears plausibly explained by the Cygnus superbubble and OB associations, the massive-star activity producing ^{26}Al

puzzled by this high fraction, since ^{26}Al decays on a time scale longer than the observable features of supernova remnants and WR winds survive. It has been suggested that the ^{26}Al from this region attributed to 'seen' sources should be multiplied by a factor of between 1 and 10 to account for 'unseen' sources. COMPTEL images show structures which suggestively align with the Cygnus superbubble and the Cyg OB1 and OB2 associations. However, it is difficult to spatially separate source regions, and in particular assess the significance of emission from the prominent group of WR stars in this region.

The Carina region ($l = 282\text{–}295°$) presents a tangential view along a spiral arm, identified through a large molecular-cloud complex $\sim$2–5 kpc away, houses the prominent $\geq 140\,\mathrm{M}_\odot$ star Eta Carinae ($l = 288°$) and shows the densest concentration of young open clusters along the plane of the Galaxy. ^{26}Al production within these clusters as part of the Car OB1 association may relate to the observed 1.809 MeV feature at $l = 286°$. This feature appears almost as a point source for the $4°$ resolution COMPTEL instrument, and thus may be spatially more confined than the expected signature from a tangential view into one of the Milky Way's spiral arms.

It is also interesting that some of the 1.809 MeV image structures which fail to align with spiral arms do coincide with directions towards nearby associations of massive stars. Patchiness in such a nucleosynthetic snapshot might be expected from the clustering of formation environments of massive stars.

If viewed from the outside, the Milky Way might also display the signs of massive star populations in the form of HII regions arranged like beads on a string along spiral arms, such as observed in M31 from H$_\alpha$ emission analysis, or in M51 from heated dust seen in infrared continuum at $\geq 15\,\mu$m. Interstellar absorption and source confusion prevents such mapping within the Milky Way. Therefore, detailed investigation of local systematic uncertainty in the COMPTEL image, that is, a quantitative limit to artificial bumpiness of the imaging algorithm, is important for such interpretation of 1.809 MeV emission. Similar concerns also apply to other instruments and future measurements of the morphology of Galactic nucleosynthesis radioactivity along the Galactic plane.

In a balloon flight, the GRIS Ge detector instrument ($\sim$3 keV energy resolution) drift scanned the Galactic Center region with its $\sim$100° field of view, and detected the 1.809 MeV line at 7σ significance and a flux of 4.8×10^{-4} photons cm^{-2} s^{-1} (Naya et al. 1996). The main surprise in this measurement is the width of the line profile: it was significantly broader than the instrumental resolution of the germanium detector and reported as ΔE = 5.4 $\pm$ 1.4 keV (Fig. 10.13). This line width is much larger than expected from Galactic rotation ($\sim$1 keV), which again dominates above the broadening from random motions in the interstellar medium. It is presently difficult to understand how such high-velocity motion could be maintained over the million year time scale of ^{26}Al decay (Chen et al. 1997). Thermal broadening by a very hot

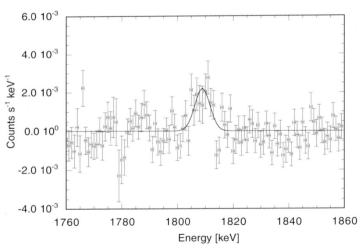

Fig. 10.13. The high-resolution study of the 1.809 MeV line from ^{26}Al may hold rich kinematic information about the source regions. Here the GRIS experiment reported significant line broadening, suggesting that ^{26}Al nuclei remain at high ($\sim$500 km s^{-1}) over 10^5 yr or more. (Reproduced with permission from Naya et al. 1996)

phase of the interstellar medium (~10^8 K) with long cooling times (~10^5 yr) or a kinetic broadening at high average velocities (~500 km s^{-1}) seems to be required. Either case requires extremely low-density phases of the interstellar medium on large spatial scales. Alternatively one may hypothesize massive, high-speed dust grains as carriers of ^{26}Al to explain the measurement. This view derives some support from our current understanding of cosmic-ray acceleration. Further observations of the line shape are required to examine any spatial variations in the line broadening. A spectral resolution of ~2 keV is required for such a study. Although the INTEGRAL mission may still have insufficient energy resolution to make complete velocity maps of the 1.809 MeV emission along the plane of the Galaxy, the brightest features can probably have their velocity centroids determined well enough to place them on the Galactic rotation curve and thus derive the distance to the features. Surveys which combine velocity information and Galactic latitude extent could then examine the existence and nature of any Galactic 'fountains' and 'chimneys' from possible 'venting' of ^{26}Al into the Galactic halo. The COMPTEL-measured latitude width may constrain the mean velocity of observed ^{26}Al below the GRIS value for an assumed young and hence narrow population of sources and isotropic expansion over 10^6 yr. In any case, the large line width measured by GRIS is inconsistent with the HEAO-C line width limit of <3 keV and needs confirmation, since it could have profound implications on our understanding of the interstellar medium in the Galaxy.

^{60}Fe in the Galaxy

Physically, ^{60}Fe should be a good discriminant of different source types generating ^{26}Al, because massive stars produce ^{26}Al and ^{60}Fe in the same regions and in roughly comparable amounts. While the ^{26}Al production occurs in the hydrogen shell and the oxygen–neon shell, ^{60}Fe is produced in He-shell burning and at the base of the oxygen–neon shell. Most important is that the bulk of both ^{26}Al and ^{60}Fe are produced, mainly during the pre-supernova evolution, at mass coordinates between 3 and 6 M$_\odot$ of typical ~20 M$_\odot$ stars. These two isotopes should have similar spatial distributions after the explosion of these stars, if supernovae from massive stars dominate both ^{26}Al and ^{60}Fe nucleosynthesis.

An estimate for the injection rate into the Milky Way is the steady-state event rate times the average mass ejected per event; taking M(^{26}Al) ~10^{-4} M$_\odot$, M(^{60}Fe) ~ 4×10^{-5} M$_\odot$, and ~2 core-collapse supernovae per century, one obtains $\dot{\text{M}}$(^{26}Al) ~ 2.0 M$_\odot$ Myr^{-1} and $\dot{\text{M}}$(^{60}Fe) ~ 0.8 M$_\odot$Myr^{-1}. More refined chemical evolution calculations suggest that Type II supernovae are responsible for a steady-state abundance of 2.2 ± 1.1 M$_\odot$ of ^{26}Al and 1.7 ± 0.9 M$_\odot$ of ^{60}Fe in the Galaxy, producing an intensity ratio in the decay γ-ray lines of ^{26}Al and ^{60}Fe of ~16% (Timmes et al. 1995).

A few recent ^{60}Fe measurements and flux ratios with ^{26}Al have been obtained so far. SMM reported an upper limit of 8.1×10^{-5} photons cm^{-2} s^{-1}

for the 1.173 MeV ^{60}Co line over the central radian of Galactic longitude, giving an upper limit of 1.7 M$_\odot$ of ^{60}Fe (Leising and Share 1994). From the GRIS balloon experiment, a more stringent upper limit of 17% has been derived on the ^{26}Al/^{60}Fe γ-ray line intensity ratio (Naya et al. 1998); this was confirmed by COMPTEL (Diehl et al. 1997). If this is the case, the initial model estimates for the total Galactic flux ratio might be too large. However, there are uncertainties of factors ∼2 in the models, from nuclear cross sections, explosion energy uncertainties (affecting the large ^{60}Fe contribution from explosive He-burning), and chemical evolution models.

Once ^{60}Fe can be detected with γ-ray telescopes, we can test the supernova origin of ^{26}Al hypothesis, since ^{60}Fe from other sources is probably negligible.[17]

Positrons in the Galaxy

Radioactive decays can generate positrons whenever the energy level of the daughter atom is below the energy level of the parent by more than the 1.022 MeV pair-production threshold. Interesting numbers of positrons are produced from the β^+ decay of ^{26}Al, ^{44}Ti, ^{56}Co, and the distinct nova products ^{13}N and ^{18}F. Annihilation produces two 511 keV photons if the positron and electron spins point in opposite directions or three photons in a continuous energy spectrum if the spins are parallel. The three-photon annihilation process usually involves formation of positronium, which then decays before being collisionally destroyed because of the low densities in interstellar space. As a result, the fraction of positronium radiation in the total annihilation signal carries information about the thermodynamic properties of the annihilation environment. A cold, neutral environment results in low positronium fractions and predominant annihilation in flight (Bussard et al. 1979), while larger positronium fractions tend to indicate annihilation in warmer environments, 5×10^3 K or hotter. The positrons produced by the interesting radioactive decays have average kinetic energies of ∼MeV, and thus are relativistic. Deceleration and thermalization are more likely than annihilation in-flight, so that the positron lifetime in interstellar space before annihilation is ∼10^5 yr (Chan and Lingenfelter 1993). The thermalization process implies an intrinsically narrow 511 keV linewidth, which is related to the annihilation environment rather than to the positron production environment. In addition to annihilation from radioactive decays, positron annihilation is expected from the disks around accreting compact remnants, from the jets caused by dynamo action of accreting compact sources, and from γ–γ reactions in strong magnetic fields.

---

[17] Thermonuclear supernovae may be important sources, if neutron-rich nucleosynthesis happens in (supposedly rare) carbon deflagration supernovae (Woosley 1997). In this case, a few ^{60}Fe hot spots from such events would be superimposed on the diffuse SNII ^{60}Fe glow.

How can these various signals be differentiated? Positrons from radioactive decays usually annihilate in the diffuse interstellar medium where the thermalization lifetime is long. Positrons produced around compact objects should annihilate near the compact objects, since the density is usually much larger in those environments. This is expected to result in a more localized and possibly time variable signal. Annihilation spectra may also contain such recognizable signatures as a high-energy tail above 511 keV from annihilation in-flight and/or bremsstrahlung, a blue-shift from positron jets moving towards the observer, a red-shift when originating from sources with a large gravitational field, or a distinctly smaller positronium fraction. Observations of the 511 keV line from positron annihilation show a steady diffuse component from the Galactic Disk, possibly superimposed upon a time-variable point source located near the Galactic Center reported in the 1980s (Ramaty et al. 1994). OSSE and SMM measurements of the annihilation radiation can be analyzed in terms of plausible spatial distribution models that aim to separate the disk component from the Galactic Bulge component [see Fig. 10.14 (Purcell et al. 1997; Milne et al. 2000)]. This decomposition implies annihilation rates of 10^{43} e$^+$ s^{-1} for the disk and 2.6×10^{43} e$^+$ s^{-1} for the bulge. Almost all of the annihilation luminosity from the Galactic Disk may be explained by radioactive sources. About $16 \pm 5\%$ is assigned to ^{26}Al, with the remainder partitioned between ^{44}Ti, ^{56}Co, and old stellar population products. The estimated positronium fraction of 0.94–1.0 for the inner Galaxy suggests that the contribution from compact sources might be small. However, this constraint strongly depends on the environment of the compact sources: for example, the entire bulge component could be explained from jet sources exemplified by the 'Great Annihilator'[18] alone if positrons are not rapidly annihilated in a target close to that compact source (Ramaty et al. 1994). Estimates of the contribution from compact sources may also be derived from simulations of classical novae. These show that the peak 511 keV emission reaches $\sim 10^{-2}$ $(D/1\,\text{kpc})^2$ photons cm^{-2} s^{-1} for a period of about 7 h after the outburst (Gomez-Gomar et al. 1998). This would make nova detections from distances as far as the Galactic Center feasible, if the timing of the observations were fortunate. Overall, however, the nova contribution to the diffuse 511 keV glow of the Galaxy is expected to be low.

Spectral decomposition of the γ-ray spectrum from the inner Galaxy is difficult, in particular for extraction of the three-photon continuum from positronium annihilation (Kinzer et al. 1999), which is crucial for extraction of a radioactivity- or nucleosynthesis-related signal. Problems arise from the

---

[18] This source 1E1740.7-2942 had been understood to inject a large number of positrons through its jets, which have been mapped by radio observations. The large amount of experimental data on the inner-Galaxy 511 keV intensity had suggested time variability on the scale of a few months. Later these measurements were explained to be consistent with time-invariable flux; differing instrumental fields of view and backgrounds explain the different measured flux values.

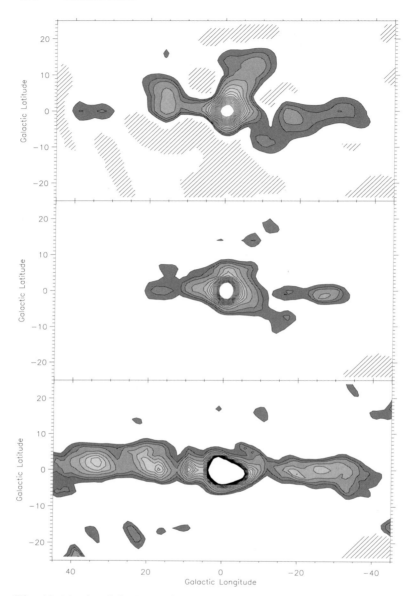

Fig. 10.14. Annihilation radiation images from the inner Galaxy region (Milne et al. 2000). The two upper maps show 511 keV line data from the nonimaging instruments OSSE and SMM, deconvolved into an intensity map with two different techniques. The lower panel shows the continuum annihilation emission measured with OSSE below the 511 keV line, and demonstrates that here the disk of the Galaxy has a larger contribution. (Reproduced with permission from Milne et al. 2000)

diffuse Galactic continuum emission (see Chap. 9), beyond the uncertain contribution from variable hard X-ray sources such as seen by the French SIGMA coded-mask X-ray telescope aboard the Russian GRANAT satellite. Therefore the uncertainty from spectral analysis alone remains large, and imaging results are needed.

The OSSE team reported a spectacular result from their efforts to map the annihilation emission of the inner Galaxy, using their many pointings during the CGRO mission (Purcell et al. 1997). According to their interpretation, a 'fountain' of positrons appears to emerge from the inner Galaxy and annihilate over a region extending several 100 pc into the northern halo. Imaging analysis is very difficult for such data from a nonimaging collimator instrument, and some care should be applied in interpretations. Nevertheless, this inner region of the Galaxy appears to have interesting peculiarities, also from other observational hints such as EGRET and SIGMA results (see also Chaps. 4 and 13), and models for a positron jet source have been advocated (Dermer and Skibo 1997). Better imaging measurements of annihilation γ-rays are needed.

10.5 Summary and Perspective

Nucleosynthesis bridges many aspects of modern astrophysics, from the basic nuclear physics processes and studies of stars and supernovae to investigations of chemical evolution throughout galaxies and the universe. Gamma-ray observations have joined this field, with a significant boost from the CGRO, through discoveries of several γ-ray lines with measurements of spatial distributions and line shapes. Although the astronomical precision of these measurements lags far behind what has been achieved in other fields, the more direct constraints set by γ-ray lines have distinct advantages over other approaches to understand the sites of nucleosynthesis in the different places in the evolving universe.

We have seen that supernova light is powered by radioactivity at very different time scales, from early ^{56}Ni to late ^{44}Ti and positrons. The puzzling observation of ^{44}Ti from exceptional supernovae may turn out to give fundamentally new insight on processes close to the compact remnant of a core collapse. We learnt that massive stars can be probed through ^{26}Al decay γ-rays throughout otherwise opaque regions. We have obtained a glimpse of the potential information carried by different line shapes of the radioactivity γ-rays. The γ-ray universe has shown once more that 'color' information educates us about the variety and detail of fundamental physical processes, in the regime of γ-rays, as in other domains of the electromagnetic spectrum.

What remains is reconciliation of these lessons with our models of the sites of nucleosynthesis, such that we can sufficiently refine those models to explain isotopic abundances as they evolved from the Big Bang to sites we observe today.

References

Anders, E., Grevesse, N., 1989, Geochim. Cosmochim. Acta, **53**, 197
Arnett, W.D., 1996, *Supernovae and Nucleosynthesis*, Princeton University Press, Princeton
Arnett, W.D. et al., 1977, Ann. NY Acad. Sci., **302**, 90
Arnould, M., Takahashi, K., 1999, Rep. Prog. Phys., **62**, 395
Aschenbach, B., 1998, Nature, **396**, 141
Aschenbach, B., Iyudin, A.F., Schönfelder, V., 1999, Astron. Astrophys. **350**, 997
Bernatowicz, T.J., Walker, R.M., 1997, Phys. Today, **12**, 26
Branch, D., 1998, Annu. Rev. Astron. Astrophys. **36**, 17
Burbidge, E.M., Burbidge, G.R., Fowler, W.A., Hoyle, F., 1957, Rev. Mod. Phys. **29**, 547
Burrows, A., 2000, Nature **403**, 727
Bussard, R.W., Ramaty, R., Brachman, R.J., 1979, Astrophys. J. **228**, 928
Cameron, A.G., W., 1957, Chalk River Report, CRL-41
Chan, K., Lingenfelter, R., 1991, Astrophys. J., **368**, 515
Chan, K., Lingenfelter, R., 1993, Astrophys. J., **405**, 614
Chen W., Diehl, R., Lehrels, N., et al., 1997, ESA-SP **382**, 105
Clayton, D.D., 1968, *Principles of Stellar Evolution and Nucleosynthesis*, Mac Graw-Hill, New York
Clayton, D.D., Colgate, S.A., Fishman, G., 1969, Astrophys. J., **220**, 353
Del Rio, E., von Ballmoos, P., Bennett, K. et al., 1996, Astron. Astrophys., **315** 237
Dermer, C.D., Skibo, J.G., 1997, Astrophys. J. **487**, L57
Diehl, R., Dupraz, C., Bennett, K. et al., 1995, Astron. Astrophys. **298**, 445
Diehl, R., Wessolowski, U., Oberlack, U. et al., 1997, AIP Conf. Proc. **410**, 1109
Diehl, R., Timmes, F.X., 1998, Publ. Astron. Soc. Pac. **110**, 748, 637
Diehl, R., Oberlack, U., Plüschke, S. et al., 1999, Astroph. Lett. Commun., **38**, 357
Dupraz, C., Bloemen, H., Bennett, K. et al., 1997, Astron. Astrophys. **324**, 683
Forestini, M., Charbonnel, C., 1997, Astron. Astrophys. S, **123**, 241
Gamow, G.Z., 1928, Z. Phys. **51**, 204
Gehrels, N., Leventhal, M., MacCallum, C.J., 1988, AIP Conf. Proc. **170**, 87
Gehrz, R., Truran, J., Williams, R., Starrfield, S., 1998, Publ. Astron. Soc. Pac. **110**, 3
Gomez-Gomar, J., Hernanz, M., Jose, J., Isern, J., 1998, Mon. Not. R. Astron. Soc. **296**, 913
Georgii, R., Plüschke, S., Diehl, R. et al., 2000, AIP Conf. Proc. **510**, 49
Harris, M.J., Leising, M.D., Share, G., 1991, Astrophys. J. **375**, 216
Hartmann, D.H., Predehl, P., Greiner, J. et al., 1997, Nucl. Phys. A **621**, 83
Hernanz, M., Gomez-Gomar, J., José, J., Isern, J., 1997, ESA-SP **382**, 47
Hix, W.R., Thielemann, F.-K., 1999, Astrophys. J. **511**, 862
Iben, I., Renzini, A., 1983, Annu. Rev. Astron. Astrophys. **21**, 271
Iwamoto, K., Brachwitz, F., Nomoto, K., Kishimoto, N., 1999, Astrophys. J.S **125**, 439
Iyudin, A.F., Diehl, R., Bloemen, H. et al., 1994, Astron. Astrophys. **284**, L1
Iyudin, A.F., Bennett, K., Bloemen, H. et al., 1995, Astron. Astrophys. **300**, 422
Iyudin, A., Schönfelder, V., Bennett, K. et al., 1998, Nature **396**, 142
José, J., Hernanz, M., 1997, Nucl. Phys. A **621**, 491

Käppeler, F., Beer, H., Wishak, K., 1989, Rep. Prog. Phys. **52**, 945
Kinzer, R.L., Purcell, W.R., Kurfess, J.D., 1999, Astrophys. J. **515**, 215
Kirsten, T., 1999, Rev. Mod. Phys. **71**, 1213
Knödlseder, J., 1999, Astrophys. J., **510**, 915
Knödlseder, J., Dixon, D.D., Diehl, R. et al., 1999a, Astron. Astrophys. **345**, 813
Knödlseder, J., Bennett, K., Bloemen, H. et al., 1999b, Astron. Astrophys. **344**, 68
Kolb, U., Politano, M., 1997, Astron. Astrophys. **319**, 909
Kurfess, J.D., Johnson, W.N., Kinzer, R.L. et al., 1992, Astrophys. J. **399**, L137
Lattanzio, J.C., Boothroyd, A.I., 1997, AIP Conf. Proc. **402**, 85
Leising, M.D., Share, G.H., 1994, Astrophys. J. **424**, 200
MacPherson, G.J., Davis, A.M., Zinner, E.K., 1995, Meteoritics **30**, 365
Merill, P.W., 1952, Science **115**, 484
Meyer, B.S., 1993, Phys. Rep. **227**, 1-5, 257
Meyer, B.S., 1994, Annu. Rev. Astron. Astrophys. **32**, 153
Meynet, G., Arnould, M., Prantzos, N., Paulus, G., 1997, Astron. Astrophys. **320**, 460
Milne, P. et al., 2000, AIP Conf. Proc. **510**, 21
Mochizuki, Y., Takahashi, K., Janka, H.-Th. et al., 1999, Astron. Astrophys. **346**, 831
Morris, D.J., Bennett, K., Bloemen, H. et al., 1995, 17th Texas Symposium, N.Y. Acad. Sci., **759**, 397
Naya, J.E., Barthelmy, S.D., Bartlett, L.M. et al., 1996, Nature **384**, 44
Naya, J.E., Barthelmy, S.D., Bartlett, L.M. et al., 1998, Astrophys. J. **499**, L169
Niemeyer, J., 1999, Astrophys. J. **523**, 57
Nomoto, K., Iwamoto, K., Kishimoto, N., 1997, Science **276**, 1378
Oberlack, U., Wessolowski, U., Diehl, R. et al., 1999, Astron. Astrophys. **353**, 715
Pagel, B.E.J., 1997, *Nucleosynthesis and Chemical Evolution in Galaxies*, Cambridge University Press, Cambridge
Prantzos, N., Diehl, R., 1996, Phys. Rep. **267**, 1
Purcell, W.R., Cheng, L.-X., Dixon, D.D. et al., 1997, Astrophys. J. **491**, 725
Phillips, M.M., Suntzeff, N.B., 2000, *SN 1987A: Ten Years After, The Fifth CTIO/ESO/LCO Workshop*, ASP, in preparation
Qian Y-Z., Woosley, S.E., 1996, Astrophys. J. **471**, 331
Ramaty, R., Skibo, J.G., Lingenfelter, R.E., 1994, Astrophys. J.S **92**, 393
Ramaty, R., Lingenfelter, R.E., 1977, Astrophys. J. **213**, L5
Ramaty, R., Kozlovsy, B., Lingenfelter, R.E., 1999, Phys. Today **4**, 30
Ritossa, C., Garcia-Berro, E., Iben, I., 1996, Astrophys. J. **460**, 489
Rohlfs, C., Rodney, W. 1988, *Cauldrons in the Cosmos*, Univ. of Chicago Press, Chicago
Schatz, H., Aprahamian, A., Görres, J. et al., 1998, Phys. Rep. **294**(4), 198
Schönfelder, V., Bloemen, H., Collmar., et al., 2000, AIP Conf. Proc. **510**, 54
Schramm, D.N., Turner, M.S., 1998, Rev. Mod. Phys. **70**, 1, 303
Shu, F.H., Adams, F.C., Lizano, F., 1987, Annu. Rev. Astron. Astrophys. **25**, 84
Simpson, J.A., Connell, J.J., 1998, Astrophys. J. **497**, L85
Suess, H.E.P., Urey, H.C., 1956, Rev. Mod. Physics **28**, 53
Stone, E.C., Cohen, C.M.S., Cook, W.R. et al., 1998, Sp. Sci. Rev. **86**, 357
The, L.-S., Diehl, R., Hartmann, D.H. et al., 2000, AIP Conf. Proc. **510**, 64
Thielemann, F.-K., Nomoto, K., Yokoi, Y., 1986, Astron. Astrophys. **158**, 17
Thielemann, F.-K., Nomoto, K., Hashimoto, M.A., 1996, Astrophys. J. **460**, 408

Timmes, F.X., Woosley, S.E., Hartmann, D.H. et al., 1995, Astrophys. J. **449**, 204
Timmes, F.X., Woosley, S.E., Hoffman, R.D., Hartmann, D.H., 1996, Astrophys. J. **464**, 332
Trimble, V., 1975, Rev. Mod. Phys. **47**, 877
Trimble, V., 1991, Astron. Astrophys. Rev. **3**, 1
Tutukov, A.V., Yungelson, L.R., Iben, I. Jr., 1992, Astrophys. J. **386**, 197
Wallerstein, G., Iben, I., Parker, P. et al., 1997, Rev. Mod. Phys. **69**, 4, 995
Woosley, S.E., Hartmann, D.H., Hoffman, R.D., Haxton, W.C., 1990, Astrophys. J. **356**, 272
Woosley, S.E., Weaver, T.A., 1995, Astrophys. J.S **101**, 181
Woosley, S.E., 1997, Astrophys. J. **476**, 801

11 Nuclear Interaction Gamma-Ray Lines

Volker Schönfelder and Andrew W. Strong

11.1 Introduction

Nuclear interaction lines are produced when energetic nuclei (typically >1 MeV nucleon^{-1}) interact with ambient matter. Such interactions have been observed from the Sun during strong solar flares, as has been extensively discussed in Chap. 5. Nuclear interaction lines are also expected from accreting compact stellar objects like white dwarfs, neutron stars, and black holes, from supernova remnants, from the nuclei of galaxies, and from the interstellar medium.

The first comprehensive calculations of expected line fluxes from the interstellar medium were performed in the 1970s (Rygg and Fishman 1973; Meneguzzi and Reeves 1975; Ramaty et al. 1979). Since most of the γ-ray line production processes have their maximum cross sections at several tens of MeV nucleon^{-1}, and since the inelastic cross sections decrease rapidly towards higher energies, energetic particles below 100 MeV nucleon^{-1} are essentially responsible for the production of nuclear interaction lines. Our knowledge of MeV nuclei in interstellar space is extremely poor, because such low-energy particles cannot penetrate from interstellar space into the solar system due to solar modulation. The observation of nuclear interaction lines therefore provides practically the only means to study the low-energy cosmic-ray component in interstellar space. If this component is sufficiently strong, it may play an important role in the heating and ionization of the interstellar medium. It is also of interest in the context of the production of the light elements lithium, beryllium and boron, which are believed to be the result of cosmic-ray interactions with gas over the lifetime of the Galaxy.

The most important nuclear reactions and their resulting line energies are described in Chap. 2. Recent displays of the excitation cross sections for the ^{12}C* state at 4.44 MeV and the ^{16}O* state at 6.13 MeV, 6.92 MeV, and 7.12 MeV are provided by Tatischeff et al. (1996) and displayed in Fig. 11.1. These cross section determinations are based on limited experimental results combined with theory. [For recent experimental measurements see also Kiener et al. (1998).]

The width of the lines depends on whether energetic protons interact with heavier nuclei in the ambient medium (e.g., carbon or oxygen nuclei), or whether heavier cosmic-ray nuclei (e.g., carbon or oxygen nuclei again)

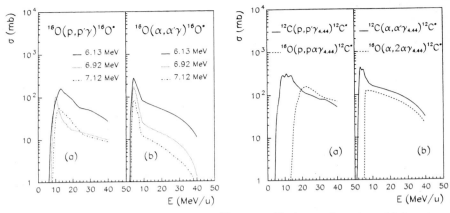

Fig. 11.1. Excitation cross sections of $^{12}C^*$ and $^{16}O^*$ as a function of laboratory energy in nuclear reactions induced by (a) protons and (b) α particles. (Reproduced with permission from Tatischeff et al. 1996)

interact with ambient hydrogen. In the first case, the recoil of the excited carbon and oxygen nuclei will be small, resulting typically in 1% broadening of the corresponding lines at 4.4 and 6.1 MeV, whereas in the latter case, the excited nuclei have much higher velocities and therefore the line broadening is of the order of 15%. If the target material is in the form of dust grains, the excited nuclei may come to rest before decaying. In this case, the line may be very sharp.

The γ-ray line flux ϕ_γ (photons cm^{-2} s^{-1}) from an astronomical object is determined by

$$\phi_\gamma = \int_V \int_e \frac{n_{\text{am}} \phi_{\text{CR}}(E_{\text{CR}}) \sigma(E_{\text{CR}})}{r^2} dE dV, \qquad (11.1)$$

where n_{am} is the number density of target nuclei in the "ambient medium," $\phi_{\text{CR}}(E)_{\text{CR}}$ is the flux (cm^{-2} s^{-1} ster^{-1}) of energetic particles, $\sigma(E_{\text{CR}})$ is the γ-ray line production cross section, and r is the distance to the emitting volume V.

For an order-of-magnitude estimate of the expected line fluxes let us calculate the flux of the 4.43 MeV line from excited ^{12}C nuclei from the Galactic Center direction. For this purpose, we assume that the line is produced by the bombardment of ambient carbon nuclei of density 4×10^{-4} cm^{-3} with 10 to 30 MeV protons of flux 1 cm^{-2} s^{-1} ster^{-1}, corresponding to an energy density of 4×10^{-2} eV cm^{-3}. With $\sigma \approx 100$ mbarn we obtain for the Galactic Center region ($r = 10$ kpc, $dV = 7 \cdot 10^{65}$ cm^3, corresponding to a radius 5 kpc and thickness 300 pc) a line flux of $\phi_\gamma \approx 3 \times 10^{-8}$ photons cm^{-2} s^{-1} from the central radian of the inner Galaxy. This intensity is extremely small – orders of magnitude below the presently achievable sensitivity of instruments.

The most comprehensive and accurate calculations of nuclear interaction lines from interstellar space are those of Ramaty et al. (1979). In these calculations the energy density of the low-energy cosmic-rays and the heavy-element abundances in the ambient matter were kept as free parameters. As an example Fig. 11.2 shows the expected nuclear interaction line spectrum from the general direction of the Galactic Center superimposed on a bremsstrahlung continuum component. A low-energy cosmic-ray density of $1\,\text{eV}\,\text{cm}^{-3}$ was assumed.

For both the ambient medium and the low-energy cosmic-ray component, abundances about 5 times higher than for the solar system were assumed for all elements heavier than helium (except for the N abundance, which was assumed to be a factor of 15 higher). The three classes of line shapes (narrow, broad and very sharp) can easily be recognized. The most intense lines are those at 4.43 MeV, 6.129 MeV, 847 keV and at 511 keV. The total fluxes of these lines (integrated over the line profiles) are summarized in Table 11.1 (from Ramaty et al. 1979).

From Table 11.1 we see that the combined contributions from nuclear interaction lines to the total 3 to 7 MeV flux from the central Galactic Plane would be typically 1.0×10^{-4} photons $\text{cm}^{-2}\,\text{s}^{-1}\,\text{rad}^{-1}$.

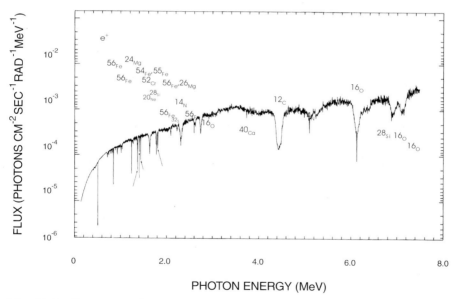

Fig. 11.2. Gamma-ray line spectrum expected (under certain assumptions, see Sect. 11.1) from nuclear reactions in interstellar space towards the general direction of the Galactic Center, superimposed on a bremsstrahlung continuum component. (Reproduced with permission from Ramaty et al. 1979). The total flux in each line is obtained by integrating over the entire line profile

Table 11.1. Expected γ-ray line intensities

Gamma-ray line energy (MeV)	γ-ray line flux (cm^{-2} s^{-1} rad^{-1})
4.43	7.2×10^{-5}
6.129	3.0×10^{-5}
0.847	9.2×10^{-6}
0.511	6.0×10^{-5}

The 4.43 MeV line flux turns out to be so much larger than in our previous order-of-magnitude estimate because of the much more optimistic assumption made in Fig. 11.2 of a higher energy density of low-energy cosmic-ray particles and, in addition, because of the enhanced heavy element composition in both ambient medium and cosmic-rays. But even these higher fluxes are only close to or still below the COMPTEL (Compton Telescope) sensitivity limit (see Chap. 3). Therefore there was little hope to detect these lines with the Compton Gamma-Ray Observatory. When the COMPTEL team, nevertheless, reported hints of nuclear interaction lines from the direction of the Orion complex (Bloemen et al. 1994, 1997a), and the Galactic Center region (Bloemen et al. 1997b), this came as a surprise and stimulated lively discussions about energetic particle production in the Galaxy in general, and near OB star associations in particular. Based on a better understanding of the COMPTEL instrumental background, an improved analysis method, and more exposure time, the Orion results were finally revised: the previously reported Orion fluxes were converted into 2σ upper limits, which are about 4 times lower than the original fluxes (Bloemen et al. 1999). At the time of writing, no firm detections of nuclear interaction lines from any astronomical object other than the Sun can be claimed.

11.2 Nuclear Interaction Lines from the Inner Galaxy

The broadband energy spectrum of continuum emission from the Milky Way is discussed extensively in Chap. 9. COMPTEL spectra with a finer energy binning from the inner two radians have revealed evidence for deviations from the smooth continuum shape. Spectral excesses seem to be present from 1.3 to 1.7 MeV and from 3 to 7 MeV. The excess emission seems to have a rather wide latitudinal distribution ($\sim 20°$ FWHM, full width at half-maximum). The flux in each of the two excesses is typically 2×10^{-4} photons cm^{-2} s^{-1} for the inner two radians (Bloemen et al. 1997b) – roughly consistent with the optimistic predictions of Fig. 11.2.

The presence of such a wide latitudinal component would, however, be difficult to understand in a scenario where energetic particles interact with the interstellar medium. If the origin of both excesses is indeed nuclear interaction lines (no individual lines can be resolved in the COMPTEL spectrum), then the most plausible explanation seems to be that the collective signal of an ensemble of unresolved objects is seen. The wide latitudinal extent could be understood if low-mass X-ray binaries, having a large scale height, play an important role. Individually the sources would be undetectable (see below), but their global effect may be visible. The line emission would be gravitationally redshifted and could extend over a larger energy range (Bloemen and Bykov 1997).

11.3 Nuclear Interaction Lines from Accretion Processes

Shvartsman (1972) first suggested that matter accreting onto neutron stars would produce observable γ-ray spectral features. The most accurate estimates of expected line fluxes from nuclear reactions involving ^{12}C, ^{14}N, and ^{16}O nuclei, and neutron capture lines (especially at 2.223 MeV) have been performed by Bildsten et al. (1992), and Bildsten et al. (1993).

The kinetic energy of a proton in free-fall onto a neutron star is

$$E = \frac{GMm_\mathrm{p}}{R} = 194 \, \frac{\mathrm{MeV}}{\mathrm{nucleon}} \left(\frac{M}{1.4\,M_\odot}\right) \times \left(\frac{10\,\mathrm{km}}{R}\right), \tag{11.2}$$

where M and R are the mass and radius of the neutron star and m_p the proton mass. These energies are near the range where cross sections for nuclear line production have their maxima. The infalling ions are slowed and thermalized by Coulomb interactions with ambient electrons and become part of the neutron-star atmosphere. In the accretion process all elements heavier than helium thermalize at higher altitudes above the neutron star than accreting protons. Bildsten et al. (1992) have self-consistently calculated the resulting composition of the neutron-star atmosphere as a function of altitude. The interplay between the rate at which the elements drift downwards through the atmosphere and the rate at which they are destroyed and excited by the proton beam determines the abundances in the atmosphere and the resulting γ-ray line fluxes. Essentially, only 4 parameters determine the atmosphere of an accreting neutron star, namely the mass, the radius, the local mass flux $\dot{M}$ (measured in $\mathrm{g\,cm^{-2}\,s^{-1}}$), and the elemental composition of the incident material. If the atmosphere has solar abundances, the expected γ-ray line fluxes from excited nuclei are several orders of magnitude below the detection threshold of existing experiments (of the order of 10^{-8} to 10^{-9} photons $\mathrm{cm^{-2}\,s^{-1}}$). Only if the accreting material consists predominantly of helium or heavier elements could these objects produce larger γ-ray fluxes that might become observable. The energetic helium nuclei in the beam of interacting particles liberate neutrons by spallation reactions, which subsequently recombine with ambient atmospheric protons and emit the 2.223 MeV neutron capture line. However, also here the flux of redshifted 2.223 MeV photons from even the brightest accreting X-ray source Scorpius X-1 ($\sim 10^{-6}$ photons cm^{-2} s^{-1}) is still a factor of 20 below the COMPTEL sensitivity (Bildsten et al. 1993).

11.4 The Observations of the Orion Complex

The Orion complex of molecular clouds constitutes one of the closest active regions (distance ~ 450 pc) of high star-forming activity, strong stellar winds and multiple expanding supernova shells. This configuration may provide an ideal environment for the production of energetic particles.

Our present knowledge about the γ-ray emission from the Orion complex is summarized in Fig. 11.3, where the flux spectrum of the complex is shown from 1 MeV to 10 GeV. The emissivities derived from the EGRET (Energetic Gamma-Ray-Experiment Telescope) measurements above 30 MeV and from COS-B above 300 MeV can be interpreted as cosmic-ray interactions with ambient gas nuclei via π^0 decay and bremsstrahlung emission at an intensity level which is consistent with cosmic rays observed near Earth (see also Chap. 9). The upper limits from COMPTEL below 30 MeV are, in principle, consistent with an extrapolation of the EGRET data. The 2σ upper limits for nuclear interaction lines at 4.4 MeV and 6.1 MeV depend on the actual width of the lines. Here we use a mean value of 1.0×10^{-5} photons cm^{-2} s^{-1} as a 2σ upper limit to any line within the 3 to 7 MeV interval.

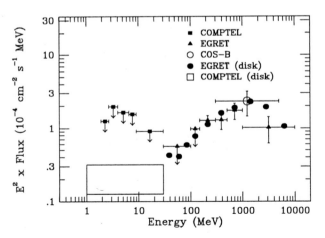

Fig. 11.3. Flux spectrum of the Orion complex from about 1 MeV to 10 GeV. Adopted from Bloemen et al. (1999). (Reproduced with permission of Bloemen et al. 1999)

The upper limits on nuclear interaction lines from Orion can be used to derive limits for the density of MeV nuclei in the complex. For the subsequent estimates we use a distance of the Orion complex of 450 pc, a total mass in the form of gas in the complex of 2×10^5 M$_\odot$, and an average gas density of 100 cm^{-3}. From the measured line limit of 1.0×10^{-5} photons cm^{-2} s^{-1} a luminosity limit of 2×10^{38} photons s^{-1} or 1.4×10^{33} erg s^{-1} ($= 1.4 \times 10^{26}$ J s^{-1}) is derived.

In order to derive limits on the energetic particle flux within the complex, we again use (11.1). If we first consider lines that are produced by energetic proton interactions, we obtain by applying the flux limit of 1.0×10^{-5} photons cm^{-2} s^{-1} to the 4.43 MeV line from ^{12}C* a proton-flux limit of ϕ_{CR} (10–30 MeV) $< 1.8 \times 10^3$ protons cm^{-2} s^{-1} ster^{-1}, by making $n_{am}\Delta V = 1.0 \times 10^{59}$ C nuclei (corresponding to a C abundance of 4×10^{-4}). The resulting proton energy density of ≈ 100 eV cm^{-3} is still about two orders of magnitude higher than the local cosmic-ray density above 1 GeV, as well as the density used for the calculations of line fluxes in Sect. 11.1. A density as high as this limit would violate the ionization rate in the interstellar medium of the complex (Cowsik and Friedlander 1995). If instead we consider energetic C nuclei interacting with ambient hydrogen, the ionization would be smaller for the same γ-ray emission.

Limits to the C and O nuclei luminosity in the 10 to 30 MeV nucleon^{-1} range can be estimated by again making use of the fact that the 4.43 MeV ^{12}C* line was not detected from Orion. The number of C nuclei is

$$N_C = 2 \times 10^{38} \tau_\gamma = \frac{2 \times 10^{38}}{n_H v_C \sigma(E_C)}. \tag{11.3}$$

For $n_H = 10^2$ atoms cm^{-3}, $v_C = 0.2c$, and $\sigma = 100$ mbarn we obtain $\tau_\gamma = 5 \times 10^5$ yr and hence $N_C < 3 \times 10^{51}$. Since these nuclei suffer energy losses by ionization with an energy loss time τ_i, the resulting luminosity limit in energetic C nuclei is N_C/τ_i, where

$$\tau_i = \frac{E_C}{\left(\frac{dE}{dt}\right)_C}, \tag{11.4}$$

with

$$E_C = \frac{1}{2} M_C v_C^2 \tag{11.5}$$

and

$$\left(\frac{dE}{dt}\right)_C = \frac{2\pi n_e Z^2 e^4}{m_e v_C} \ln \Lambda, \tag{11.6}$$

where M_C is the mass and $Z = 6$ the charge of the energetic C nuclei, m_e is the electron mass, n_e the electron density, $\Lambda = \frac{2 M_C v_C^2}{I_0}$ and hence $\ln(\Lambda_n) \approx 10$ (I_0 is the ionization potential).

For carbon nuclei of 20 MeV nucleon^{-1} in a medium with $n_e = 10^2$ cm^{-3} we obtain, from (11.4–6), $\tau_i = 5700$ yr, and hence for the nuclei a luminosity limit $N_C/\tau_i = 2 \times 10^{40}$ nuclei s^{-1} corresponding to an energy luminosity limit of 7×10^{36} erg s^{-1} (7×10^{29} J s^{-1}).

The exact value of the luminosity limit depends on the shape of the nuclei spectrum and the cut-off energy of the nuclei; luminosity limits are of the order of 10^{37}–10^{38} erg s^{-1}. These limits are extremely high, namely only a factor of 1000 lower than the cosmic-ray luminosity (>1 GeV) of the entire Galaxy, although the volume of the Orion complex is at least 10^7 times smaller than that of the Galaxy. One may, therefore, speculate that the MeV cosmic-ray luminosity in Orion is actually much lower than such upper limits.

As was first pointed out by Dogiel et al. (1997), due to the extremely low efficiency of the γ-ray line production ($\sim 10^{-5}$), one can expect that the energetic particles will produce other effects as well, which can be used to trace nuclear line production sites by other means. One such effect is the production of X-ray emission. X-rays can be produced in three different ways by the same energetic particles that produce the γ-ray lines: first, by inverse bremsstrahlung with ambient electrons, second by bremsstrahlung of knock-on electrons (produced by collisions with the energetic particles), and third by X-ray line emission stimulated either by electron capture of the energetic nuclei from the ambient medium or by interactions of the energetic nuclei with ambient ions leading to inner-shell vacancy creation (Tatischeff et al. 1998). The spatial distribution of the X-ray production should exactly follow that of the γ-ray line production. The observable X-ray fluxes will, however, be affected by photo electric absorption along the line of sight (3×10^{21} H atoms cm^{-2} towards Orion) for X-rays which have energies below 2 keV. The expected continuum X-ray intensities are too low to give useful limits based on observations from present X-ray instruments.

References

Bildsten, L., Salpeter, E.E., Wasserman, I., 1992, Astrophys. J. **384**, 143
Bildsten, L., Salpeter, E.E., Wasserman, I., 1993, Astrophys. J. **408**, 615
Bloemen, H., Wijnands, R., Bennett, K. et al., 1994, Astron. Astrophys. **281**, L5
Bloemen, H., Bykov, A.M., 1997, Proc. of the 4th Compton Symposium, Williamsburg, AIP Conf. Proc. **410**, ed. by Dermer, C.D., Strickman, M.S., Kurfess, J.D., p. 249
Bloemen, H., Bykov, A.M., Bozhokin, S.V. et al., 1997a, Astrophys. J. **475**, L25
Bloemen, H., Bykov, A.M., Diehl, R. et al., 1997b, Proc. of the 4th Compton Symposium, Williamsburg, AIP Conf. Proc. **410**, ed. by Dermer, C.D., Strickman, M.S., Kurfess, J.D., p. 1074
Bloemen, H., Morris, D., Knödlseder, J. et al., 1999, Astrophys. J. Letters, **521**, L137
Cowsik, R. Friedlander, M.W., 1995, Astrophys. J. Letters **444**, L29

Dogiel, V.A., Freyberg, M.J., Morfill, G.E., Schönfelder, V., 1997, Proc. of 25th Int. Cosmic Ray Conf., Durban, South Africa, ed. by Potgieter M.S., Raubenheimer B.C., van der Walt D.J., Vol. **3**, 133
Kiener, W., Berheide, M., Achouri, N.L., 1998, Phys. Rev. **C 58**, 2175
Meneguzzi, M. Reeves, H., 1975, Astron. Astrophys. **40**, 91
Ramaty, R., Kozlovski, B., Lingenfelter, R.E., 1979, Astrophys. J. Suppl. **40**, 487
Rygg, T.A. Fishman, G.J., 1973, Proc. of 13th Int. Cosmic Ray Conf., Denver, Univ. of Denver, ed. by Chasson, R.L. **1**, 472
Shvartsman, V.F., 1972, Astrophysics **6**, 56
Tatischeff, V., Cassé, M., Kiener. J. et al., 1996, Astrophys. J. **472**, 205
Tatischeff, V., Ramaty, R., Kozlovsky, B., 1998, Astrophys. J. **504**, 874

12 Gamma-Ray Emission of Active Galaxies

Werner Collmar

12.1 The Nature of Active Galaxies

12.1.1 Differences Between Normal and Active Galaxies

The observable universe consists of billions of galaxies, many of them containing 100 billion or more stars, like our "Milky Way" for example. On a morphological basis these galaxies are grouped into three main categories: spirals, ellipticals, and irregulars.

In 1943 the US astronomer Carl Seyfert noted a distinct class of galaxies with unresolved bright cores (starlike) that emit broad emission lines. Galaxies showing these two characteristics were subsequently named Seyfert galaxies and mark today a subclass of the so-called active galaxies. Among active galaxies subclasses have been named quasars (quasistellar radio sources), QSOs (quasistellar objects), radio galaxies, and blazars for example. The names of the different subclasses of active galaxies developed historically according to the name of the discoverer (e.g., Seyferts), morphological characteristics (QSOs), or from other properties, such as radio galaxies from strong radio emission.

There are distinct observational differences between ordinary or normal galaxies and active galaxies. In contrast to the images of normal galaxies, which are basically an assembly of stars, the images of active galaxies show bright nuclei. Their optical spectra typically show broadened emission lines instead of absorption lines (Fig. 12.1). Their overall spectra show luminosity maxima in IR, UV, X-rays or even γ-rays, and are often dominated by nonthermal emission, while normal galaxies radiate most of their energy in the optical band, this simply being the sum of the starlight. Another major difference between normal and active galaxies is their variability. While normal galaxies always look the same – apart from when a supernova shines occasionally for a few months – the emission of active galaxies changes significantly on short time scales, down to days or even less, which gave rise to the term active. This short-term variability points to small sizes of the emission regions which are consistent with the bright nuclei. In the particular case of quasars or QSOs the nucleus outshines the rest of the galaxy by far, resulting in a starlike image. On the basis of light-travel arguments, the emission region radius is estimated from

$$r < c\Delta t/(1+z), \qquad (12.1)$$

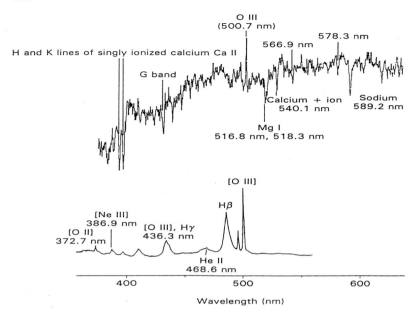

Fig. 12.1. Comparison of optical spectra of a normal (spiral) galaxy above and an active (Seyfert) galaxy below. The spectrum of the spiral galaxy shows mainly absorption lines, while the spectrum of the Seyfert galaxy is dominated by strong and broad emission lines. (From Robson 1997, who adapted the data from "The Cambridge Atlas of Astronomy", reproduced with permission)

with Δt the observed time variability, z the redshift of the source, and c the speed of light. For a Δt of 1 day, as is observed in many quasars, and a redshift z of 1, for example, the emission region has to be smaller than $\sim 1.3 \times 10^{10}$ km, which is roughly the size of the solar system. So, the active part in active galaxies is in fact the nucleus, which led to the commonly used term Active Galactic Nuclei (AGN) for this class of celestial objects. Some active galaxies or AGN show jets – a collimated plasma flow coming straight out from the central region – as in the famous quasar 3C 273, which in fact was the first quasar ever detected (Schmidt 1963), or in strong radio galaxies like Cygnus A or Centaurus A.

12.1.2 The Source of Energy for Active Galaxies

After the radio source with the number 273 of the 3rd Cambridge catalogue of radio sources ("3C 273") had been identified as a starlike celestial object, an optical spectrum was taken. It showed strong and bright emission lines, which, however, are redshifted by a factor of about 0.16 (see Schmidt 1963), suggesting an extragalactic nature for the object. If this redshift is interpreted to be a sign of cosmological expansion, a source distance of the order of 1 Gpc ($H_0 = 50\,\mathrm{km\,s^{-1}\,Mpc^{-1}}$) is determined by applying Hubble's law to convert recession velocity into distance. This huge distance together with the measured optical magnitude indicates a luminosity of about 5×10^{12} solar luminosities ($1.9 \times 10^{46}\,\mathrm{erg\,s^{-1}}$), which is, of course, far beyond the most luminous stars and more than a factor of 10 larger than the luminosity of the most luminous galaxies. Such a luminosity, together with the above estimates on the source regions, indicates an incredible scenario: in such sources a luminosity of the order of 10^{12} Suns is generated in a volume as small as the solar system! This fact immediately leads to the question about the nature of the energy source. The most efficient process known is the release of gravitational energy in a deep gravitational potential:

$$L_{\mathrm{acc}} = \epsilon \cdot \dot{m}_{\mathrm{acc}} c^2, \qquad (12.2)$$

with L_{acc} being the source luminosity, ϵ the efficiency for converting gravitational energy into radiation, $\dot{m}_{acc}$ the mass accretion rate, and c the speed of light. This process, which has an efficiency $\sim 10\%$ of converting the rest-mass energy into radiation (nuclear fusion reaches only $\sim 0.7\%$) fuels the Galactic X-ray binary sources (see Chap. 7) and is thought to be at work in AGN as well. However, there is an upper limit on the mass accretion rate ($\dot{m}_{\mathrm{acc}}$), and therefore on the luminosity (L_{acc}) resulting from the accretion process: the so-called Eddington limit. It is derived from the balance between the gravitational force and the radiation pressure on the accreting material. If the radiation pressure dominates, the accretion stops. For spherical accretion and Compton scattering in the Thomson regime the Eddington limit for the luminosity of a source as function of the central mass, M, is given by

$$L_{\mathrm{Edd}} = \frac{4\pi G m_{\mathrm{p}} c}{\sigma_{\mathrm{T}}} M = 1.3 \times 10^{38} \frac{M}{M_\odot} (\mathrm{erg\,s^{-1}}), \qquad (12.3)$$

where G is the gravitational constant, m_{p} the proton mass, c the speed of light, and σ_{T} the Thomson cross section. If a source is radiating at its Eddington limit with a luminosity of $10^{46}\,\mathrm{erg\,s^{-1}}$, its mass is of the order of $10^8\,M_\odot$. Therefore nuclei of active galaxies should have masses in the range between $10^6\,M_\odot$ and $10^{10}\,M_\odot$. If the source is radiating below the Eddington limit, even larger masses would result. In summary, the energy source which ultimately powers AGN is believed to be the release of gravitational energy in mass accretion onto a supermassive ($>10^6\,M_\odot$) black hole. To sustain, for example, an AGN luminosity of $10^{46}\,\mathrm{erg\,s^{-1}}$ generated by accretion, a mass accretion rate of the order of $1.7\,M_\odot\,\mathrm{yr}^{-1}$ is necessary, assuming a conversion efficiency of 10% (see 12.2).

12.1.3 Unification Schemes for Active Galaxies

Roughly 3% of all galaxies are classified as being active. Although there is no overall agreement amongst astronomers on a precise definition of active galaxies, one common criterion is the generation of large luminosities in small core regions. In many cases the active and bright core outshines by far the remaining galaxy.

As described above there exist a large number of classes and subclasses of AGN, which developed historically and sometimes overlap in their defining parameters. Therefore they are often confusing for physicists and even astrophysicists not working in AGN research. For example, the most luminous Seyferts are brighter than the low-luminosity quasars. If a Seyfert galaxy were put much further away, only the bright core would remain visible, and, subsequently, it would be classified as a QSO. Thus, in a first step towards unification, it is suggested that some Seyferts are simply "nearby" QSOs.

Observationally, AGN are classified on the basis of three parameters: the strength of their radio emission, their emission line properties in the optical and UV range, and their luminosity. However, the question has arisen as to whether there exists some common underlying physics or physical structure. One of the first steps towards unification concerned the difference between the two subclasses of Seyfert galaxies, simply called Type 1 and Type 2. While the Type 1 Seyferts show broad ($\Delta v > 2000\,\mathrm{km\,s^{-1}}$) *and* narrow emission lines ($\Delta v < 2000\,\mathrm{km\,s^{-1}}$), the Type 2 sources show *only* narrow emission lines. A key observation by Antonucci and Miller (1985) revealed broad emission lines at a weak level in the polarized light from the Seyfert Type 2 galaxy NGC 1068. Because scattered light is polarized, this observation was interpreted to imply that NGC 1068 also generates broad emission lines, which, however, are somehow hidden from direct observation. From such observational facts a simplified classification scheme evolved recently, which is shown in Table 12.1. The different classes of AGN are ordered according to two parameters only: their "optical emission line properties" and their "radio loudness." With respect to their line properties, AGN are subdivided into so-called Type 2 sources, showing only narrow emission lines; Type 1 sources, also showing broad emission lines; and Type 0 sources, showing weak or even no emission lines at all. AGN are considered to be radio-loud if the ratio of the radio flux at 5 GHz to the flux in the optical B band ("blue") is larger than 10 ($f_{5\,\mathrm{GHz}}/f_B \geq 10$). According to this definition roughly 10% to 15% of all AGN are radio-loud.

The main AGN types are sorted according to this two-dimensional scheme in Table 12.1. It is believed that the horizontal axis reflects an orientation effect, namely the decreasing angle of the line of sight towards the jet (plasma outflow) of the active galaxy (see Fig. 12.2). There is no commonly accepted interpretation of why some AGN are radio-loud and others not. Whether this has to do with the host galaxy – radio-loud sources are mainly found in elliptical galaxies, radio-quiet ones in spirals – or with the spin of the putative central black hole or something else remains unclear to date. This question is currently one of the major unresolved problems in AGN research. Fig. 12.2 shows the "standard model" for AGN, which explains the different types as

Table 12.1. This table [adapted from Urry and Padovani (1995)] shows the simplified classification of AGN according to the two parameters "radio–loudness" (vertical direction) and "optical emission line properties" (horizontal direction). While the distinctions in the horizontal direction are thought to be due to a viewing effect ("decreasing angle between jet and observers line of sight"), the cause of distinction in the vertical direction, i.e., radio-loud or radio-quiet, is still unknown (indicated by question marks). For more discussion of this scheme see Sect. 12.1.3 or, for example, Urry and Padovani (1995). The main types of AGN are sorted in the table. NLRG, BLRG: narrow and broad line radio galaxies; SSRQ, FSRQ: steep and flat spectrum radio quasars; FR I, II: Fanaroff–Riley Type 1 and 2 radio galaxies; and BL Lac: BL Lacertae objects. The BL Lacs and the FSRQs form the class of blazars

Radio loudness	Optical emission line properties		
	Type 2 (narrow lines)	Type 1 (broad lines)	Type 0 (weak/absent)
Radio-quiet (85–90%)	Seyfert 2	Seyfert 1 QSO	
Radio-loud (10–15%)	NLRG (FR I, FR II)	BLRG SSRQ, FSRQ	Blazars (BL Lac, FSRQ)
	Decreasing jet angle to line of sight →		

simply being an orientational effect. Despite the fact that many details are not yet understood, this scenario – which is described in the following – is commonly accepted because it explains the major observational facts reasonably well and consistently. The central object is thought to be a supermassive black hole with masses of the order of $\sim 10^6$ to $\sim 10^{10} \, M_\odot$. The Schwarzschild radius,

$$R_\mathrm{s} = \frac{2GM_\mathrm{BH}}{c^2} \tag{12.4}$$

of an $10^8 \, M_\odot$ black hole is $\sim 3 \times 10^8$ km, which is approximately 2 Astronomical Units (AU) or $\sim 10^{-5}$ pc. The black hole is surrounded by an accretion disc consisting of ionized material reaching out to several hundreds of Schwarzschild radii. The central region of the active galaxy is surrounded by an extended molecular torus, with an inner diameter of ~ 1.5 pc and an outer diameter of the order of ~ 30 pc. Within the molecular torus and near the center of the active galaxy fast moving ($v \geq 2000 \, \mathrm{km\,s^{-1}}$) gas clouds (dark spots in Fig. 12.2) exist which are ionized by the accretion disk radiation and which emit the observed broad emission lines. These fast-moving clouds

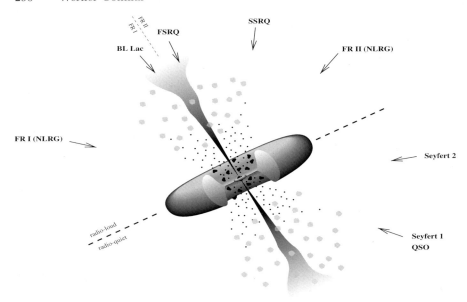

Fig. 12.2. The "standard model" for radio-loud AGN. For an explanation see Sect. 12.1.3. Adapted from Urry and Padovani (1995). The distinctions between radio-loud and radio-quiet objects, and FR I and FR II sources, are added. In the radio-quiet (lower) part of the figure no jet or only a weak one is required.

are located within $\sim(2\text{--}20)\times10^{16}$ cm and mark the so-called broad-line region (BLR). Further out such clouds (grey spots in Fig. 12.2) move slower ($v \leq 2000\,\mathrm{km/s}$) and therefore give rise to the observed narrow emission lines. This so-called narrow-line region (NLR) extends roughly from 10^{18} cm to 10^{20} cm. In addition to these clouds a hot electron corona (dark dots in Fig. 12.2) populates the inner region; this can scatter some continuum and emission from the broad-line region, as has been observed in NGC 1068. In a radio-loud AGN a strong jet of relativistic particles emanates perpendicular to the plane of the accretion disc. The generation of such jets is still not understood; however it is believed that strong magnetic fields play a fundamental role. Radio jets have been observed on scales from 10^{17} cm to $\sim 10^{24}$ cm, which is significantly larger than the largest galaxies.

The "unification-by-orientation" scenario assumes such a general structure for all AGN. Depending on the spatial orientation with respect to our line of sight we observe the different types of AGN. If we look towards the "central engine" along the jet axis ($\lesssim 10°$) for a radio-loud object, we observe a BL Lacertae object or a quasar which both show a flat ($S \propto \nu^\alpha$, $\alpha > -0.5$) radio spectrum. A steep-spectrum radio quasar is observed at offset angles of the order of 30°. A typical radio galaxy, showing two strong oppositely aligned jets, is observed at viewing angles approximately perpendicular to the jet axis. For the radio-quiet objects, a Seyfert Type 1 galaxy or a QSO is seen for viewing angles at which both the narrow- and broad-line regions are visible. At larger angular offsets the broad-line region will be hidden by this extended molecular torus, giving rise to Seyfert Type 2 galaxies.

This scenario is very intriguing and is widely accepted among astronomers. However, there are still several unresolved questions, such as "Where are the radio-quiet narrow-line (Type 2) QSOs?", and – as already mentioned above – "Which physics causes the difference between radio-loud and radio-quiet AGN?". In addition, this scenario neglects evolution, which is probably another important parameter for active galaxies.

For recent comprehensive reviews about all aspects of AGN I refer the reader to, for example, the books "Active Galactic Nuclei" by Robson (1997), "An Introduction to Active Galactic Nuclei" by Peterson (1997), and "Active galactic nuclei: from the central black hole to the galactic environment" by Krolik (1999). In the following I shall focus on the high-energy emission of AGN with emphasis on γ-rays. I shall outline the major developments in this research area in recent years.

12.2 Gamma-Ray Blazars

12.2.1 Observational Status on AGN before CGRO

Prior to the launch of the Compton Gamma-Ray Observatory (CGRO) in April 1991 the extragalactic γ-ray sky was largely unknown. Only the quasar 3C 273 had been detected significantly as an emitter of γ-ray radiation at energies above 50 MeV by the COS-B satellite (Swanenburg et al. 1978). For three more sources – the radio galaxy Centaurus A (von Ballmoos et al. 1987) and the two Seyfert galaxies NGC 4151 (Perotti et al. 1981a) and MCG-8-11-11 (Perotti et al. 1981b) – the detection of γ-rays up to $\sim$20 MeV had been reported. However, these measurements resulted from short-duration balloon flights and contained large systematic uncertainties. In view of these reports and considering the greatly improved sensitivities of the CGRO instruments compared to previous experiments, and the capability for long-term source exposures provided by the satellite mission, it was expected prior to the launch that CGRO would detect several quasars, Seyfert galaxies, and radio galaxies. In addition, there was hope for new and unexpected discoveries, which are often made when new instruments come into operation.

12.2.2 CGRO Source Detections

Shortly after the launch, CGRO was pointed towards the quasar 3C 273. A strong γ-ray signal was immediately observed by EGRET (Energetic Gamma-Ray Experiment Telescope), which, however, was not at the expected location. It was offset by $\sim 8°$ from the position of 3C 273, thereby causing some confusion within the EGRET and satellite operations teams. After the satellite pointing and the EGRET data analysis was confirmed to be correct, the strong γ-ray source was identified with the blazar-type quasar 3C 279. 3C 279, known to be a radio source and an optically violently variable (OVV) quasar, was the most "dramatic" object inside the EGRET location error box. Further observations of other sky regions often showed fairly bright γ-ray sources in the EGRET field of view, and their location error boxes always contained a blazar-type AGN, supporting the identification of 3C 279. Therefore the conclusion that blazars can – at least occasionally – be strong emitters of γ-ray radiation became firm after some time of CGRO operation. Blazars are radio-loud quasars or BL Lac objects which show a flat radio spectrum, strong and rapid variability in both optical and radio bands, and strong optical polarization. In addition, many blazars show superluminal motion of radio "knots" resolved by very long baseline interferometry (VLBI). The term blazar derives from merging of the designations *BL* Lac and qu*asar* (with a "z" for the "s" to emphasize the "blazing" luminosities of these sources) and was coined after it became apparent that OVV quasars and BL Lac objects share so many observational properties and therefore might belong to a distinct category.

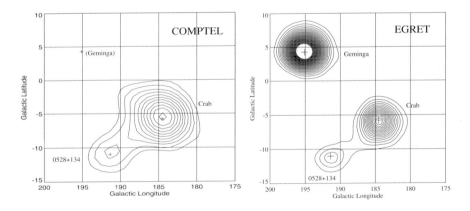

Fig. 12.3. Comparison of COMPTEL (10–30 MeV) and EGRET (above 100 MeV) detection-significance maps of the Galactic Anticenter. The contour lines indicate detection-significance levels. Three significant sources are seen by EGRET: the previously known Galactic γ-ray pulsars Crab and Geminga and the newly discovered γ-ray blazar PKS 0528+134 . The absence of Geminga in the COMPTEL maps is due to its hard γ-ray spectrum, which causes it to fall below the COMPTEL detection threshold despite its prominence in the EGRET data. (From Hartman et al. 1997, reproduced with permission of The American Institute of Physics)

12.2.3 Observational Properties

After seven years of operation, roughly 90 blazars had been detected by EGRET at energies above 100 MeV (e.g., Hartman et al. 1999). Roughly ten of them were also observed at lower-energy γ-rays (1 to 30 MeV) by COMPTEL (Compton Telescope) (Collmar et al. 1999) and also about ten by the OSSE experiment (Oriented Scintillation Spectrometer Experiment) (Kurfess 1996), mainly at hard X-ray energies (∼50 keV to ∼500 keV). Figure 12.3, as an example, shows simultaneous COMPTEL and EGRET maps of the Galactic Anticenter region. The evidence for the newly discovered γ-ray blazar PKS 0528+134 is obvious in both maps. Gamma-ray blazars have been detected with redshifts ranging from ∼0.03 to ∼2.3. They show a similar redshift distribution to flat-spectrum radio quasars, which additionally supports their association.

The γ-ray fluxes of blazars are observed to be highly variable, with variability time scales (e.g., doubling of the flux) as short as one day and even less. For example, during a major outburst of PKS 1622-297 flux variations up to a factor of 80 with flux doubling time scales of a few hours were observed by EGRET (Mattox et al. 1997). Another prominent example for γ-ray variability of blazars is shown in Fig. 12.4: the long-term γ-ray light curve of 3C 279 – covering ∼5 yr in time – as observed by EGRET at energies above

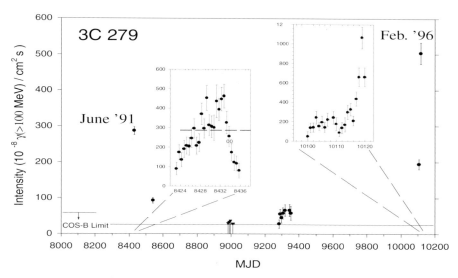

Fig. 12.4. Long-term γ-ray flux history of the blazar 3C 279. For details see Sect. 12.2.3. The two flaring events in June 1991 and February 1996 are shown in the insets with improved time resolutions of 0.5 days and 1 day, respectively. Variability on a daily time scale is clearly visible. (From Hartman et al. 1997, reproduced with permission of The American Institute of Physics.)

100 MeV. Short-term variability (∼days) is clearly seen in the insets, which resolve the two major γ-ray flares observed from this blazar.

The energy spectra measured by the different CGRO instruments are consistent with power-law shapes. At energies above 100 MeV, EGRET measures power-law shapes with photon indices ranging from 1.4 to 2.8 with a Gaussian mean value of 2.16 and a standard deviation of 0.31 (Mukherjee et al. 1997). EGRET results indicate two spectral trends: (a) the tendency for spectral hardening at high flux levels and (b) that BL Lac objects have slightly harder spectra than EGRET-detected quasars. COMPTEL, being a pioneering instrument at MeV energies, has detected ∼10 of these blazars (Collmar et al. 1999). Because the COMPTEL detections often occur near threshold and are significant only in parts of its energy band, our knowledge of the MeV spectra of blazars is incomplete. Nevertheless some trends are apparent. In time-averaged analyses the spectra are described well by power-law shapes with photon indices of the order of two. However, during flaring periods reported by EGRET above 100 MeV, the MeV shapes tend to be harder. The OSSE experiment detects blazars most significantly at hard X-ray energies between 50 keV and 150 keV. The spectra are consistent with power-law shapes with photon indices around 1.6 (McNaron-Brown et al. 1995), thereby being significantly harder than those observed by EGRET. If spectra of the

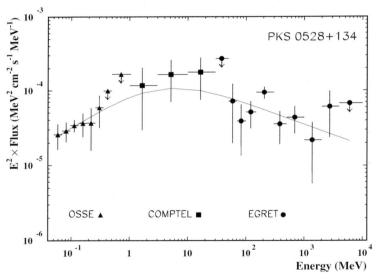

Fig. 12.5. Simultaneous OSSE, COMPTEL, and EGRET spectrum during a γ-ray high state of the blazar PKS 0528+134 . The *solid-line* represents the best-fitting function, which assumes a smooth turnover between two different power-law shapes. The spectral turnover at MeV energies is evident (Collmar, unpublished). The *down-arrows* represent upper flux limits

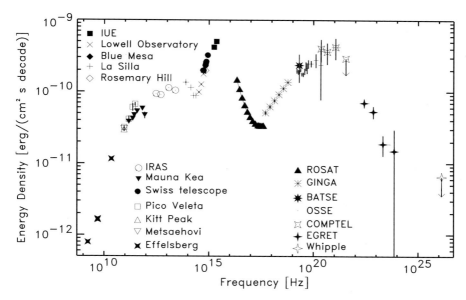

Fig. 12.6. Quasisimultaneous broadband spectrum of the blazar 3C 273. In this representation, the radiated power in the different bands can be directly compared. It is obvious that 3C 273 radiates most of its power in the UV (10^{15}–10^{16} Hz) and MeV γ-ray band (10^{20}–10^{21} Hz). The spectrum shows at least three, probably four, maxima indicating different emission components and mechanisms. (From Lichti et al. 1995, reproduced with permission)

different CGRO experiments are combined, a spectral turnover at MeV energies becomes evident (Fig. 12.5) for a number of sources, which indicates that the MeV range is, at least for some blazars, a spectral transition region (see also Fig. 12.6).

Integrating the measured γ-ray spectra (>1 MeV) and assuming isotropic emission yields blazar luminosities typically of the order of $\sim 10^{48}$ erg s^{-1} with maximum values up to several 10^{49} erg s^{-1}. Keeping in mind that the bolometric luminosity of our Milky Way is about 10^{44} erg s^{-1}, of which only a tiny fraction of $\sim 10^{-5}$ (about 10^{39} erg s^{-1}) is emitted at γ-ray energies, blazars are spectacular objects which, during flaring periods, are in γ-rays alone as luminous as thousands of galaxies, and – even more spectacular – can switch the total from this luminosity on and off on timescales of a day or even less (see Fig. 12.4).

COMPTEL has found indications for an unexpected class of "MeV blazars" (e.g., Bloemen et al. 1995; Blom et al. 1995), which are exceptionally bright in the 1–10 MeV range compared to their fluxes measured simultaneously with EGRET at >30 MeV energies. Two prime objects, GRO J0516-609 (which has PKS 0506-612 as the probable counterpart) and PKS 0208-512, have been found to exhibit this behavior on occasion. In contrast to normal blazars, where the COMPTEL measurements require a spectral softening between the COMPTEL and EGRET energy ranges, spectral "bumps" occur at MeV energies (∼1 to 10 MeV) in the spectra of these so-called MeV blazars. A simple extrapolation of the simultaneous EGRET spectrum passes well below the COMPTEL points. This strong MeV excess is a time-variable phenomenon.

12.2.4 Broadband Emission of CGRO-Detected Blazars

The CGRO observations of γ-ray blazars have shown that these sources can be enormously luminous at γ-ray energies and at the same time show short-term variability. Simultaneous multi-wavelength spectra and especially multi-wavelength monitoring have the potential to significantly increase our knowledge about the physics of these spectacular objects. In spectra measured simultaneously from radio to γ-ray energies, the luminosity of the objects in different energy bands can be directly compared, providing the possibility to identify the major emission components and processes. Especially valuable would be multi-frequency monitoring because it would provide clues about correlated/uncorrelated time variability in different frequency bands and thereby reveal crucial information about their connection and possibly about their physical emission sites. For example, the information as to whether an X-ray or optical flare occurs at the same time as, earlier than, later than, or is even absent during a γ-ray flare would strongly constrain blazar models. To utilize this possibility several multi-frequency campaigns have been organized around CGRO blazar observations. While the individual observations are easy to carry out, simultaneous observations of a particular source are difficult to achieve. Because telescope time is precious, especially on space experiments, severe pointing constraints exist, and scientific review panels have to be convinced, so only a few examples of such campaigns have been carried out with good coverage from the radio to γ-ray bands.

To date probably the best results have been derived for the blazars 3C 273 and 3C 279. 3C 273 is one of the best studied celestial sources because it is the closest ($z = 0.158$) blazar-type quasar and is bright in all wavelength bands. Figure 12.6 shows a quasi-simultaneous multi-frequency spectrum of 3C 273 from the radio to TeV γ-ray bands. The ordinate represents the radiated power per natural logarithmic frequency interval, so the energy release – the power – in the different wavelength bands can be directly compared. First of all the spectrum shows that the high-energy emission (X- to γ-rays) is a significant part of the bolometric luminosity. The spectrum shows (probably) four maxima indicating different emission components and processes, which are interpreted as synchrotron emission from relativistic electrons in the radio and far-IR band, thermal emission from a dust (molecular) torus in the IR and from an accretion disk in the UV (blue bump), and inverse-Compton radiation in X- and γ-rays generated by relativistic electrons and

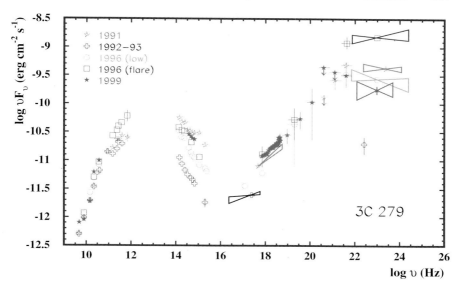

Fig. 12.7. Broadband spectra of 3C 279 during different γ-ray states in brightness. The 1991 to 1996 data are adapted from Wehrle et al. (1998). The variability at γ-rays is much larger than at lower energies (e.g., $\sim 10^{14}$ Hz). (From Collmar et al. 2000, reproduced with permission)

soft photons. For comparison see the unified AGN model in Fig. 12.2. In summary, the spectrum consists of a mixture of thermal and nonthermal emission components.

Figure 12.7 shows simultaneous multi-frequency spectra of the blazar 3C 279 observed at different times; with a redshift of $z = 0.58$ this object is much more distant than 3C 273. These spectra show only two maxima. One is in the radio/IR region and one in the γ-ray band, which during γ-ray flares clearly dominates the overall power output. It is also obvious that there is less variability at frequencies below 10^{15} Hz compared to the γ-rays where the source is variable by about two orders of magnitude. The spectrum is interpreted to be completely of nonthermal origin with only two visible emission mechanisms at work: synchrotron emission from the radio to the UV/soft X-ray band and inverse-Compton emission from X-rays to γ-rays. The thermal signatures, which are probably weak due to the large source distance, are completely outshone by the nonthermal emission.

Correlated monitoring observations have suggested that detectable γ-ray emission in blazars is generally present in coincidence with enhanced radio activity. Sometimes correlated activity between γ-rays and other wavelength bands (e.g., X-rays, optical) has been observed. However, since the emission of these sources is continuously variable in all wavelength bands, these activities cannot be correlated unambiguously due to the sparse sampling in γ-rays.

12.2.5 TeV Emission from Blazars

Recently a significant improvement in detector sensitivity has been achieved at ground-based γ-ray detectors. This is especially true for the Atmospheric Cherenkov Telescopes (ACTs) operating in the so-called VHE (very high-energy) range at photon energies between $\sim$300 GeV and 50 TeV. Sensitivity improvements have also been made for the particle detectors operating in the UHE (ultrahigh-energy) range, at photon energies between $\sim$50 TeV and $\sim$10 PeV. For more details on these γ-ray detectors see Chap. 3. Previous to the CGRO era no convincing AGN detections had been reported at these energies. However, with the increased sensitivities, five blazars have now been reported to emit at least occasionally in the VHE range. Mkn 421 (Punch et al. 1992) and Mkn 501 (Quinn et al. 1996), both discovered by the Whipple collaboration, are the most prominent sources. Both have been significantly re-detected many times by the Whipple telescope and confirmed by other VHE groups such as the HEGRA (High-Energy Gamma-Ray Astronomy) collaboration for example. Evidence for the three further sources, 1ES 2344+514 (Catanese et al. 1998), with the main signal (6σ) coming from a 1.5 h observation in 1995, 3C 66A (Neshpor et al. 1998) with evidence at the 5σ level for energies above 900 GeV, and very recently PKS 2155-304 (Chadwick et al. 1999), has been obtained by a single group only. These sources are all members of the BL Lac subclass of blazars; in fact they all belong to the so-called X-ray selected BL Lacs, which are more prominent in X-rays than at radio energies. Apart from 3C 66A ($z = 0.444$), which is also an EGRET source (Hartman et al. 1999), the sources are among the closest BL Lac objects; Mkn 421 ($z = 0.031$) is the closest one and Mkn 501 ($z = 0.034$) is the second closest BL Lac object. Despite extensive searches, no blazar has yet been found in the UHE range.

Due to their multiple detections in recent years Mkn 421 and Mkn 501 are reasonably studied at TeV energies. Both are time variable and show strong flaring activity with flux-doubling times of the order of $\sim$1 h or even below. While for Mkn 421 the flux was found to be typically 30–50% of the Crab Nebula – the standard candle at these energies – the flux of Mkn 501 exhibits intensity variations up to a factor of $\sim$50, between 0.1 and 5 times the Crab flux. During its most intense outbursts Mkn 501 was the brightest VHE source known in the sky. Figure 12.8 shows the 1996 to 1999 VHE light curve of this blazar as measured by the HEGRA collaboration. VHE energy spectra have been derived for Mkn 421 and Mkn 501. The spectrum of Mkn 421 shows no obvious cut-off either at the lower or at the upper end and is therefore well fitted by a simple power-law model with a photon index of about 2.6 ($E^{-\alpha_{\rm ph}}$) in the energy band between $\sim$300 GeV and $\sim$8 TeV (e.g., Krennrich et al. 1999). A recent analysis of the VHE spectrum of Mkn 501 during γ-ray flares in 1997 indicates curvature (e.g., Samuelson et al. 1998), which is consistent with the simultaneous upper limits derived by EGRET at

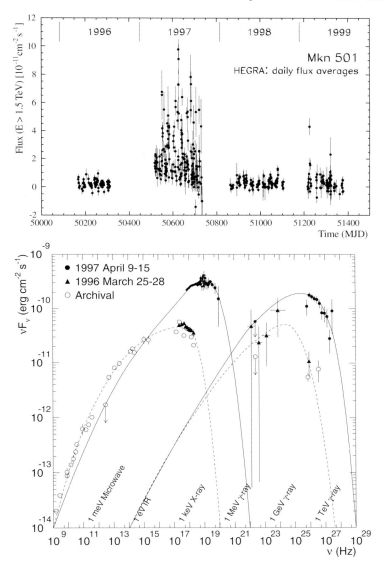

Fig. 12.8. The upper panel shows the long-term light curve of the TeV emission (>1.5 TeV) of the blazar Mkn 501. The daily flux averages as measured by the HEGRA telescope during the years 1996 to 1999 are shown. Strong flaring activity is visible during the observation period in 1997. (Reproduced with permission from the HEGRA Collaboration.) The lower panel shows multi-frequency spectra of Mkn 501 for two observation periods (TeV-quiet and flaring), which are supplemented by archival data. The *curves* in the figure are spline functions and do not represent proper model fits. They are meant to guide the eye for the anticipated broadband spectrum of Mkn 501. It is obvious that during the TeV flaring period in April 1997 the TeV flux as well as the X-ray flux increased significantly compared to the TeV quiescent period in March 1996. (From Catanese and Weekes (2000). Copyright 2000, Astronomical Society of the Pacific; reproduced with permission)

GeV energies. In the meantime EGRET has detected Mkn 501 in its highest energy range (Kataoka et al. 1999).

Multi-frequency observations of both Mkn 421 and Mkn 501 indicated a connection between the TeV and X-ray bands (e.g., Buckley et al. 1996; Catanese et al. 1997). Flares seem to occur simultaneously, but with larger amplitudes at TeV energies. Surprisingly, both sources are weak γ-ray emitters in the MeV/GeV range. This is well illustrated by broadband spectra, which during TeV flares (making the source observable at these energies) show the same two-peak spectra as the usual EGRET-type blazars. However, both maxima are shifted to higher energies (Fig. 12.8). One appears in the X-ray or occasionally even in the hard X-ray band and the other in the TeV range. EGRET and COMPTEL typically cover the "flux valley" between the two peaks.

A recent review of the TeV emission of AGN is provided in the article by Catanese and Weekes (2000), which contains further details and the references to individual papers.

12.2.6 Interpretation

Since the CGRO discovery of γ-ray blazars, the origin of their γ-ray emission has been widely discussed. The two most intriguing observational results are the short-term variability and the inferred huge γ-ray luminosities. Both facts together imply a highly compact γ-ray-emitting region, where large luminosities are generated in a small (solar-system size) volume. If the photon density in this region, however, is above a certain threshold value, the γ-rays cannot escape. Instead they interact with each other and generate electron–positron pairs by the pair-production process, which can schematically be written as

$$\gamma + \gamma \rightarrow e^- + e^+ . \tag{12.5}$$

For photons above the threshold energy $m_e c^2 = 511$ keV, the optical depth for pair production of a source of size R is given by

$$\tau_{\gamma\gamma} = n_\gamma \sigma_{\gamma\gamma} R , \tag{12.6}$$

where n_γ is the photon density and $\sigma_{\gamma\gamma}$ the cross section for pair production. Around 1 MeV (just above the threshold for this process) the cross section for pair creation reaches its maximum values, which are close to the Thomson cross section, σ_T. The number density of photons (n_γ) can be estimated from their energy density ($\sim L/4\pi R^2 c$) divided by their mean energy ($\sim m_e c^2$) (for more details see, e.g., Peterson 1997). Therefore, approximately, a source will be optically thick for pair production if

$$\tau_{\gamma\gamma} \approx \left(\frac{L_\gamma}{4\pi R^2 m_e c^3}\right) \sigma_T R = \frac{\sigma_T}{4\pi m_e c^3} \frac{L_\gamma}{R} = 2 \times 10^{-30} \frac{\text{cm s}}{\text{erg}} \frac{L_\gamma}{R} \geq 1 , \tag{12.7}$$

where σ_T is the Thomson cross section, L_γ the γ-ray luminosity, and R the source radius. For higher-energy γ-rays (e.g., >100 MeV) this is an order of magnitude estimate because it neglects the energy dependence of $\sigma_{\gamma\gamma}$. For typical isotropic blazar luminosities of 10^{48} erg s^{-1} and radii of a few light days, estimated from variability arguments, typical values for the optical depth for pair production of

$$\tau_{\gamma\gamma} \sim 200 \gg 1 \tag{12.8}$$

are derived. Such values simply imply that the γ-rays cannot escape from the source region without generating electron–positron pairs. Nevertheless, these γ-rays are observed, implying an intrinsic contradiction. To resolve it, beamed emission from a relativistic jet is considered instead of the central isotropic emission which is assumed in the calculations above. A jet origin of the observed γ-rays is consistent with the facts that for many blazars superluminal motion has been observed, which is indicative of jets pointing at small offset angles to our line of sight, and with the redshift distribution of these sources showing that the source distance is not the critical parameter for their detection. Gamma-ray blazars are observed far out into the universe up to redshifts of 2.3. Furthermore, this hypothesis is consistent with the unified AGN scenario (see Sect. 12.1). A relativistic jet origin of the γ-rays would imply beamed emission, which in turn implies that the observed γ-radiation is Doppler boosted. The Doppler factor, D,

$$D = \frac{1}{\Gamma(1-\beta \cos\Theta)}, \quad \beta = v_{\text{jet}}/c, \quad \Gamma = (1-\beta^2)^{-1/2} \tag{12.9}$$

is determined by the jet's bulk speed β and its orientation angle Θ with respect to our line of sight. The apparent γ-ray luminosity, $L_{\gamma,\text{app}}$, is enhanced with respect to the intrinsic γ-ray luminosity, $L_{\gamma,\text{int}}$, by the factor

$$L_{\gamma,\text{app}} = D^n \times L_{\gamma,\text{int}}, \tag{12.10}$$

where n is a model-dependent factor which for current emission models has values between 3 and 4. For a Doppler factor of 10, as is estimated in 3C 279 for example, the intrinsically generated γ-ray luminosities are a factor of 10^3 to 10^4 smaller than determined without taking relativistic beaming into account. The luminosity reduction in γ-rays results directly in the reduction of the optical depth for γ–γ pair creation by the same factor [see (12.7), which assumed isotropic, unboosted emission]. With values of the order of 10^{-1} for $\tau_{\gamma\gamma}$, the γ-ray-emitting region is optically thin, and therefore the generated γ-rays can escape, resolving the above-mentioned contradiction.

On the basis of this more general argument the detailed models have been developed within the unified AGN model and the relativistic jet scenario, which assumes that we are viewing almost along the axis of a relativistically outflowing plasma jet which is ejected from an accreting supermassive black

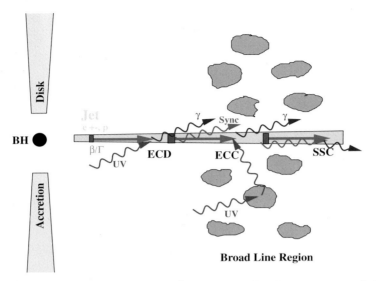

Fig. 12.9. Sketch (not to scale) of a model for the inner part of a radio-loud active galaxy. The γ-ray emission (*blue*) is thought to be generated in the inner-jet region, a few hundreds of Schwarzschild radii from the central supermassive black hole (BH), where blobs of relativistic particles are injected and accelerated in the jet. Several leptonic emission scenarios are indicated (see Sect. 12.2.6). The abbreviations have the following meaning: UV: UV-radiation from the accretion disk, Sync: synchrotron emission from the jet, ECC: Comptonization of accretion disk photons which are scattered into the the jet by the broad-line region clouds, ECD: Comptonization of direct accretion disk photons, and SSC: Comptonization of self-generated jet synchrotron photons

hole. The broadband radiation is thought to be produced by nonthermal electron synchrotron and inverse-Compton radiation in outflowing plasma blobs or shocked jet regions. Figure 12.9 sketches the central part of a radio-loud AGN, where according to current models the γ-ray emission is generated. Two classes of models have been proposed to explain the blazar γ radiation, with either leptons or hadrons as the primary accelerated particles, which then radiate directly or through the production of secondary particles, which in turn emit photons.

Leptonic Models

In leptonic models, leptons (electrons and positrons) are assumed to be the primary accelerated particles in the jet, reaching relativistic energies. The γ-ray emission of blazars is then produced by these nonthermal relativistic leptons which scatter soft photons to γ-ray energies via the inverse-Compton (IC) process. These leptonic models come in two flavors, depending on the nature of the soft photons. In the synchrotron self-Compton (SSC) process (e.g., Maraschi et al. 1992; Bloom and Marscher 1996) the relativistic electrons moving along the magnetized jet generate synchrotron photons. Their frequency is given by

$$\nu_s \propto B \times E_e^2, \qquad (12.11)$$

where B is the magnetic field strength in the jet and E_e the electron energy. Some of these synchrotron photons are boosted by the same relativistic electron population to γ-ray energies via the IC process:

$$\nu_{IC} \approx \nu_s \times E_e^2 \propto B \times E_e^4, \qquad (12.12)$$

where ν_{IC} is the frequency of the IC photon. The IC spectrum follows to the first order the shape of the (curved) synchrotron spectrum (just shifted to higher energies), which explains the observed spectral turnover at MeV energies in the γ-ray spectra of several blazars. For more details on the synchrotron and IC radiation mechanisms see Chaps. 2 and 9.

The so-called external Compton scattering (ECS) models consider a different origin for the soft photons. They assume that UV and soft X-ray photons which are radiated from the accretion disk directly into the jet (e.g., Dermer and Schlickeiser 1993) or are scattered into the jet by the broad-line region clouds (e.g., Sikora et al. 1994), are the soft target photons for the relativistic jet electrons [see (12.12)]. The ECS models can also reproduce the broadband nonthermal spectral shape of the γ-ray blazars. These models explain the spectral bending at MeV energies by the so-called incomplete Compton cooling of the electrons. When a blob of relativistic electrons is injected into the jet or is generated inside the jet by energization of leptons in a shocked region, a power-law-shaped IC spectrum is generated with low- and high-energy cutoffs corresponding to the low- and high-energy cutoffs in the electron spectrum. Because the high-energy electrons cool first, the high-energy cutoff in the IC spectrum moves towards lower energies with time. The electron cooling by the IC process stops when the blob has moved out into regions where the photon field becomes too thin to maintain this process. Integrating these spectra over time generates the spectral turnover at MeV energies observed by CGRO in time-averaged (days to weeks) γ-ray spectra. Because the above-mentioned leptonic processes, SSC and ECS, are physically expected, they all should – or at least could – be operating in blazars at some level, depending on the actual local conditions. Consequently modelling of the observed data has recently started by assuming that several leptonic processes contribute significantly to the observed high-energy spectra (e.g., Böttcher and Collmar 1998; Mukherjee et al. 1999). An example of such modelling is shown in Fig. 12.10, where a simultaneously measured multi-frequency spectrum of 3C 279 which was obtained during a γ-ray flaring period is modelled by several leptonic components (Hartman et al. 2001). According to this model the X-rays are dominated by SSC emission, while

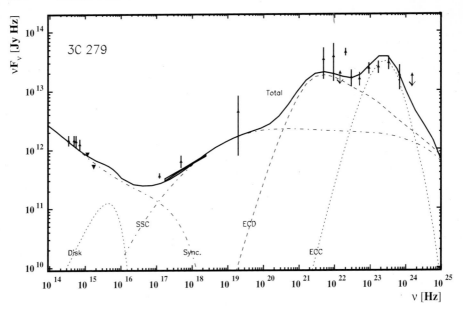

Fig. 12.10. Fit to a simultaneously measured optical to γ-ray multi-frequency spectrum of the blazar 3C 279 during a γ-ray flaring state. A leptonic emission scenario (see Sect. 12.2.6) with several emitting components is assumed. The sum ("Total", *solid line*) and the contributions of individual emitting components are shown; the latter are indicated by the following abbreviations: Sync.: synchrotron emission component, and Disk: thermal emission from the inner part of the accretion disk. For the abbreviations SSC, ECD, and ECC see Fig. 12.9. According to this model the γ-rays (>1 MeV) are dominated by Comptonization of photons external to the jet, while the X-ray part is dominated by the SSC component, i.e., Comptonization of photons generated within the jet. (Reproduced with permission from Hartman et al. 2001)

the γ-rays (>1 MeV) are dominated by ECS emission. However, such an approach is not commonly accepted yet, and the data quality of current γ-ray instruments is not sufficient to discriminate among the different models.

Hadronic Models

Models have also been proposed in which accelerated hadronic particles (mainly protons) carry the bulk of the energy (e.g., Mannheim and Biermann 1992; Bednarek 1993). Because protons do not suffer severe radiation losses, they can be accelerated up to energies of 10^{20} eV, reaching the threshold for photopion production. In this process the protons transfer energy to pairs, photons, and neutrinos via pion production:

$$p + \gamma \to \pi^0 + p, \, \pi^+ + n \tag{12.13}$$

and

$$p + \gamma \to e^- + e^+ + p\,. \tag{12.14}$$

These secondary pairs and pions are considered to be the origin of the high-energy jet emission. The pairs immediately lose their energy via synchroton emission, and the pions decay into γ-rays ($\pi^0 \to \gamma+\gamma$) and generate additional positrons via the decay chain

$$\pi^+ \to \mu^+ + \nu_\mu,\ \mu^+ \to e^+ + \nu_\mu + \nu_e\,. \tag{12.15}$$

Because of the high optical depth for pair production, the γ-rays are absorbed, generating pairs which in turn produce further γ-rays via the IC process. This reprocessing stops when after the n th cycle the energy of the γ-rays is below the threshold energy E_{thres}, for which $\tau_{\gamma\gamma}(E_{\text{thres}}) = 1$, i.e., the γ-rays can escape. Because the generation of neutrinos is a by-product of this process, the detection of a strong neutrino flux from blazars would definitively identify hadrons as the primary accelerated particles, finally supplying the energy for the γ-rays. In the leptonic and hadronic models the nature of the accelerated primary particles is different; however, the γ-ray production process (inverse Comptonization of soft photons by relativistic leptons) is the same.

Interpretation of TeV Emission

The emission of TeV blazars, for which we have only two well-studied sources, is interpreted in the same relativistic jet scenario as is the MeV/GeV emission of the regular CGRO-type blazars. It is thought to be IC emission from a relativistic jet, however with the peak of the emission shifted to higher energies. Instead of being in the MeV range (see e.g., Fig. 12.6), it is located somewhere in the GeV to TeV range (see Fig. 12.8). This explanation is clearly suggested by simultaneous multi-frequency spectra of these sources. The TeV flare spectrum of Mkn 501 (Fig. 12.8) clearly shows two humps. The lower-energy one, peaking at hard X-rays, is interpreted to be the synchrotron emission peak, while the high-energy one, peaking at TeV energies, is considered to be due to IC emission. The global shift in energy suggests – assuming a leptonic model – that during such time periods the leptons – being responsible for the emission in both peaks – are accelerated to larger maximum energies. The reason for that is as yet unknown. One possibility might be that in BL Lacs, which have less matter in their central regions (indicated by the weak or absent emission lines) than quasars, lepton cooling is generally on a lower level. Simultaneous multi-frequency observations have shown that in X-rays and TeV γ-rays the sources vary simultaneously, with a larger amplitude at the high energies. Because this behavior is expected in the SSC models, this process is currently favored. Since there are only two sources convincingly detected, which also seem to have intrinsically different spectra, final conclusions cannot yet be drawn. If the hadronic model turns out to be the relevant one for these TeV blazars, one could also consider them as being sources of cosmic-rays, whose origin is – decades after their discovery – still heavily debated.

Interpretation of MeV Blazar Signature

The MeV enhancement observed with COMPTEL in the MeV blazars (see Sect. 12.2.3) cannot be simply explained by the Compton-scattering models, because they predict spectra with breaks. To account for bumps, Doppler-boosted $e^{\pm}$ annihilation radiation emitted by a relativistically moving electron–positron cloud in a jet has been proposed (e.g., Henri et al. 1993; Roland and Hermsen 1995). Roland and Hermsen (1995) assumed that the blazar nuclei eject two-plasma fluids consisting of a mildly relativistic e^{-}-proton jet, which accounts for the jet formation, hot spots and extended radio lobes, and a relativistic $e^{\pm}$ beam, with a bulk Lorentz factor between 3 and 10, which produces the γ-ray emission. Applying their model to the γ-ray spectrum of PKS 0506-612, they derived a value for the jet plasma density. Recently Skibo et al. (1997) explained the γ-ray bump of MeV blazars by the same general scenario of Doppler-boosted annihilation radiation, although they propose that the MeV blazar phenomenon is an orientation effect. The beaming pattern of Doppler-boosted thermal annihilation radiation is much broader than that of IC emission from the scattering of external photons by jet electrons. Gamma-ray emission consisting of these two components would naturally lead to the signature of the MeV blazar class of sources when viewed at moderate angles ($\sim 15°$) with respect to the axis of the outflowing jet. At these angles the annihilation radiation is still observable, while the flux from the scattering of external photons by jet electrons is not sufficiently Doppler boosted.

12.2.7 Summary and Open Questions

A dramatic increase in our knowledge on blazars, phenomenologically as well as with respect to source physics, occurred during the CGRO era. For recent reviews including more details and further references see, for example, Hartman et al. (1997) for the MeV/GeV range and Catanese and Weekes (2000) for the TeV range. Blazars were found to be occasional strong emitters of MeV/GeV γ-ray radiation, some nearby BL Lac objects emitting even up to TeV energies. The most remarkable characteristics are (a) that for some sources the γ-ray luminosity by far dominates the bolometric power, showing the importance of γ-rays in considering the energetics and the modelling of blazars and (b) that several sources show time variability on the order of days and even less. The inferred large isotropic luminosities in combination with the short-term variability provide evidence for relativistic beaming of the γ-ray emission based on pair-production attenuation of γ-rays in dense emission regions. Therefore the γ-ray emission is thought to be IC emission from a relativistic jet which points towards us within $\sim 10°$ and is enhanced in intensity due to Doppler boosting. The intrinsically generated luminosities are then expected to be a factor of the order 1000 to 10 000 smaller. According to the models, the γ-ray radiation is generated in the inner-jet region, at distances of only several hundreds of Schwarzschild radii from the putative supermassive black hole, and so probes the jet physics in the innermost AGN region. Although the general picture seems to be well established now, the details are not understood yet. For example, there is an ongoing debate about the nature – hadrons or leptons – of the primary accelerated particles. If leptons are considered, opinions differ as to whether internally generated synchrotron photons or photons from the accretion disk are the dominant target photons, or whether a combination of both is involved. Other open issues are the duty cycles of γ-ray emission and whether there is a quiescent γ-ray emission level or whether the emission completely ceases outside γ-ray flaring periods. These latter two issues are crucial for estimating the blazar contribution to the diffuse extragalactic γ-ray background (see Chap. 14 for more details). Another open issue is the relation of the γ-ray emission to the emission in other wavelength bands. Although many individual hints for possible correlations have been reported, no generally accepted overall picture has emerged yet. Since these objects are continuously variable at all frequencies, only continuous monitoring, which is currently not feasible in γ-rays, would unambiguously reveal such relationships.

The recent detections of blazars have finally opened the field of extragalactic γ-ray astronomy. However, as usual in science, this unexpected detection has provided more questions (unexpected ten years ago) than answers.

12.3 Seyfert and Radio Galaxies

Prior to CGRO, the AGN subclasses Seyfert and radio galaxies had been considered to be promising candidates for γ-ray emission detectable by the CGRO instruments. They should be bright γ-ray sources if their measured power-law spectra in hard X-rays continue this shape into the γ-ray band. The main observational difference between these two types of AGN is that Seyfert galaxies are usually radio-quiet while radio galaxies are radio-loud (see Fig. 12.1), showing strong radio lobes emanating from their centers.

12.3.1 Seyfert Galaxies

OSSE has detected about 25 out of roughly 40 observed Seyferts at the $>3\sigma$ level, typically at energies between 50 and 150 keV (Johnson et al. 1997). This is consistent with the results of BATSE (Burst and Transient Source Experiment), which has detected ~30 objects at the same energies by dedicated searches of several Seyfert samples (Bassani et al. 1996; Maliza et al. 1997). No object has been detected yet at MeV energies by COMPTEL (e.g., Maisack et al. 1995) or at energies above 100 MeV by EGRET (e.g., Lin et al.

1993), despite systematic searches of large samples of Seyfert galaxies. These results clearly show that Seyfert galaxies are rather X-ray and hard X-ray sources than γ-ray sources.

Although no Seyfert galaxy has yet been "seen" at energies around 1 MeV and above, the significant detection of many sources by OSSE and BATSE has provided new and interesting insights into these objects. Especially the knowledge of their high-energy spectra, on the relationship between different source classes (e.g., Seyferts Type 1 and Type 2), and of their emission mechanisms has greatly increased. Because roughly 30 Seyferts have been detected, different subclasses can be studied and compared: the radio-quiet Seyfert Type 1 (Seyfert 1s) and Type 2 (Seyfert 2s) sources, and the class of radio-loud Seyferts, which also belong to the class of broad-line radio galaxies (see Tab. 12.1). The latter class, however, is populated by only a few sources, which show jets and one – 3C 120 – even superluminal motion (Walker et al. 1988). Because many individual source detections are weak, detailed information has been derived either from investigating the strongest sources or by averaging over several sources. The best-studied object is NGC 4151, which at a distance of $\sim$20 Mpc is the closest and brightest Seyfert galaxy and which is considered to be the model Seyfert 1 galaxy. NGC 4151 was found by OSSE and BATSE to be relatively constant in flux in the 20 keV to $\sim$150 keV band. The long-term average flux has varied by no more than a factor of 2 during 5.5 yr. Day-to-day variability of the order of 25% was observed. Its spectrum is described well by a power law which cuts off exponentially at higher energies (Fig. 12.11):

$$dN/dE \propto E^{-\alpha} \times e^{-\frac{E}{E_C}}, \tag{12.16}$$

where α is the photon index and E_C is the cut-off energy. Such a mathematical shape approximates a thermal Comptonization spectrum. Best-fit values of α of $\sim$1.6 and E_C of $\sim$100 keV are derived. Simple power-law models, as fitted in the pre-CGRO era, are not consistent with these newly measured spectra, due to the observed cut-off at high energies. In particular, no 511 keV pair-annihilation emission is found, which severely restricts emission models for Seyferts. In summary, the OSSE data on NGC 4151 suggest that its high-energy emission is described well by thermal processes. With an upper limit of 15% for their contribution, nonthermal processes can probably be neglected. The detailed spectral results on NGC 4151 are consistent with the average results on complete Seyfert classes, which are also best modelled by a thermal Comptonization spectrum.

Investigating the emission of different Seyfert subclasses provides the possibility to check their relationship, i.e., whether they are intrinsically similar or not. This is particularly important in view of the unification model of AGN, according to which Seyfert 1s and 2s only differ in their orientation angle towards us (see Sect. 12.1.3). Average spectra of the different subclasses have been generated by summing up the emission of the individual

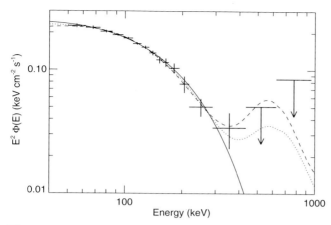

Fig. 12.11. The average OSSE spectrum of the radio-quiet Seyfert galaxy NGC 4151, which is the closest and brightest member of its class. The spectral cut-off around 100 keV is visible. There is no indication for an annihilation feature around ∼500 keV. The *solid curve* represents the best-fit for the thermal Comptonization plus Compton-reflection model. The *dashed* and *dotted* curves give the best fit of hybrid, thermal/nonthermal, models. Because the thermal model fits well, contributions from nonthermal emission is not required. (From Johnson et al. 1997, reproduced with permission of The American Institute of Physics)

class members (Fig. 12.12), but leaving out the brightest representatives (e.g., NGC 4151) of each class. This procedure ensures that these averaged spectra are not dominated by single sources. The main result is that the hard X-ray spectra of radio-quiet Seyfert 1 and 2 sources are very similar (Johnson et

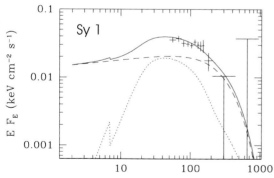

Fig. 12.12. Average OSSE spectrum of the Seyfert 1 class together with the best-fit model and the individual model components. The *dashed curve* represents the best-fit thermal Comptonization spectrum and the *dotted curve* the reflected component. The *solid curve* is the sum of both. (From Johnson et al. 1997, reproduced with permission of The American Institute of Physics)

al. 1997). Both are well fitted by an exponentially cut-off power-law model. Single power-law shapes do not provide an adequate representation for both types. The plasma parameters of the Comptonizing region are found to be very similar: $kT \sim 100$ keV and $\tau \sim 1$. For both classes no significant emission above 200 keV is found, which indicates that thermal processes dominate in all radio-quiet Seyferts. Whether there is any low-level nonthermal emission remains an open question to be answered by future γ-ray instruments. Despite the difference in the optical spectra, the hard X-ray emission of both classes is the same. This is expected if the sources are intrinsically similar, because hard X-rays are not affected by intervening material between their production site and the observer. Therefore, the similarity between the two classes is in accord with the unification model and hence supports it.

Although radio-quiet Seyfert 1s and 2s are similar, OSSE has provided evidence that the radio-loud Seyferts differ somewhat. The spectra of the latter can be sufficiently well modelled by two simple power-law functions, a harder one breaking to a softer one around ~100 keV, thus showing substantial emission above ~200 keV and no exponential cut-off (Wozniak et al. 1998). This suggests that in these sources nonthermal processes might operate as well, and therefore they might be a link between the "normal" radio-quiet Seyferts and blazars, in the sense that probably a mixture of thermal and nonthermal processes is visible. However, the spectra above about 200 keV are too poorly known to derive safe conclusions to date.

To test emission models for the high-energy emission of Seyferts, broadband X-ray to soft γ-ray spectra have been compiled. The spectra of the radio-quiet Seyferts are described well by an X-ray power-law shape, which is modified by the presence of nearby matter. This primary emission is visible between ~2 keV and ~6 keV, showing a power-law shape with index ~1.9. Below 2 keV it is absorbed due to intrinsic matter and intervening matter between source and observer. At 6.4 keV an Fe $K\alpha$ line is visible. At energies above this iron line a spectral hardening occurs (index ~ 1.7) which is attributed to a Compton-reflection component on top of the primary spectrum. This spectral component peaks at ~30 keV and cuts off around 100 keV. The primary spectrum cuts off at ~150 keV. A typical spectrum together with a model fit to the high-energy data, also showing the individual emission components, is given in Fig. 12.12. This spectral shape is the same for Seyfert 1s and 2s. Spectral differences between the two classes only show up at energies below 2 keV. There Seyfert 2s are more strongly absorbed than Seyfert 1s, indicating a larger amount of intervening matter, which is in accord with the unification scheme for these objects. Radio-loud Seyferts (broad-line radio galaxies) show strong Fe $K\alpha$ emission but weak Compton reflection. As mentioned earlier, their hard X-ray spectra have to steepen above ~100 keV, but an exponential cut-off is not required. They could break to a second power-law shape with a softer index.

Prior to CGRO the high-energy emission of Seyferts was thought to be the result of nonthermal acceleration of particles to relativistic energies with a subsequent pair cascade and nonthermal Compton scattering. This model predicted power-law spectra which reach up to several hundreds of keV before cutting off, and also a strong electron–positron annihilation line near 511 keV. The recent CGRO results definitely rule out this model, because (a) the spectra cut-off around 100 keV and (b) no annihilation line is observed. Instead, models assuming thermal Comptonization of soft seed photons – probably UV photons from the accretion disk – in a hot plasma successfully reproduce the observed spectra. To explain the Fe $K\alpha$ line and the Compton-reflection component, an optically thick and cold medium, which could be a cold accretion disk, is needed in the vicinity of the hot emitting plasma. Because the modelling of the reflection component often indicates the subtended solid angle of the cold medium to be less than 2π (e.g., Zdziarski et al. 1996), a simple slab-like emitting region is unlikely. A possible emission scenario could be the following: Above a cold ($\sim$1 keV) accretion disk exists a hot and inhomogeneous ("patchy") accretion-disk corona, which consists of isolated active regions of hot plasma, which dissipate most of the energy released in the accretion process (see e.g., Johnson et al. 1997, and references therein).

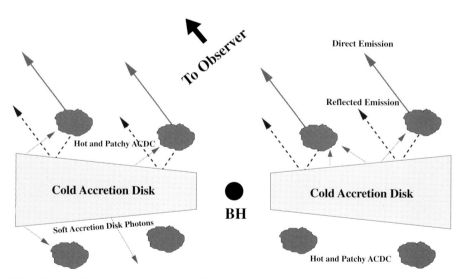

Fig. 12.13. Sketch (not to scale) of a possible model of the central and emitting region of radio-quiet Seyfert galaxies. The two emission components (direct and reflected) at hard X-ray energies from the anticipated hot and patchy accretion disk corona (ACDC) are indicated. The ACDC could consist of isolated active regions (e.g., hot plasma blobs), which are responsible for the emission. However other emission scenarios are currently still possible. For more details see Sect. 12.3.1

These hot plasma "blobs" ($kT \sim 100\,\text{keV}$), consisting of electrons or pairs, Comptonize UV photons from the accretion disk to generate the hard X-ray to γ-ray spectrum. Some photons reach the observer not directly, but after reflection on the cold accretion disk, giving rise to the observed reflection component and the iron line. Such a picture (Fig. 12.13) would satisfy all observational constraints: the spectral cutoff at $\sim 100\,\text{keV}$, the Fe $K\alpha$ line, and the strength of the Compton-reflection component. However, other plasma geometries are possible as well, such as a hot inner disk overlapping with a cold outer disk (see Fig. 7.7), which is a popular emission scenario for X-ray binaries.

12.3.2 Centaurus A

The radio galaxy Centaurus A (Cen A) is the closest active galaxy ($\sim 3.5\,\text{Mpc}$) and the brightest source for OSSE in the hard X-ray band. Cen A is a peculiar object: it is a Seyfert 2-like galaxy. However, a classification is difficult because of the strong optical absorption due to the band of dust. The source was found by BATSE to be variable by a factor of 5 on a long time scale, and, on a short time scale, OSSE observed an intensity drop by $\sim 25\%$ within 12 h. In contrast to the regular Seyferts discussed in the previous section, Cen A is detected at MeV energies by COMPTEL (Steinle et al. 1998) and probably also above $\sim 100\,\text{MeV}$ by EGRET (Nolan et al. 1996), although source confusion is not yet definitely ruled out for this source. The broadband X-ray to γ-ray spectrum can be described by different power-law shapes, showing a spectral steepening towards higher energies (Fig. 12.14). In hard X-rays, the spectrum has the canonical hard power-law shape with photon index 1.7, like the Seyferts. It breaks at $\sim 150\,\text{keV}$ to a softer power-law shape ($\alpha_{\text{ph}} \sim 2.3$) which holds up to $\sim 30\,\text{MeV}$. At higher energies, it has to break again ($\alpha_{\text{ph}} \sim 3.3$) to match the weak EGRET flux above $\sim 100\,\text{MeV}$ (Steinle et al. 1998). Since there is no exponential cut-off visible in hard X-rays, the spectrum resembles that of the radio-loud Seyferts, again indicating that thermal as well as nonthermal processes are operating in this object. The emission around 1 MeV and above is considered to be of nonthermal origin. Because Cen A is viewed from the "side" – the angle between its jet axis and our line of sight is $\sim 70°$ (e.g., Jones et al. 1996) – the hypothesis of a "misdirected" blazar is considered. We might observe the intrinsically (*not* Doppler-boosted) generated γ-ray luminosity, which is of the order of $10^{42}\,\text{erg}\,\text{s}^{-1}$ at MeV energies. This MeV luminosity – if viewed "jet on-axis" and subsequently Doppler boosted by a factor of $\sim 10^3$–10^4 – would place Cen A in luminosity near the weaker blazars detected by COMPTEL. According to this picture this source is detectable by current γ-ray instruments only because of its proximity.

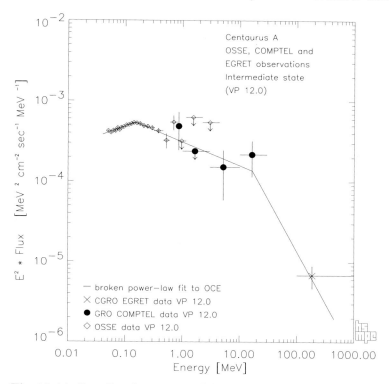

Fig. 12.14. Broadband spectrum of Centaurus A from X-rays to GeV γ-rays. The break of the hard X-ray power-law spectrum at $\sim 150\,\mathrm{keV}$ to a softer one up to $\sim 30\,\mathrm{MeV}$ is evident. To match the EGRET flux, a second spectral break is necessary. (From Steinle et al. 1998, reproduced with permission)

12.3.3 Summary

CGRO observations have revealed that Seyfert galaxies do not emit detectable γ-rays around 1 MeV and above. Instead their spectra cut off near $\sim 100\,\mathrm{keV}$, showing the shape of a thermally Comptonized spectrum with plasma parameters of $kT \sim 100\,\mathrm{keV}$ and $\tau \sim 1$. Therefore Seyferts are rather to be considered as X-ray and hard X-ray emitters than γ-ray emitters. No distinct spectral differences between Seyfert 1s and 2s are found at these energies, which supports (or at least is consistent with) the unified model of AGN. The spectra of the radio-loud objects seem to be harder and show, at energies above $\sim 150\,\mathrm{keV}$, rather a break to a softer power-law shape than an exponential cut-off. This might indicate that in these objects nonthermal processes operate as well, which is consistent with the presence of radio jets in these objects. Although detailed modelling is still in an early stage, the observed emission can be understood by a scheme in which it is generated in a dynamical (e.g., hot and patchy) accretion-disk corona above a colder disk.

Another possible scenario would be a hot inner disk overlapping with a colder outer one, like in X-ray binaries (see Chap. 7). A peculiar case is the radio galaxy Cen A, the closest active galaxy. It shows emission probably up to energies of 100 MeV and above, indicating a mixture of thermal and nonthermal processes. In hard X-rays its spectrum resembles that of the radio-loud Seyferts but maintains its emission well into the γ-ray range, where nonthermal processes dominate. No Seyfert or radio galaxy has yet been detected by any ground-based Cherenkov telescope or UHE air-shower array.

A recent review of the hard X-ray and γ-ray emission of Seyfert and radio galaxies is provided by Johnson et al. (1997) and – with respect to their modelling – by, for example, Zdziarski (1999), which contain further details and references to individual papers.

12.4 Summary and Conclusions

In the study of the high-energy (> 50 keV) emission of AGN large progress has been made in recent years. At the beginning of the 1990s, before CGRO was launched, only four AGN had been reported to emit detectable γ-rays around 1 MeV and above. This small number of sources, belonging to several AGN subclasses, led to a very vague and patchy knowledge about their hard X-ray to γ-ray emission. It was dominated by reports on a few individual sources, for which detections as well as nondetections were obtained, and therefore an overview picture did not exist at all.

Now, at the end of this decade, about 100 sources have been significantly detected by the CGRO experiments from hard X-ray (≥ 20 keV) to GeV γ-ray energies, providing reliable new information in this energy band. Even at VHE γ-rays at least two sources have been convincingly detected. During these years a consistent general picture has emerged; however, it is not understood in detail yet, and, as usual, also provided many new questions. The current overall picture is that radio-loud AGN, especially blazars, can occasionally be emitters of strong, nonthermal γ-ray radiation. This nonthermal emission, which most likely emerges from a jet via inverse Comptonization, maintains the emission up to GeV or even TeV γ-rays. In contrast, no radio-quiet AGN has been yet detected around 1 MeV and above. Radio-quiet sources show the signs of thermally Comptonized spectra, which cut off around $\sim$100 keV and which are thought to be generated in the central region, i.e., in the vicinity of the supermassive black hole. The basic result is shown in Fig. 12.15, which shows the high-energy luminosity spectra (assuming isotropic emission) of different sources belonging to different source classes as function of their rest-frame energy. This figure, in fact, allows to compare the amounts of radiated energy or power. The different spectral behavior of blazars, having nonthermal jet emission, and Seyferts, showing thermal emission from the accretion disk, is obvious. Additionally, the figure shows that some BL Lac-

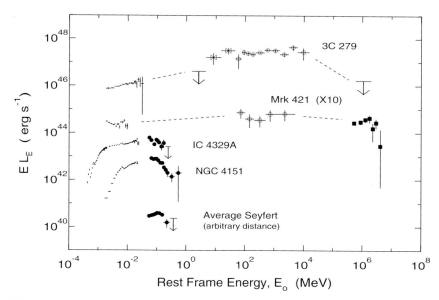

Fig. 12.15. Multi-wavelength power spectra for the "average" Seyfert galaxy, the bright Seyferts NGC 4151 and IC 4329A, the BL Lac object Markarian 421, and the blazar 3C 279. It is clearly shown that in contrast to both blazars, the high-energy Seyfert spectra cut off around 100 keV, indicating thermal instead of a nonthermal emission processes. (From Dermer and Gehrels 1995, reproduced with permission)

type blazars even maintain their emission into the TeV range, although they are "weak" GeV sources.

The recent progress in this field has not only provided new results and answers, as usual it also posed many new questions. For example, "What causes the γ-ray flares?", "What are the duty cycles of the γ-ray emission in blazars?", and "What is the the level of emission in quiescent (or off) γ-ray states?" are important ones. The answers of the latter two questions will have a big impact on the – so far unknown – level of blazar contribution to the diffuse extragalactic γ-ray background. It is generally believed that the γ-rays in blazars arise from accelerated jet particles. However it is still an open issue whether hadrons or leptons are the primary accelerated particles, being responsible for the bulk of the energy which is converted to γ-rays. More fundamentally, it is still not understood how and where in the jet these particles are accelerated, and even more fundamentally, how such a jet is generated and stabilized in the first place. Other important remaining issues are how the spectra of the radio-loud Seyferts continue towards higher energies, whether these sources link the radio-quiet Seyferts and the radio-loud blazars, and whether there is low-level nonthermal emission in radio-quiet Seyferts.

The prospects to further increase our phenomenological knowledge and subsequently the physical understanding of these spectacular sources are promising for the foreseeable future. The GLAST (Gamma-ray Large Area Telescope) experiment, measuring at energies between ~15 MeV and 100 GeV with improved sensitivity and an enlarged field of view with respect to EGRET, is expected to detect a significantly larger number of blazars than EGRET did. This will result in important information on their duty cycles for γ-ray emission and will provide the possibility to generate a much more accurate $\log N$–$\log S$ distribution, which in turn will provide more reliable estimates about the blazar contribution to the diffuse extragalactic γ-ray background. The INTEGRAL mission, spanning the hard X-ray (>20 keV) to γ-ray (~ 8 MeV) energy range, will provide detailed spectral measurements in this band, which has been found to be a spectral transition region for many blazars. This energy range is also important to allow us to follow the spectra of the radio-loud Seyferts towards higher energies. The ground-based experiments measuring VHE and UHE γ-rays are being continuously improved. Cherenkov telescopes with improved sensitivities and larger fields of view are already under construction. They will extend their bandwidth towards lower energies, even to as low as 20 to 40 GeV, thereby closing the gap in energy coverage with the space-based γ-ray measurements. Of course, these improvements will result in the detection of more sources, which in turn will lead to a better overall picture of AGN emission in the TeV band, where today the knowledge is still dominated by the data on only two sources. The detection of further sources at larger distances may provide an estimate of the extragalactic IR-radiation field, by measuring spectral cut-offs as a function of distance, and in a turn would have further implications on cosmology.

In summary, the field of extragalactic γ-ray astronomy was really opened in the 1990s, mainly by the CGRO experiments, and is expected to grow significantly in the near future with the upcoming improved instruments.

References

Antonucci, R.R.J., Miller, J.S., 1985, Astrophys. J. **297**, 621
Bednarek, W., 1993, Astrophys. J. **402**, L29
Bassani, L., Malaguti, G., Paciesas, W.S. et al., 1996, Astron. Astrophys. Suppl. **120**, 559
Bloemen, H., Bennett, K., Blom, J.J. et al., 1995, Astron. Astrophys. **293**, L1
Blom, J.J., Bennett, K., Bloemen, H. et al., 1995, Astron. Astrophys. **298**, 33
Bloom, S.D., Marscher, A.P., 1996, Astrophys. J. **461**, 657
Böttcher, M. and Collmar, W., 1998, Astron. Astrophys. **329**, L57
Buckley, J.H., Akerlof, C.W., Biller, S. et al., 1996, Astrophys. J. **472**, L9
Catanese, M. and Weekes, T.C., 2000, PASP **111**, 1193
Catanese, M., Bradbury, S.M., Breslin, A.C. et al., 1997, Astrophys. J. **487**, L143
Catanese, M., Akerlof, C.W., Badran, H.M. et al., 1998, Astrophys. J. **501**, 616
Chadwick, P.M., Lyons, K., McComb, T.J.L. et al., 1999, Astrophys. J. **513**, 161

Collmar, W., Bennett, K., Bloemen, H. et al., 1999, *Proc. of the 3rd INTEGRAL Workshop*, Taormina, Italy 1998, ed. by Palumbo, G., Bazzano, A., Winkler, C., Astrophys. Letters and Communications, 38, 445 1998)
Collmar, W., Benlloch, S., Grove, J.E. et al., 2000, in *Proc. of 5th Compton Symposium*, Portsmouth, AIP Conference Proceedings **510**; ed. by McConnell, M.L., Ryan, J.M., p. 303
Dermer, C.D., Gehrels, N., 1995, Astrophys. J. **447**, 103
Dermer, C.D., Schlickeiser, R., 1993, Astrophys. J. **416**, 458
Hartman, R.C., Collmar, W., v. Montigny, C., Dermer, C.D., 1997, in *Proc. of the fourth Compton Symposium*, AIP Conference Proceedings **410**; ed. by Dermer, C., Strickman, M., Kurfess, J., p. 307
Hartman, R.C., Bertsch, D.L., Bloom, S.D. et al., 1999, Astrophys. J. **123**, 203
Hartman, R.C., Böttcher, M., Aldering, G., et al., 2001, Astrophys. J. **553**, in press
Henri, G., Pelletier, G., Roland, J., 1993, Astrophys. J. **404**, L41
Johnson, W.N., Zdziarski, A.A., Madejski, G.M. et al., 1997, in *Proc. of the fourth Compton Symposium*, AIP Conference Proceedings **410**; ed. by Dermer, C., Strickman, M., Kurfess, J., p. 283
Jones, D.L., Tingay, S.J., Murphy, D.W. et al., 1996, Astrophys. J. **466**, L63
Kurfess, J., 1996, Astron. Astrophys. Suppl. **120**, 5
Kataoka, J., Mattox, J.R., Quinn, J. et al., 1999, Astrophys. J. **514**, 138
Krennrich, F., Biller, S.D., Bond, I.H. et al., 1999, Astrophys. J. **511**, 149
Krolik, J.H., 1999, *Active galactic nuclei: from the central black hole to the galactic environment*, Princeton, N.J., Princeton University Press
Lichti, G.G., Balonek, T., Courvoisier, T.J.-L. et al., 1995, Astron. Astrophys. **298**, 711
Lin, Y.C., Bertsch, D.L., Dingus, B.L. et al., 1993, Astrophys. J. **416**, L53
Maisack, M., Collmar, W., Barr, P. et al., 1995, Astron. Astrophys. **298**, 400
Maliza, A., Bassani, L., Malaguti, G. et al., 1997, *Proc. of 2nd INTEGRAL Workshop*, The Transparent Universe, St. Malo, France, ed. by Winkler, C., Courvoisier, T.J.-L., Durouchoux, P., p. 439
Mannheim, K., Biermann, P.L., 1992, Astron. Astrophys. **253**, L21
Maraschi, L., Ghisellini, G., Celotti, A., 1992, Astrophys. J. **397**, L5
Mattox, J., Wagner, S.J., Malkan, M. et al., 1997, Astrophys. J. **476**, 842
McNaron-Brown, K., Johnson, W.N., Jung, G.V. et al., 1995, Astrophys. J. **451**, 692
Mukherjee, R., Bertsch, D.L., Bloom, S.D. et al., 1997, Astrophys. J. **490**, 116
Mukherjee, R., Böttcher, M., Hartman, R.C. et al., 1999, Astrophys. J. **527**, 132
Neshpor, Y.I., Stepanyan, A.A., Kalekin, O.R. et al., 1998, Astron. Letts., **24**, 132
Nolan, P., Bertsch, D.L., Chiang, J. et al., 1996, Astrophys. J. **159**, 100
Perotti, F., Della Ventura, A., Villa, G. et al., 1981a, Astrophys. J. **247**, L63
Perotti, F., Della Ventura, A., Villa, G. et al., 1981b, Nature **292**, 133
Peterson, B.M., 1997, *An introduction to active galactic nuclei* Cambridge, Cambridge University Press
Punch, M., Akerlof, C.W., Cawley, M.F. et al., 1992, Nature **358**, 477
Quinn, J., Akerlof, C.W., Biller, S.D. et al., 1996, Astrophys. J. **456**, L83
Robson, I., 1997, *Active Galactic Nuclei* Chichester, Wiley
Roland, J., Hermsen, W., 1995, Astron. Astrophys. **297**, L9
Samuelson F.W., Biller, S.D., Bond, I.H., et al., 1998, Astrophys. J. **501**, L17
Schmidt, M., 1963, Nature 197, 104

Sikora, M., Begelman, M.C., Rees, M.J., 1994, Astrophys. J. **421**, 153
Skibo, J.G., Dermer, C.D., Schlickeiser, R., 1997, Astrophys. J. **483**, 56
Steinle, H., Bennett, K., Bloemen, H. et al., 1998, Astron. Astrophys. **330**, 97
Swanenburg, B.N., Hermsen, W., Bennett, K. et al., 1978, Nature **275**, 298
Urry, M., Padovani, P., 1995, PASP**107**, 803
von Ballmoos, P., Diehl, R., Schönfelder, V., 1993, Astrophys. J. **312**, 134
Walker, R.C., Walker, M.A., Benson, J.M., 1988, Astrophys. J. **335**, 668
Wehrle, A.E., Pian, E., Urry, C.M. et al., 1998, Astrophys. J. **497**, 178
Wozniak, P.R., Zdziarski, A.A., Smith, D., et al., 1998 MNRAS **299**, 499
Zdziarski, A.A., 1999, in *High Energy processes in Accreting Black Holes*, ASP Conference Series 161, ed. by Poutanen, J., Svensson, R., p. 16
Zdziarski, A.A., Johnson, W.N., Magdziarz, P., 1996, MNRAS **283**, 193

13 Unidentified Gamma-Ray Sources

Olaf Reimer

13.1 Objective

After γ-ray measurements opened a new window to observational astronomy, the question arose as to what kind of objects and which physical processes produce the observed emission. Because the early balloon experiments suffered from inadequate resolution and sky coverage, the first phenomenon associated with unidentified γ-ray-emitting sources was γ-ray bursts. Later on it became clear from the results of the SAS-2 (Small Astronomy Satellite) and COS-B satellite missions that the observed γ-ray emission consists of a diffuse component as well as point-like sources. Since their discovery the identification of γ-ray point sources has remained a fundamental key to understanding the nature of these sources and is an ongoing topic of contemporary high-energy γ-ray astrophysics. Especially obstructive is the fact that a source appears not as a single point in the sky, but rather as a region with different probabilities for the source location. This is due to instrumental techniques and the resulting width of the point spread function (PSF), thus limiting the angular resolution of any current γ-ray instrumentation. Although the classes of uniquely identified sources like pulsars, Active Galactic Nuclei (AGN), and X-ray binaries are described in separate chapters (see Chaps. 6, 12, and 9, respectively) I will here discuss the collective properties and individual characteristics of unidentified γ-ray point sources. Special classes of unidentified γ-ray sources such as γ-ray bursts and regions of enhanced diffuse emission are discussed separately within this book (see Chaps. 14 and 9, respectively). The choice of unidentified γ-ray point sources as the subject of this chapter means that no already known unique counterpart exists at other wavelengths which is able to account for the observed γ-ray emission. Therefore a short review of γ-ray source identifications will be given to introduce the subject of unidentified γ-ray sources.

13.1.1 Gamma-Ray Source Detection History

With the identification of the Crab pulsar (Hillier et al. 1970) as a γ-ray emitter it became clear that apart from diffuse γ-ray emission features, localized regions could well be γ-ray sources. With the satellite experiments SAS-2 and COS-B the existence of individual point-like sources of γ-ray emission became evident (Fichtel et al. 1975, Hermsen et al. 1977). With the publication of 25 discrete γ-ray sources (Swanenburg et al. 1981) the identification problem arose. Only 4 COS-B sources could be identified with a proper counterpart – two pulsars by their timing signature, one AGN due to the remarkable positional coincidence between the radio position and the γ-ray source, and a molecular cloud complex. For the remaining 21 sources no unambiguous counterparts appeared to exist at other wavelengths and the concept of a class of unidentified sources was introduced. Later several of these γ-ray excesses were explained as concentrations of interstellar gas irradiated by cosmic-rays (Mayer-Hasselwander and Simpson 1990). However some new sources were discovered also. More than a dozen of the COS-B sources still remained unidentified.

In the 1980s various claims of TeV emission were announced from ground-based observations, primarily X-ray binaries (Cyg X-3, Her X-1, Vela X-1, Cen X-3, SCO X-1, SMC X-1, 4U 1145-619), pulsars (Crab/Crab Nebula, 1E 2259+586, 4U 0115, PSR 1957+20, PSR 0355+54), but also extragalactic objects (Cen A, 3C273, 3C279, M87). For a review see Fegan (1990). Although most of the claims still need verification, no unidentified excess emission has been reported from ground-based γ-ray telescopes so far. This is due to the observational techniques for ground-based γ-ray detectors and their pointing strategy to check candidate objects for very high-energy γ-ray emission within a very small field of view. With the start of operation of ground-based Cherenkov telescopes the identifications of TeV sources were raised to a new level of significance, immediately resulting in high-significance detections of γ-ray emission from Mkn 421, and Mkn 501, the confirmation of the Crab Nebula, Vela, and X-ray binaries like Cen X-3, and Vela X-1, as well as nonconfirmation of various sources reported earlier as detections. Up to now no γ-ray point source above 100 GeV remains unidentified, mainly because no sensitive large-area survey is yet feasible with Cherenkov telescopes.

13.1.2 Compton Gamma-Ray Observatory Era

The situation developed dramatically with the launch of the Compton Gamma-Ray Observatory (CGRO) in 1991. All instruments (BATSE [Burst And Transient Source Experiment], OSSE [Oriented Scintillation Spectrometer Experiment], COMPTEL [Compton Telescope], and EGRET [Energetic Gamma-Ray-Experiment Telescope]) discovered new γ-ray sources in the corresponding energy regimes. COMPTEL (Schönfelder et al. 2000) and EGRET (Fichtel et al. 1994) performed the first all-sky surveys in high-energy γ-rays, resulting in all-sky source catalogs. One of the most striking results was the detection and identification of blazar-type AGN as a class of γ-ray point sources (von Montigny et al. 1995). With CGRO at least seven pulsars were identified as high-energy γ-ray emitters by observing their characteristic light curves (Thompson et al. 1997). OSSE discovered Seyfert-type AGN as γ-ray emitters up to a few hundred keV (Johnson et al. 1994) but not at higher energies. One of our neighboring galaxies, the Large Magellanic Cloud was seen by EGRET, as reported by Sreekumar et al. (1992). Despite the many source identifications, the majority of the high-energy γ-ray sources still remains unidentified. This fact is perhaps the biggest mystery in contemporary γ-ray astronomy. So far none of these sources has been definitively identified with an object at other wavelengths, although various objects or object classes have been studied in order to investigate whether they are likely sources of the observed high-energy γ-ray emission or not. A dozen sources discovered during the COS-B mission still await identification, while in the early CGRO era about 100 sources are listed in the EGRET source catalogs (2EG: Thompson et al. 1995, 2EGS: Thompson et al. 1996) without identification. The interest in identifying these sources is based on several facts. First, and probably most disturbing, is that still as many as half of the observed high-energy γ-ray emitters are of unknown origin. Second, there is a certain chance to discover something new or unexpected from these sources. Even if these sources belong to classes of already known objects, their identification could significantly improve the understanding of their collective properties.

13.2 The Observations

Discussing the characteristics of unidentified sources means basically to accumulate all available information about their properties, such as locations, flux histories, spectral features, timing information, collective classifications, counterpart studies. In order to understand these collective properties, a few remarks with respect to observational aspects and instrumental effects of the EGRET telescope and their consequences for the results need to be given. Due to the limited field of view of the EGRET telescope all observations are pointings towards selected regions of the sky. By co-adding these pointings all-sky surveys were derived as the basis for the compilation of the point source catalogs. As a consequence, regions of the sky are covered with nonuniform instrumental exposure, directly affecting the point source sensitivity. Furthermore, the diffuse γ-ray background is highly structured, strongest along the Galactic Plane, thereby reducing the sensitivity towards low Galactic latitudes. A point-source detection depends strongly on the exposure and diffuse background of the region surrounding the γ-ray source, as well as on its flux. Also some peculiarities in the EGRET point-source catalogs have to be considered in the following discussion of source distributions. All EGRET catalogs are compiled using two different detection significance thresholds, determined by means of a likelihood method: $\geq 5\sigma$ for $|b| < 10°$, and $\geq 4\sigma$ for $|b| > 10°$. This results in a nonuniform detection sensitivity function ("step" at $10°$). In addition, the criterion for including a source in the EGRET catalogs is the fulfilment of the detection significance criterion in either a single viewing period, a combination of single viewing periods or the total superposition of all viewing periods. Therefore there are cases where a source is considered as detected if it fulfils the detectability criterion in one single viewing period, although being well beneath that criterion in the analysis of the superposition of all viewing periods. This becomes an essential point in any attempt to balance the uneven sky coverage of EGRET by means of an exposure correction. Furthermore the number of identifications depends on the completeness required for the catalogs. Effects such as the "zone of avoidance" for extragalactic objects at low Galactic latitudes caused by interstellar absorption and high background intensity affect radio and optical surveys, therefore reducing the chance to find counterparts.

The third catalog of EGRET-detected γ-ray point sources (3EG: Hartman et al., 1999) (see Table 4.1) lists about 270 sources. A total of 100 sources are identified (66 AGN with high confidence, 27 AGN with lower confidence, 5 pulsars and the Large Magellanic Cloud), but 170 sources remain unidentified. Concerning the identifications of AGN in the 2EG catalog and its supplement 2EGS, some of the marginal detections are controversial, but some of the unidentified sources are likely associated with blazar-like flat-spectrum radio sources (Mattox et al. 1997). Figure 13.1 shows the class distributions of the sources listed in the 3EG catalog. Note that the total of all catalog sources as well as a more uniform subsample suitable for source distribution studies is sketched here. The subsample of catalog sources shown is corrected for the biases from nonuniform source detection criteria within the catalogs, as mentioned above.

The histograms clearly indicate differences between the source distributions of high-energy γ-ray emitters. Moreover, before going into details, one has to consider the EGRET detection significance function, shown along with its constituents in a Galactic latitude projection in Fig. 13.2.

The significance s of a detection for an isolated point source can be expressed as $s \sim f\sqrt{e/bg}$, where f is the flux, e the instrumental exposure and bg the diffuse γ-ray background (Mattox et al. 1996). Clearly an increase of the detection probability towards high Galactic latitudes is expected; it is seen especially towards positive latitudes. This is due to the uneven sky coverage of EGRET pointed observations. The particular exposure enhancement at positive Galactic latitudes is caused by long observation runs pointing to the Virgo region. The lowest chance to detect equally bright sources on the sky with EGRET is directly in the Galactic Plane, due to the pronounced diffuse Galactic γ-ray background. With this knowledge one can describe the derived source distributions qualitatively. The identified AGN show an isotropic sky distribution. The excess towards high positive Galactic latitudes can be explained by the corresponding detection significance for this region. In contrast, the six EGRET-detected pulsars show, although rather statistically limited, a pure Galactic distribution – as expected for these Galactic sources. The distribution of the unidentified sources consists of at least two components. While a distinct excess in the Galactic equator and the Galactic

13 Unidentified Gamma-Ray Sources 323

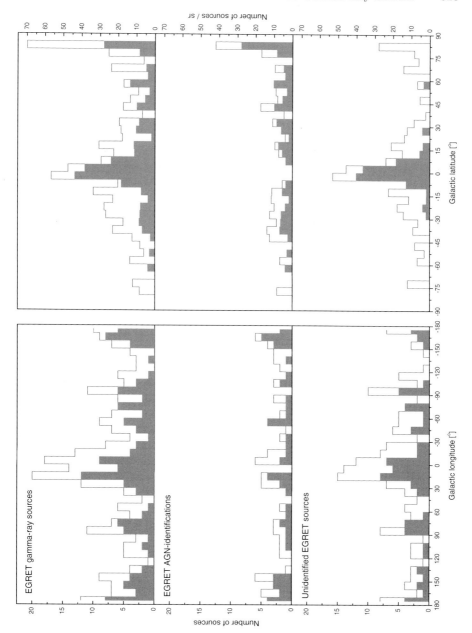

Fig. 13.1. Class distributions of EGRET-detected γ-ray sources in Galactic longitude und latitude. The *thin line* represents the complete 3EG catalog, whereas the *shaded regions* represent more uniform source selections suitable for correlation studies

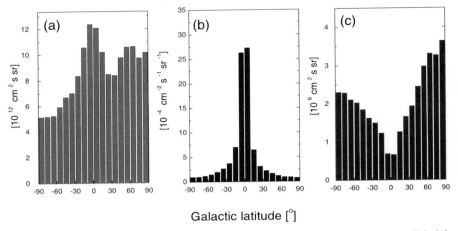

Fig. 13.2. Galactic latitude distribution of the (**a**) instrumental exposure, (**b**) diffuse background, and (**c**) the detection significance function for equally bright γ-ray point sources. See Sect. 13.2 for more details

Center indicates a strong Galactic component, a second, apparently isotropic component seems to be superimposed on the dominant Galactic distribution (see also Fig. 13.1). The noticeable enhancement of sources in mid-latitude regions between $5° < |b| < 30°$ of the complete sample appears to be strongly suppressed compared to the distribution determined from the more uniform data sample. Obviously the uniform data set suppresses highly variable objects or badly exposed sources predominantly in the mid-latitude regions. The compilations of EGRET-detected sources above 1 GeV (Lamb and Macomb 1997, Reimer et al. 1997a) amplify even more the characteristics of the source distributions as described above. Whereas the AGN population still appears isotropically distributed above 1 GeV, the latitude distribution of the unidentified sources shows clustering in the Galactic Plane. High-latitude unidentified sources above 1 GeV are extremely rare. Among these the absence of bright sources compared to low Galactic latitude unidentified sources is especially striking. Another aspect of high-energy γ-ray sources is their characterization in terms of their flux variability. Variability is addressed here on time scales of EGRET viewing periods, which typically range from 2 to 3 weeks. Variability on shorter time scales is a typical feature of AGN and discussed elsewhere (see Chap. 12). A systematic study of γ-ray flux variability (McLaughlin et al. 1996) for all sources within the second EGRET catalog of γ-ray point sources has shown that under the variability criteria V defined therein pulsars are indeed nonvariable sources (see Fig. 13.3). In contrast most AGN are indeed variable. More strictly, the ones with high average γ-ray fluxes appear without exception to be variable. High-energy γ-ray sources can also be discussed in two other categories: variable and nonvariable sources. Both are concentrated towards low Galactic latitudes. It

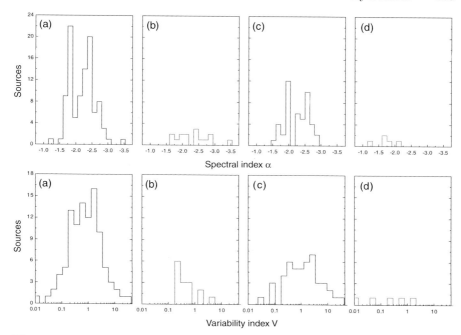

Fig. 13.3. Spectral index and variability for different source classes. From left to right: the complete catalog sample, the high-latitude unidentified sources, the AGN, and the pulsars are shown

has turned out that the low-latitude, variable γ-ray sources cannot be easily related to any known source population on the basis of their variability. In contrast, blazars are known to be highly variable, but could be occasionally in a quiescent state, preventing any straightforward identification. Short-time variability analysis techniques on time scales of days rather than weeks improve the situation here (Wallace et al. 2000).

Further indication for the existence of more than two classes of γ-ray sources is given by the observations of GRO J1838-04 and 2CG135+01 (Kniffen et al., 1997, Tavani et al., 1997, 1998). The transient character of GRO J1838-04 leads to the conclusion that objects other than AGN and pulsars might account for these particular sources, and perhaps other unidentified sources too. 2CG135+01 seems to be the first candidate of such an association (see Sect. 13.3.2).

Table 13.1 attempts to compare unidentified sources not only by means of their latitudinal arrangement, but also by addressing their flux variability. Hence, this scheme is suitable for the majority of the unidentified sources, but is expected to be imperfect in some individual cases.

The EGRET point-source catalogs above 100 MeV and above 1 GeV indicate a concentration of unidentified sources towards the central part of

Table 13.1. This scheme arranges the unidentified sources with respect to their Galactic coordinates as well as their variability characteristics. The separation between low-latitude and high-latitude sources is primarily a qualitative way to separate obviously different populations. It is even less stringent to distinguish between variable and nonvariable sources by value. However, the most likely astronomical objects accounting for the unidentified sources arrange rather well into individual cells

Variability	Predominant occurrence	
	Low Galactic latitudes	High Galactic latitudes
Nonvariable	Galactic objects (more distant, higher luminosity)	Galactic objects (nearby, lower luminosity)
	Pulsars, OB-star associations, and other young star-forming regions	Possible pulsars (radio quiet ?), Gould-Belt sources
Variable	Galactic objects (likely with some contermination from not yet identified AGN)	Extragalactic objects (but maybe a few nearby objects too)
	Compact objects, black holes, colliding stellar winds	AGN (but currently not identifiable as such)

the galaxy. Whereas the latitudinal distribution clearly peaks in the Galactic Plane, the longitudinal distribution does not show a similar enhancement within ~30° of the Galactic Center. The longitudinal distribution suggests that the majority of these sources are Galactic objects with an upper limit of about 6 kpc for their typical distance (Mukherjee et al. 1995). This is well below the distance to the Galactic Center. Although the luminosities of unidentified sources cannot be determined on an individual basis due to their unknown distances, an average luminosity may be estimated from the limit on their typical distances. With a typical flux of $5 \times 10^{-7}\,\mathrm{cm^{-2}\,s^{-1}}$ for photon energies above 100 MeV and an E^{-2} power law, the luminosity of these sources ranges from 6×10^{34} to $4 \times 10^{35}\,\mathrm{erg\,s^{-1}}$ for assumed distances between 40 pc and 1.4 kpc and a beaming into 1 sr. This is compatible with the observed longitudinal and latitudinal distribution of low-latitude unidentified sources (Kanbach et al. 1996). Also similarities in the spectral characteristics between low-latitude unidentified high-energy γ-ray sources and identified γ-ray emitters have been examined (Merck et al. 1996). It was found that some of the unidentified sources show the characteristic spectral feature known from γ-ray pulsars: a very hard spectrum as well as a turnover at GeV energies. But most of the Galactic sources differ significantly from these – indicating either the presence of a yet unknown source class among the Galactic γ-ray sources or pointing towards a complexity among pulsars higher than we expect from the few γ-ray pulsars currently known.

In the case of high-latitude sources a rather isotropic latitudinal and longitudinal distribution is observed. A significant fraction of these sources has been already identified to be AGN, so far the dominant type of identified γ-ray objects among all EGRET-detected γ-ray sources. In order to derive conclusions from the AGN population detected at high latitudes, $\log N$–$\log S$ distributions were determined (see Fig. 13.4) for the identified AGNs as well as the unidentified high-latitude sources (Özel and Thompson 1996). Above a lower limit, which reflects the instrumental detection threshold, a linear fit to the data appears to be consistent with the expected $S^{3/2}$ dependence for an isotropic/spatial uniform distribution of underlying objects for a Euclidean universe. The only difference here is that AGN extend to higher flux levels compared to the unidentified sources. However an enhancement of unidentified sources exists at intermediate galactic latitudes ($5° < |b| < 30°$). Recent studies show evidence (Gehrels et al. 2000) that a distinction between weak unidentified sources at mid-latitudes and bright unidentified sources close to the Galactic Plane could be made. Thus, steady but weak unidentified sources at mid-latitudes represent a distinct population, likely associated with the Gould Belt of massive and late-type stars, molecular clouds and expanding interstellar gas. The high-latitude source population can be explained

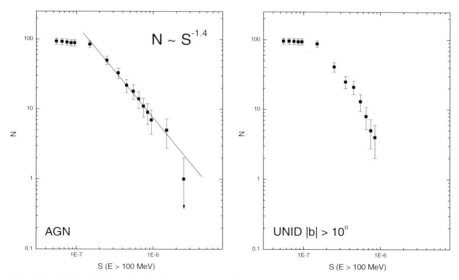

Fig. 13.4. $\log N$–$\log S$ plot for the identified AGN and the high-latitude unidentified EGRET sources. The flattening of the observed logN–logS dependence towards low sensitivities is an instrumental effect due to the limited sensitivity of the EGRET telescope

with an isotropic extragalactic and at least one local Galactic component. When comparing the spectra of AGN and unidentified EGRET high-latitude sources (Reimer et al. 1997b), a similar variety of spectral indices is found (see Fig. 13.3). In contrast to observed pulsars with predominately hard spectra this gives additional confirmation that AGN are the dominant source class at high Galactic latitudes. This is true for already identified objects as well as for the objects still awaiting identification. Therefore it is expected that a certain fraction of yet unidentified high-latitude sources are AGN, which has already been confirmed by the identification of formerly unidentified sources with AGN, i.e., 2EG J0432+2910 with 0430+2859 (Dingus et al. 1996) and 2EG J0852-1257 with 0850-1213 (Bloom et al. 1997).

13.3 Correlation Studies

13.3.1 Correlations with Source Populations

To obtain information on the identity of the unidentified γ-ray sources, object class correlation studies are useful and applicable. In this approach one examines classes of objects known from other wavelengths which are supposed to be able to account for the observed γ-ray emission. We ask whether their spatial, temporal and physical characteristics match the properties of the unclassified γ-ray sources. One of the first correlation studies was performed for the COS-B γ-ray sources by Montmerle (1979), who investigated the relation between unidentified γ-ray sources, supernova remnants (SNRs), hot emission-line star associations (spectral type O and B [OB-star associations]), and HII regions. It showed that not all known bright SNRs are detectable as γ-ray sources, but that γ-ray sources are closely linked with young objects. In cases of spatial proximity between SNRs, OB stars, and molecular cloud complexes (e.g., HII regions) a significant correlation with (Galactic) γ-ray sources was noticed. Such "SNOBs" (SNRs and OB stars in association with a molecular cloud) would be sources of diffuse emission, but with the rather poor angular resolution of γ-ray telescopes and the size of the interaction and emitting region they would not be detectable as truly diffuse sources.

Because the only unique identifications of γ-ray sources in the early era of γ-ray astronomy were pulsars, the dominant interest concentrated on the relation between the rather well-studied radio pulsar population and recently discovered γ-ray sources. A pulsar in γ-rays is detected by its pulsed emission or by the point-like intensity excess at the position of a known pulsar. Whereas a pulsar is unambiguously identified by its light curve, the confidence region for the location of a γ-ray source often contains more than one counterpart. Therefore the class of radio pulsars (Taylor et al. 1993) is a good candidate for correlation studies. In a series of EGRET-related pulsar study publications (Thompson et al. 1994, Fierro et al. 1995, Nel et al. 1996) young spin-powered pulsars, ms pulsars, and finally all catalogued pulsars were studied in detail. Except for the six significantly detected high-energy γ-ray pulsars no additional γ-ray source turned out to exhibit pulsed emission. Only positional coincidences were noticed for a few radio pulsars. None of them, however, show any evidence for identification on the basis of their light curve. The most promising candidates for further γ-ray pulsars based on coincidences with cataloged radio pulsars and measured (but still not with high statistical significance) light curves are PSR B1046-58 (Kaspi et al. 2000) (positionally coincident with 3EG J1048-5840), PSR J0218+4232 (Kuiper et al. 2000) (positionally coincident with the low energy emission of 3EG J0222+4253), PSR J2229+6114 (Halpern et al. 2001) (positionally coincident with 3EG J2227+6122), PSR B0656+14 (Ramanamurthy et al. 1996), and PSR B0355+54 (Thompson et al. 1994). However, this negative result led to the expectation that we might see only the top-ranked pulsars, according to $\dot{E}/d^2$, where $\dot{E}$ is the spin-down energy and d the distance. A class of radio-quiet pulsars with properties similar to radio-loud pulsars has been suggested, initiated by the existence of Geminga as a γ-ray and X-ray pulsar while not detectable as a strong radio pulsar. Speculations have been made as to how many of these objects might be in the γ-ray sky (Helfand 1994). Finally it was concluded that unidentified EGRET sources could well be represented by young, radio-quiet pulsars with high luminosity (Mukherjee et al. 1995), although estimates of the supernova birthrate limit this to a fraction of the observed γ-ray sources only. Later an estimate of the γ-ray beaming factor and the pulsar luminosity dependence with age were taken into account as well (Yadigaroglu and Romani 1995), based on a γ-ray pulsar emission model. The results in terms of mean distances and luminosities were compared with γ-ray observations of sources close to the Galactic Plane. It was found that indeed many of the Galactic unidentified sources are pulsar candidates. This has been confirmed by comparing results from an outer-gap model (see Chap. 6) with a consistent parameter set of observed properties from unidentified sources (Cheng and Zhang 1998, Zhang and Cheng 1998). This interpretation addresses individual γ-ray sources as pulsar candidates, a case where a source population study allows conclusions about individual γ-ray sources. In the framework of a polar-cap pulsar model (see Chap. 6), a similar conclusion was reached with respect to unidentified γ-ray source distributions (Sturner and Dermer 1996). For distance, period, age, magnetic field strength, and γ-ray flux distributions the model and observed γ-ray pulsars are in statistical agreement, when radio selection processes are included. A fraction of the low-latitude unidentified γ-ray sources are predicted to be radio-quiet pulsars, especially when radio and γ-ray beams are strongly misaligned. A rather different approach concerning the correlation of radio pulsars and unidentified γ-ray sources has been made by radio observers (Nice and Sayer 1997). Here, deep radio searches for pulsed emission towards unidentified γ-ray sources were performed. Although no new radio pulsar at these positions was found, two known radio pulsars were rediscovered, underlining the positional coincidence of individual γ-ray sources and radio pulsars. No conclusive identification of an unidentified γ-ray source could yet be achieved using a similar approach.

Since young pulsars are promising candidates for unidentified γ-ray sources, a revisit of the SNOB association between the low-latitude unidentified γ-ray sources and interacting supernova remnants with molecular clouds in OB-star associations is desirable. For the unidentified low-latitude EGRET sources the correlation with SNRs has been reviewed by Sturner and Dermer (1995), Sturner et al. (1996), and Esposito et al. (1996). Meanwhile 11 unidentified EGRET sources were noted to be positionally coincident with known SNRs, confirming the indication of such associations already known from COS-B observations. The correlation of unidentified γ-ray sources and populations of young stellar objects has also been re-investigated. A statistically significant association with OB-star associations has been reported by Kaaret and Cottam (1996) (see Fig. 13.5). On the basis of the 2EG catalog, 16 correlations were found: four sources well within an OB-star association, five OB-star associations within the source location contour of an unidentified EGRET source and seven positionally coincident within 1°. When comparing the luminosity distribution obtained by using OB-star association distances, all luminosities are compatible with detected γ-ray pulsars, also being consistent with the longitudinal, latitudinal, and flux distribution of unidentified Galactic EGRET sources as determined by Kanbach et al. (1996). A more general approach has been made by including all possible tracers of recent massive star formation and death to correlate with unidentified Galactic EGRET sources (Yadigaroglu and Romani 1997). Here objects from most recent catalogs of OB-star associations, young open clusters, HII regions, young radio pulsars and SNRs were tested for coincidence with individual unidentified EGRET sources. Clearly all but young open clusters and local HII regions could be considered as counterparts of Galactic γ-ray sources. From individual coincidences conclusions for the entire class are predicted by modelling our local neighborhood, constructed by means of known distances to the objects considered. Following an interpretation in an outer-gap pulsar

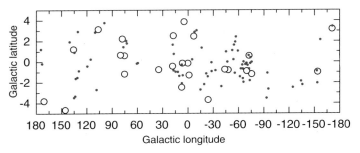

Fig. 13.5. Positions of unidentified EGRET point sources (*open circles*) and OB-star associations (*dots*), based on the 2EG-catalog

emission model, the number, spatial distributions, and luminosities of unidentified Galactic EGRET sources appear consistent with the hypothesis of all sources being pulsars. However, a mixture of these sources with a new class of γ-ray sources could still not be ruled out. In addition, a problem arises when considering the SNOB hypothesis with recent EGRET data: if pulsars are indeed sources of the γ-ray emission observed from unidentified sources, then the sample of SNR counterparts is biased towards those in OB-star associations. This is due to selection effects in the radio identification of pulsars, which may be either within OB-star associations or outside them, and also by the occurrence of Type Ia explosions leaving no neutron star. However it is argued on the basis of the expected contribution of γ-ray pulsars to the Galactic diffuse emission above 1 GeV that only a minor fraction of either young pulsars or radio-quiet pulsars could be hidden among the unidentified γ-ray sources (Pohl et al. 1997).

Another class of objects tested for positional coincidence with unidentified γ-ray sources are Wolf–Rayet (WR) stars with strong stellar winds (Kaul and Mitra 1997). These objects might be capable of producing γ-ray emission up to 100 MeV through nonthermal processes. The comparison of a pure chance coincidence of unidentified EGRET sources with 160 catalogued WR stars suggest these objects as possible counterparts too. More recently Romero et al. (1999) found, by comparing WR stars on the basis of the 3EG unidentified γ-ray sources, that only a few individual WR stars are positionally coincident with unidentified EGRET sources, but a class correlation turned out to be only marginal. However, theoretical predictions suggest that WR stars could generate under favourable cases luminosities up to only $10^{34} \mathrm{erg\,s^{-1}}$, which is not sufficient to account for the observed luminosity spectrum of the unidentified sources.

Finally, accreting black holes have been suggested as possible counterparts for unidentified EGRET sources by Dermer et al. (1997). A population of isolated black holes which accrete from the interstellar medium could account for a fraction of low-latitude unidentified sources and mid-/high-latitude sources as well. However, because accreting black holes are expected to be prominent X-ray sources, the unidentified EGRET sources should be coincident with bright X-ray sources, which is not the case.

When mid-latitude sources are treated separately because of their composite character as a Galactic and extragalactic component, tests against known or assumed structural features might also provide information on their nature. One test compared the observed sources (unidentified EGRET sources within $|b| > 2.5°$) versus modelled sources based on a 5° by 5° pattern (Grenier 1997). The simulated distributions were linear combinations of isotropically distributed extragalactic sources and different Galactic components: a spherical Galactic Halo ($r = 20$ kpc), local Galactic Disks (with either a density gradient falling towards the Galactic Plane or with a scale height of 0.4 kpc and radial scale length of ∼9 kpc in the Galactic Plane), and finally a local population. It was suggested that a structure in the local interstellar medium – the Gould Belt – correlates rather well with the observed distribution of unidentified EGRET sources. Proposed candidates for this correlation are young pulsars or old neutron stars wandering through the galaxy, accreting gas on their way through the interstellar medium. However there are already sensitive and conflicting limits in the literature from X-ray observations on old neutron stars accreting from the interstellar medium, which do not fit easily within such a scenario.

The prime candidate source class for the high Galactic latitudes are AGN, especially blazars (BL Lac objects and flat spectrum radio quasars [FSRQ]), since these are known γ-ray sources. Based on positional considerations the EGRET catalog AGN identifications were examined statistically (Mattox et al. 1997). Many of the EGRET source catalog identifications were clearly verified, although some AGN assignments were found questionable, and alternative identifications were suggested. Furthermore, a correlation between unidentified EGRET sources and radio-loud AGN has been investigated, resulting in new AGN identification proposals. It appears that only flat-spectrum, radio-loud quasars (blazars) could account for high-latitude γ-ray sources. Neither steep-spectrum radio sources nor radio-quiet AGNs are likely explanations for the observed γ-ray sources, either observationally or theoretically. It is expected that a certain fraction of the radio-loud AGN also account for low- and mid-latitude sources, as demonstrated recently with the identification of 3EG J2016+3657 with B2013+370 (Mukherjee et al. 2000). However due to the reduced detectability at low Galactic latitudes selection effects for extragalactic sources exist.

When comparing the γ-ray sources with the most-complete sample in a neighbouring wavelength regime, a search for a correlation with the ROSAT All-Sky Survey (RASS) (Voges 1993) is of interest. The main problem is that the RASS consists of various types of X-ray-emitting objects, which are only identified to about 30%. The density of RASS sources and often their lack of identification will make any conclusive source class correlation difficult. Many chance coincidences are expected and certain types of X-ray emitters (stars, distant normal galaxies, etc.) are not expected sources of high-energy γ-ray emission. Rather than complete object class correlations, deep X-ray imaging can provide essential information for identifying individual γ-ray sources.

13.3.2 Prominent Individual Unidentified Sources

Unlike correlation studies of entire object classes, for a few individual sources the multi-frequency approach of identification is preferred. Deep observations, often carried out contemporaneously, provide detailed information about the region of the γ-ray source at different wavelengths. Most useful for such identification attempts are X-rays due to some similarities between X-ray- and γ-ray-emitting objects (pulsars, AGN, X-ray binaries, SNRs); radio wavelengths, due to the excellent angular resolution and the nonthermal character of their emission; and optical wavelengths, due to the complete coverage at sufficient sensitivity. Typically, optical identification is needed in order to further reduce all candidate X-ray sources from X-ray bright objects which are unable to account for the γ-ray emission. Accurate positions from radio observations can verify high-energy source locations and are essential for correlated flux observations. Prominent individual unidentified sources and the attempts to identify them will be described in the following. Historically, some AGN among the unidentified EGRET sources were identified by means of multifrequency campaigns on the basis of correlated flux variations during flare phases. Such sources will not be discussed further in this chapter.

A point source coincident with the Galactic Center position has been observed since the beginning of the EGRET mission, clarifying the earlier known enhancement assigned to only diffuse emission features. The source, 3EG J1744-3039, is also a strong emitter above 1 GeV, and therefore the most accurate source location is provided by an analysis above 1 GeV, yielding to the appearance of this source in the GeV catalogs as GEV J1746-2854. The observed intensity spectrum appears to be very hard, with a spectral index of -1.3, and a spectral turnover towards a softer spectrum above 2 GeV (Mayer-Hasselwander et al. 1998). This clearly discriminates the source from the diffuse emission from the Galactic Center region (see Fig. 13.6). There has been no strong evidence for flux variation within the several years of the CGRO mission. Comparing the observed source characteristics with various models intended to explain its high-energy γ-ray emission, three scenarios are favoured so far: a single or a few young pulsars on the line of sight to the Galactic Center, interactions of the observed radio arc structure near the

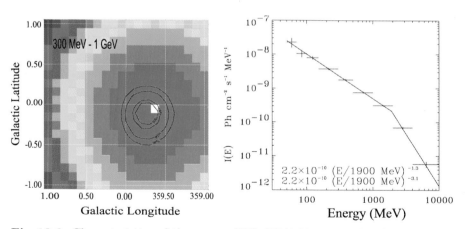

Fig. 13.6. Characteristics of the source 3EG J1744-3039 near the Galactic Center. *Left*: the position determined between 300 MeV and 1 GeV including a source location likelihood contour; *right*: the γ-ray spectrum with a double power-law fit

Galactic Center with the far IR radiation field, or γ-rays produced through decay channels for neutralino annihilation.

The brightest high-latitude unidentified source in the γ-ray sky so far, 3EG J1835+5918 (GEV J1835+59), is also an excellent target for a multifrequency identification. With half the flux of the Crab above 1 GeV, an extremely accurate source location with positional error of 8 arc min in radius at the 95% confidence contour is determined. Located 25° off the Galactic Plane, Galactic diffuse emission does not smear or confuse this discrete source. Indications of flux variations are ambiguous due to many observations at large instrument inclinations. The spectrum is described by a rather hard single power law. Especially interesting has been the long-standing absence of any obvious counterpart at other wavelengths (Nolan et al. 1994) – a unique case for such a bright γ-ray source. Therefore, the hypotheses on this source range from it being a radio-quiet pulsar, a blazar with a highly collimated relativistic jet without accompanying strong radio emission, to a representative of an entirely new class of γ-ray emitters. Only recently, overwhelming evidence indeed established 3EG 1835+5918 as being a radio-quiet isolated neutron star (Mirabal et al. 2001; Reimer et al. 2001).

A potential candidate for an entirely new class of γ-ray emitting objects is the transient source 3EG J1837-0423. This source was discovered serendipitously with EGRET due to an extremely high peak flux of 4×10^{-6} cm^{-2} s^{-1} for $E > 100$ MeV during only 3.5 days, but was never detected before or later with high statistical significance (Tavani et al. 1997). This transient soure is located only 1° off the Galactic Plane, and the spectrum can be described by a power law with a spectral index of -2.1. The obvious hypothesis is of an object similar to already identified EGRET sources with comparable characteristics: blazar-type AGN shining through the Galactic Disk. Nevertheless, this scenario of an extragalactic origin appears to be very unlikely due to the lack of an appropriate counterpart and statistical limits on the detectibility of strong blazar-type sources. Following a Galactic origin hypothesis both binary and isolated compact objects might account for the highly variable γ-ray emission. However, the γ-ray emission was not accompanied by strong X-ray emission as expected for a Galactic object of this kind. So far, no conclusive identification could be found for this source, leading to the possibility of it being an object of a yet unknown class.

An unidentified source with a long observational history is the COS-B source 2CG 135+01 (3EG J0241+6103, GEV J0241+6102), most recently reviewed by Kniffen et al. (1997) and Tavani et al. (1998). Lying only about 1° from the Galactic Plane, the source location is excellently determined due to the observed γ-ray emission above 1 GeV. The EGRET intensity spectrum is described by a power law with an spectral index of about -2.1, leading to a follow-up detection between 0.75 and 30 MeV by COMPTEL. Weak evidence for short-time flux variability has been found by EGRET at energies above 100 MeV. The most interesting counterpart is GT 0236+610, a nonthermal radio source identified with the eclipsing high-mass binary system LSI+61°303. However, the established flare periodicity in radio waves of 26.5 days has not been observed in γ-rays, leaving the suggested identification questionable.

The prominent X-ray binary Cygnus X-3 has been suggested as the counterpart of the EGRET source 2EG 2033+4112 based on positional coincidence only (Mori et al. 1997). Any conclusive identification is extremely difficult in this case: the source is located in a region where at least four γ-ray point sources and enhanced diffuse emission are observed. The known 4.79 h periodicity of Cyg X-3 has not been found in the EGRET γ-ray data so far, prohibiting definite identification.

Following the hypothesis that the majority of the unidentified Galactic γ-ray sources are young pulsars, deep X-ray investigations around γ-ray sources, close to or in SNRs, are very promising. Focussing only on bright sources which show some characteristics of a possible pulsar origin in γ-rays (hard power-law spectrum, indication for high-energy spectral turnover, steady γ-ray flux) and a well-determined source position, X-ray sources were suggested as counterparts for 3EG J2020+4017 in the γ-Cyg-SNR (Brazier et al. 1996) and 3EG J0010+7309 in SNR CTA1 (Brazier et al. 1998) (see Fig. 13.7). Each of the suggested associations is a radio-quiet object with a similar F_γ/F_X ratio to Geminga, which is higher than all other sources apart from pulsars. The X-ray objects are always inside the 68% confidence contour of the corresponding GeV source, and show a constant flux. Further-

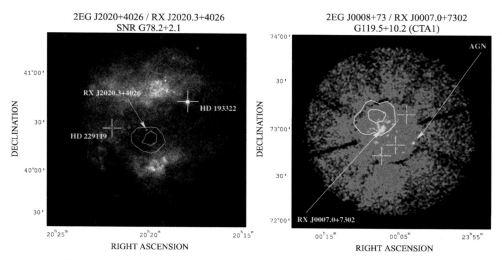

Fig. 13.7. X-ray point sources in SNRs (*images*) coincident with unidentified γ-ray sources (*contours*). In both γ-Cyg and CTA1 point sources with pulsar-like characteristics in γ-rays and X-rays were found which are likely candidates for radio-quiet pulsars

more, they could not be identified with known objects at other wavelengths, and do not appear on deep optical images. The statistics in X-rays as well as in γ-rays in conjunction with the lack of the advantage of a known radio ephemeris does not allow the final identification of these objects with pulsars, although they are already theoretically modelled and found to be consistent with the experimental suggestion of Geminga-like radio-quiet γ-ray pulsars. The same approach has been made for other objects, e.g., 3EG J1418-6038 (GEV J1417-6100) is proposed by Roberts and Romani (1998) to also be a young energetic pulsar. Perhaps an entire population of these objects is just waiting to be discovered (Brazier and Johnston 1999).

Finally, in some cases positionally coincident sources were observed by the EGRET and COMPTEL instruments. Because of the adjacent energy coverage, in a few cases joint γ-ray intensity spectra could be determined (e.g., GRO J2227+61: Iyudin et al. 1997). In conjunction with contemporaneous source flux coverage, these wide-band γ-ray observations are a vital link in their identification.

13.4 Conclusion

First of all, about two thirds of the EGRET high-energy γ-ray sources still remain unidentified. Understanding the nature of unidentified γ-ray sources has become a subject of increasing importance. Although the number of newly discovered γ-ray sources is still increasing, the fraction of unidentified point sources compared to unambiguously identified objects remains nearly the same. No new class of γ-ray-emitting objects apart from pulsars and AGN has been discovered despite a rather long observational history. The unidentified sources can be separated into a dominant Galactic component and a smaller, but more isotropic extragalactic component. Whereas for the Galactic component various explanations and suggestions have been made, the extragalactic component is widely believed to consist of still unrecognized AGN. Several statistical and observational arguments point towards an association of Galactic unidentified sources with OB-star associations and SNRs. However, until already identified point sources in SNRs are unambiguously associated with a pulsar origin, it cannot be decided whether or not these sources can solely account for the observed γ-ray emission, or whether interactions between newly accelerated cosmic-rays and the surrounding interstellar medium of the SNR itself might be an explanation. The Gould Belt, a nearby structure likely harbouring unidentified γ-ray sources, naturally supports the potential origin in massive stars or cosmic-rays interacting with enhanced gas concentrations. Finally, the possibility of the discovery of an entire new population of γ-ray-emitting objects cannot be ruled out. The unique γ-ray characteristics observed in a few individual unidentified sources in contrast to common features in the characteristics of known γ-ray sources supports this hypothesis. Beyond the approach of studying the collective properties of the unidentified sources, multi-frequency campaigns are being carried out more and more successfully, where joint, often time-correlated, deep observations at different wavelengths and with different scientific instruments are performed. Clearly one of the most important tasks for future experiments in γ-ray astronomy is to improve the source location accuracy and to enhance the sensitivity to be able to perform detailed timing analyses. Only this can provide results with sufficient statistical significance for sources other than the dominant γ-ray bright objects. Up to now, the unidentified γ-ray sources, especially those discovered by EGRET, remain one of the biggest mysteries and most intriguing issues of today's γ-ray astronomy. Determining the nature of these sources is the goal for future γ-ray instruments. GLAST (Gamma-Ray Large-Area Space Telescope), as the next major high-energy γ-ray mission, INTEGRAL (International Gamma-Ray Astrophysics Laboratory), as the upcoming low-energy γ-ray mission, and maybe ground-based Cherenkov telescopes with significant lower energy thresholds could help to obtain a more detailed picture of the many enigmatic unidentified γ-ray sources.

References

Bloom, S.D. et al. 1997, Astrophys. J. Lett. **488**, L23
Brazier, K.T.S., Johnston, S., 1999, Mon. Not. R. Astron. Soc. **305**, 671
Brazier, K.T.S. et al., 1996, Mon. Not. R. Astron. Soc. **281**, 1033
Brazier, K.T.S. et al., 1998, Mon. Not. R. Astron. Soc. **295**, 819
Cheng, K.S., Zhang, L., 1998, Astrophys. J. **498**, 327
Dermer, C.D., 1997, in Proc. 4th Compton Symposium ed. C.D.Dermer et al., AIP 410, 1275
Dingus, B.L. et al., 1996, Astrophys. J. **467** , 589
Esposito, J.A. et al., 1996, Astrophys. J. **461**, 820
Fegan, D.J., 1990, Proc. 21st ICRC, Vol.11, 23
Fichtel, C.E. et al., 1975, Astrophys. J. **198**, 163
Fichtel, C.E. et al., 1994, Astrophys. J. Suppl. **94**, 551
Fierro, J.M. et al., 1995, Astrophys. J. **447**, 807
Gehrels, N. et al., 2000, Nature **404**, 363
Grenier, I., 1997, in Proc. 2nd INTEGRAL workshop, ESA SP-382, 187
Halpern et al., 2001, Astrophys. J. **547**, 323
Hartman, R.C. et al., 1999, Astrophys. J. Suppl. 123
Helfand, D.J., 1994, Mon. Not. R. Astron. Soc. **267**, 490
Hermsen, W. et al., 1977, Nature **269**, 494
Hillier, R.R. et al., 1970, Astrophys. J. Lett. **162**, L177
Iyudin, A.F. et al., 1997, in Proc. 25th ICRC, Vol.3, 89
Johnson, W.N. et al., in Proc. 2nd Compton Symposium ed. Fichtel, C.E. et al., 1994, AIP **304**, 515
Kaaret, P., Cottam, 1996, J., Astrophys. J. Lett. **462**, L35
Kanbach, G. et al., 1996, Astron. Astrophys. Suppl. **120**, 461
Kaspi, V.M. et al., 2000, Astrophys. J. **528**, 445

Kaul, R.K., Mitra, A.K., in Proc. 4th Compton Symposium ed. C.D.Dermer et al., 1997, AIP **410**, 1271
Kniffen, D.A. et al., 1997, Astrophys. J. **486**, 126
Kuiper, L. et al., 2000, Astron. Astrophys. **359**, 615
Lamb, R.C., Macomb, D.J., 1997, Astrophys. J. **488**, 872
Mattox, J.R. et al., 1996, Astrophys. J. **461**, 396
Mattox, J.R. et al., 1997, Astrophys. J. **481**, 95
Mayer-Hasselwander, H.A., Simpson, G., 1990, Adv. Space Res. **10**, 89
Mayer-Hasselwander, H.A. et al., 1998, Astron. Astrophys. **335**, 161
McLaughlin, M.A. et al., 1996, Astrophys. J. **473**, 763
Merck, M. et al., 1996, Astron. Astrophys. Suppl. **120**, 465
Mirabal et al., 2001, Astrophys. J. Lett. **547**, L137
Montmerle, T., 1979, Astrophys. J. **231**, 95
Mori, M. et al., 1997, Astrophys. J. **476**, 842
Mukherjee, R. et al., 1995, Astrophys. J. Lett. **441**, L61
Mukherjee, R. et al., 2000, Astrophys. J. **542**, 740
Nel, H.I. et al., 1996, Astrophys. J. **465**, 898
Nice, D.J., Sayer, R.W., 1997, Astrophys. J. **476**, 261
Nolan, P.L. et al., in Proc. 2nd Compton Symposium ed. C.E.Fichtel et al., 1994, AIP **304**, 360
Özel, M.E., Thompson, D.J., 1996, Astrophys. J. **463**, 105
Pohl, M. et al., 1997, Astrophys. J. **491**, 159
Ramanamurthy, P.V. et al., 1996, Astrophys. J. **458**, 755
Reimer, O., Dingus, B.L., Nolan, P.L., 1997, in Proc. 25th ICRC, Vol.3, 97
Reimer, O. et al., in Proc. 4th Compton Symposium ed. C.D.Dermer et al., 1997, AIP **410**, 1248
Reimer, O., et al., 2001, Mon. Not. R. Astron. Soc. in press
Roberts, M.S.E., Romani, R.W., 1998, Astrophys. J. **496**, 827
Romero, G.E., Benaglia, P., Torres, D.F., 1999, Astron. Astrophys. **348**, 868
Schönfelder, V. et al., 2000, Astron. Astrophys. Suppl. **143**, 145
Sreekumar, P. et al., 1992, Astrophys. J. **400**, 67
Sturner, S.J., Dermer, C.D., 1995, Astron. Astrophys. **293**, L17
Sturner, S.J., Dermer, C.D., 1996, Astrophys. J. **461**, 872
Sturner, S.J., Dermer, C.D., Mattox, J.R., 1996, Astron. Astrophys. Suppl. **120**, 445
Swanenburg et al., 1981, Astrophys. J. **243**, L69
Tavani, M. et al., 1997, Astrophys. J. **479**, L109
Tavani, M. et al., 1998, Astrophys. J. **497**, L89
Taylor, J.H., Manchester, R.N., Lyne, A.G., 1993, Astrophys. J. Suppl. **88**, 529
Thompson, D.J. et al., 1994, Astrophys. J. **436**, 229
Thompson, D.J. et al., 1995, Astrophys. J. Suppl. **101**, 259
Thompson, D.J. et al., 1996, Astrophys. J. Suppl. **107**, 227
Thompson, D.J. et al., in Proc. 4th Compton Symposium ed. Dermer, C.D. et al., 1997, AIP **410**, 39
Voges, W., 1993, Adv. Space Res. **131**, 391
von Montigny, C. et al., 1995, Astrophys. J. **440**, 525
Wallace, P.M. et al., 2000, Astrophys. J. **540**, 184
Yadigaroglu, I.-A., Romani, R.W., 1995, Astrophys. J. **449**, 211
Yadigaroglu, I.-A., Romani, R.W., 1997, Astrophys. J. **476**, 347
Zhang, L., Cheng, K.S., 1998, Astron. Astrophys. **335**, 324

14 The Extragalactic Gamma-Ray Background

Georg Weidenspointner and Martin Varendorff

14.1 Introduction

The extragalactic background radiation (EBR) provides a unique window on a variety of fundamental topics in cosmology, astrophysics, and particle physics. These topics include the origin of the universe, the formation of structure and the evolution of galaxies, the formation of stars and the production of metals, gas and dust, and the properties of (exotic) elementary particles. Opening this window is, however, an exceedingly difficult task in almost all observable wavebands, as it requires the absolute measurement of the intensity of a presumably isotropic radiation field – one of the most challenging measurements in astronomy. A determination of the extragalactic background intensity involves a complete elimination of all instrumental background in addition to accounting for the emission from point sources and extended foreground emission, in particular Galactic diffuse emission.

There are two distinct possibilities for the origin of the EBR: it may result from the superposition of unresolved point sources or originate from truly diffuse mechanisms. The EBR may also originate from a combination of diffuse sources and point sources, with different origins dominating in different portions of the electromagnetic spectrum. An answer to the question of the origin of the EBR can be found only in an iterative process of observation and modelling of the radiative properties of various extended and discrete sources in the sky. Progress in this endeavour depends crucially on the capabilities of the telescopes and detectors available in each waveband. If the EBR is due to discrete sources (such as various types of galaxies), instruments with high angular resolution are needed to resolve and study the large number of expected faint sources. If the EBR has a truly diffuse origin it is crucial to determine its energy spectrum as accurately as possible in order to reveal its physical production process. Even if the EBR arises solely from discrete sources, a precise measurement of the energy spectrum of the total diffuse emission is needed to determine the absolute EBR intensity in the presence of appreciable Galactic diffuse foreground emission by spectral fitting. A further important diagnostic for the origin of the EBR is the spatial fluctuation of its intensity on large and small angular scales. The detection of large-scale isotropy of the EBR is a crucial test for its cosmological origin and can be exploited to separate the EBR from Galactic diffuse emission. The

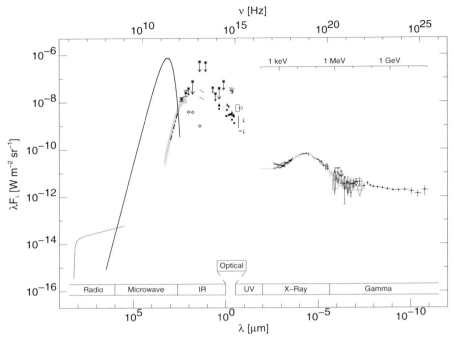

Fig. 14.1. The multi-wavelength spectrum of the extragalactic background radiation from radio wavelengths to the γ-ray regime as it is known today. (Adapted from Weidenspointner 1999)

angular-fluctuation spectrum of the EBR intensity on fine angular scales can be compared to the expected angular fluctuations due to different sources of radiation, which allows one to assess the EBR contributions from different source populations. In the following, a brief overview[1] is given on the observations of the EBR from radio wavelengths to the γ-ray regime, and on their relevance for the above-mentioned issues in cosmology, astrophysics and particle physics.

The multitude of objects and processes that manifest themselves in the EBR can only be disentangled by observing the EBR at all accessible wavelengths. The full multi-wavelength spectrum of the EBR from radio wavelengths to γ-ray energies as it is known today is depicted in Fig. 14.1. From an astrophysical point of view, the wavebands divide up naturally into those in which the physics involves "thermal" processes and those which are probably dominated by high-energy astrophysical processes, often referred to as "nonthermal" processes (Longair 1994).

---

[1] For further reading, the following proceedings are recommended: Bowyer and Leinert (1990), Sandage et al. (1994), and Calzetti et al. (1995).

The microwave, infrared, optical and ultraviolet wavebands are probably dominated by thermal processes. Considering the full background spectrum, the cosmic microwave background (CMB) is unique in a number of ways. The CMB is the only truly "cosmic" EBR component containing information on the very distant past of the cosmos, as this radiation is the cooled thermal spectrum of the hot early stages of the universe as a whole. In addition, the CMB seems to be the only EBR component of truly diffuse origin. The spectrum of the CMB is the most perfect example of a blackbody spectrum that has been found in nature to date (see e.g. Mather et al. 1994; Fixsen et al. 1996). The almost perfect blackbody shape of the CMB spectrum places severe constraints on the thermal history of the universe, even before recombination, thereby ruling out, for example, a significant contribution of a hot intergalactic medium to the extragalactic X-ray background. The anisotropy of the CMB offers a unique opportunity to probe the initial density fluctuations that were the seeds for subsequent structure formation, as well as to determine cosmological parameters such as the Hubble constant H_0, the cosmological constant Λ, and the density parameter Ω.

The EBR at infrared, optical and ultraviolet wavelengths is most likely dominated by thermal emission from galaxies and thus ultimately by thermal emission of stars and dust. The EBR in these wavebands allows us to probe the formation and evolution of galaxies as well as the formation of stars and the production of gas and dust. It is of utmost importance for this endeavour to observe the EBR at all wavelengths from far IR to near UV as the intensities in each of the individual wavebands are intimately connected. For example, studies of the star formation rate have to take into account that a significant fraction of the primary stellar emission at optical and ultraviolet wavelengths is absorbed and re-radiated at far-IR wavelengths. In this context it has to be noted that star formation activity may also reveal itself in conspicuous edges or steps in the EBR at MeV energies due to γ-ray line emission from Type Ia supernovae (see Sect. 14.4.1). The current EBR detections and limits already impose severe constraints on models for the formation and evolution of galaxies, requiring the models to include the chemical evolution of galaxies and the production of dust. The IR to UV EBR is much less suited to discriminate different cosmological models than the CMB, as the intensity in the former is influenced much more by galaxy evolution than by the specific cosmological model.

The wavelengths involving high-energy astrophysical processes are the radio, X-ray and γ-ray regions of the electromagnetic spectrum. In these wavebands, the EBR seems to be dominated by the nonthermal emission from various classes of active galaxies (Active Galactic Nuclei [AGN]; see Chap. 12). In contrast to the emission from normal galaxies, which is powered by nuclear fusion, the ultimate energy source of AGN is gravity, specifically the accretion of matter onto a massive black hole. In all models of galaxy evolution mergers and tidal interactions have a major impact not only on normal galax-

Table 14.1. The energy densities and photon densities of the EBR in different regions of the electromagnetic spectrum at the current epoch. Numbers are only given for those wavebands in which the EBR has been positively detected. Note that the quoted densities are usually rough estimates and, for any precise calculation, integrations should be carried out over the relevant regions of the spectrum. The table is adapted from Table 1.3 on p. 347 in Longair (1994) except for the numbers for the γ-ray waveband, which were re-calculated based on the current results on the EGB

Waveband	Energy density of radiation (eV m^{-3})	Number density of photons (m^{-3})
Radio	$\sim 5 \times 10^{-2}$	$\sim 10^6$
Microwave	3×10^5	5×10^8
Infrared	–	–
Optical	$\sim 2 \times 10^3$	$\sim 10^3$
Ultraviolet	–	–
X-ray	75	3×10^{-3}
γ-ray	8.2	1.3×10^{-6}

ies, but also on the formation of AGN. Unravelling the contributions of the various AGN classes to the nonthermal wavebands of the EBR will therefore provide valuable constraints not only on the evolution of AGN, but also on the evolution of normal galaxies dominating the thermal EBR wavebands.

To conclude this introduction, typical energy and photon number densities in each waveband in which the EBR has been positively detected are listed in Table 14.1, which is based on Table 1.3 in Longair (1994). These numbers are usually rough estimates and, for any precise calculation, integrations should be carried out over the relevant regions of the spectrum, as was done to redetermine the densities in the γ-ray waveband from the current results on the EBR. These figures are often useful for making crude estimates of the importance of physical processes in a cosmological setting.

As can be seen in Fig. 14.1 and Table 14.1, the majority of the EBR photons are due to the CMB, which also contains the bulk of the EBR energy density. The integrated EBR energy density in the infrared, optical and ultraviolet wavebands is most likely less than that in the CMB, although the EBR intensity in the IR still has not been positively detected except at the longest wavelengths. The EBR photon number density in the radio waveband is rather large, although the corresponding energy density is minute. Finally, in the X-ray and in particular the γ-ray wavebands the photon densities are exceedingly low, and the energy densities are very small compared to those in the thermal regions of the EBR.

14.2 History

The history of observations of the extragalactic γ-ray background (EGB), i.e. the EBR at energies above about 100 keV, is intimately tied to the history of γ-ray astronomy as a whole. The first signal detected by γ-ray astronomers was the EGB, which has remained of prime interest ever since. As described in more detail below, it was soon possible to separate an extragalactic and a galactic component in the γ-ray sky at energies above about 50 MeV. Since such a separation has only recently become feasible at MeV energies, the low-energy EGB in the ~ 0.1–10 MeV band is still customarily referred to as cosmic diffuse γ-ray background.

The first attempts to directly address the question of whether a detectable intensity of cosmic γ-rays exists were conducted around the end of the first half of the last century; all returned upper limits only. As described in more detail in Chap. 4, the first detection of celestial γ radiation around 1 MeV was claimed by Metzger et al. (1964) using omnidirectionally sensitive CsI scintillators on board the lunar probes Ranger 3 and 5. As we know today, however, these early measurements of the EGB at MeV energies were heavily contaminated by instrumental background. The first evidence for cosmic γ radiation above 100 MeV was provided by Kraushaar et al. (1965) based on data from an instrument on board the satellite Explorer 11. This small data set represents one of the first important successes of γ-ray astronomy, as it allowed Kraushaar et al. (1965) to effectively refute the steady-state cosmology based on a severe upper limit on the proton–antiproton creation rate. The presence of an isotropic component was confirmed by Kraushaar et al. (1972) using a γ-ray counter telescope on board the third Orbiting Solar Observatory (OSO-3). In addition, this experiment provided the first evidence for the existence of a galactic component concentrated in a band around the Galactic equator with a broad maximum towards the Galactic Center. The spectrum of the extragalactic component was found to be softer than that of the Galactic component.

The situation improved dramatically with the launch of the second Small Astronomy Satellite (SAS-2) in 1972, carrying a wire spark-chamber γ-ray telescope sensitive to γ-rays above about 35 MeV. The detector design of SAS-2 resulted in unprecedentedly low instrumental background. Since the direction of the incident γ-ray photons was determined rather accurately, all photons that did not originate from the telescope aperture, and may thus have been due to instrumental background, could effectively be excluded from the data. The only source of instrumental background was therefore production of secondary γ-ray photons in passive material above the veto system and within the aperture, which was kept to a minimum. With its much improved sensitivity and angular response compared to that of the telescope on board OSO-3, SAS-2 clearly detected the Galactic γ radiation, and found an apparently uniform background radiation at high Galactic latitudes. Using galaxy counts as a tracer of Galactic matter to subtract the Galactic diffuse emission, a spectral index of $-2.35^{+0.3}_{-0.4}$ was derived for the power-law spectrum of the EGB above 35 MeV (Thompson and Fichtel 1982).[2]

In the meantime progress in the MeV range was slow due to the presence of additional instrumental background from induced radioactivity and due to the difficulty in building detectors with even moderate spatial resolution. A variety of experiments similar to those flown on the Ranger spacecraft have been performed on various satellites. Of particular interest are the results of Trombka et al. (1977) obtained with an omnidirectional NaI(Tl) spectrometer crystal with plastic anticoincidence shield on an extensible boom flown on the Moon missions Apollo 15, 16 and 17. In parallel to satellite-borne experiments, considerable effort was devoted to determining the spectrum of the low-energy EGB from balloons. Among the first to attempt such a measurement were Vedrenne et al. (1971) and Daniel et al. (1972), using omnidirectional organic stilbene or NaI(Tl) scintillators with 4π anticoincidence shielding, respectively. More sophisticated balloon-borne experiments with a restricted field of view were the Compton telescopes employed by Schönfelder and Lichti (1974); White et al. (1977); Schönfelder et al. (1977, 1980) and Lockwood et al. (1981). A collimated detector system consisting of a central NaI(Tl) crystal surrounded by a cylindrical NaI(Tl) counter and an active CsI(Tl) aperture shutter was used by Fukada et al. (1975).

While instrumental background is of little concern for high-energy γ-ray telescopes such as SAS-2, its elimination is the major challenge for experiments at MeV energies. Most of the above-mentioned measurements of the low-energy EGB exploited, among other background characteristics, the fact that the background (instrumental background and/or atmospheric γ radiation), in contrast to the sought-after signal, is variable. For example, in the analysis of Mazets et al. (1975) the atmospheric γ-ray background was removed using its dependence on the geomagnetic cut-off rigidity. The modulation of the albedo γ-ray continuum was modelled by the readily observable modulation of the atmospheric 511 keV line, which was approximated by an exponential in rigidity. Both atmospheric components were then eliminated by an extrapolation to infinite rigidity. A major instrumental background component in the analysis of Trombka et al. (1977) was γ-ray photons due to interactions of primary cosmic rays and secondary particles in the Apollo spacecraft as well as primordial radioactivity of spacecraft materials. These backgrounds were extracted from the data using their dependence on the length of the extensible boom carrying the detector. Finally,

---

[2] In 1975 the COS-B satellite was launched, carrying a spark chamber γ-ray telescope similar to that on SAS-2. Unlike the nearly circular, low-Earth orbit of SAS-2, COS-B was placed in a highly eccentric orbit, resulting in an incident cosmic-ray intensity that was an order of magnitude higher. Together with an unfavourable design of the telescope (see Chap. 3) the consequence was a rather large background of secondary γ-rays produced in the detector which precluded measurement of the EGB.

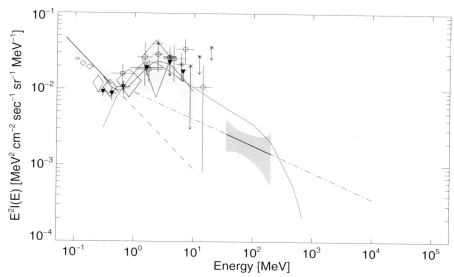

Fig. 14.2. An illustration of the status of EGB measurements above about 100 keV in 1982 (for comparison, the current results on the EGB spectrum are depicted in Fig. 14.3). The following results are plotted: Fukada et al. (1975): *diamonds*, Daniel et al. (1972) and Daniel and Lavakare (1975): *solid triangles*, Mazets et al. (1975): *solid line* up to 400 keV (the results at higher energies are not plotted as they are fully consistent with the MeV-bump), White et al. (1977): *asterisks*, Trombka et al. (1977): *light grey-shaded band*, Schönfelder et al. (1980): *circles*, Lockwood et al. (1981): *squares*, Thompson and Fichtel (1982): *grey-shaded bowtie*. The *dashed line* illustrates the extrapolation of the hard X-ray spectrum to higher energies; the *dash-dotted line* is an extrapolation of the SAS-2 result to lower and higher energies. The MeV-bump, an excess of emission at MeV energies above the extrapolations of the spectra at lower and higher energies, is clearly present. The depicted compilation of experimental results is not exhaustive, but only intended to give the qualitative and quantitative picture at that time. The *curved solid line* represents the predicted γ-ray background from matter–antimatter annihilation in a baryon-symmetric universe (Stecker 1989, see Sect. 14.4.2). For comparison, current models for the integrated emission of different point-source populations are depicted in Fig. 14.4

the intense atmospheric γ-ray background at balloon altitudes was separated using a growth-curve technique: close to the top of the atmosphere (above the Pfotzer maximum) the flux of vertically downward-moving photons is expected to increase linearly with atmospheric depth, which is the amount of residual atmosphere above the balloon in units of g cm^{-2}, while the cosmic γ-ray photons are exponentially attenuated with increasing atmospheric depth (see, e.g., Schönfelder et al. 1980, and references therein). This atmospheric growth-curve technique is particularly suited to Compton telescopes, since these instruments are capable of selecting downward-moving photons, in contrast to omnidirectional detectors. Being aware of this drawback, Fukada et al. (1975) therefore relied on artificially modulating the count rate of their omnidirectional detector with an aperture shutter to eliminate the instrumental and atmospheric backgrounds from outside the entrance aperture.

An illustration of the status of measurements of the EGB above about 100 keV in 1982 is depicted in Fig. 14.2. This compilation of experimental results is not exhaustive, but only intended to give a qualitative and quantitative picture of the EGB at that time. Of particular interest is the so-called MeV-bump, an excess of emission at MeV energies above the extrapolations of the spectra at lower and higher energies. A popular physical interpretation of the MeV-bump was to attribute it to redshifted γ radiation from matter–antimatter annihilation at an early epoch of the universe (see Sect. 14.4.2). It was also proposed that supernovae (see Sect. 14.4.1) may contribute to this excess emission below a few MeV. However, the existence of the MeV-bump was regarded with scepticism for two reasons: the energy region 0.5–30 MeV is experimentally most difficult because the cross section for the interaction of photons with matter reaches its absolute minimum, and in addition the instrumental background due to locally produced nuclear γ-rays is most important from energies of ~ 0.5 MeV to a few MeV. On the other hand, the fact that the MeV-bump was measured rather consistently with a variety of different experiments operated in different environments producing different backgrounds provided considerable support for the claimed existence of this excess emission. In the end, some doubt remained regarding the existence of the MeV-bump, and it was felt that new measurements with improved instrumentation were needed to clarify the situation.

At hard X-ray energies the HEAO-A (or HEAO-1) mission launched in 1977 was of prime importance, as it carried two experiments, A2 and A4, whose primary goal was the measurement of the EBR from keV to MeV energies. The results obtained with these detectors have important implications on the interpretation of the results at γ-ray energies (see Sect. 14.5).

14.3 Current Results

Since about 1980, numerous γ-ray observations have been performed with balloon-borne and, in particular, satellite-borne experiments. Only two satellite missions, however, delivered data that allowed astronomers to determine the EGB: the Solar Maximum Mission (SMM) and the Compton Gamma-Ray Observatory (CGRO), launched in 1980 and 1991 respectively. SMM carried, among other detectors, a wide-field NaI γ-ray spectrometer (GRS). The SMM/GRS operated stably for more than nine years. It was pointed at the Sun for most of this time, thereby scanning the ecliptic. Due to its large field of view of 140° FWHM (full width at half maximum) the SMM/GRS monitored a considerable fraction of the sky. CGRO is described in more detail in Chap. 3. Two of its instruments, namely the imaging Compton telescope COMPTEL and the spark-chamber telescope EGRET (Energetic Gamma-Ray Experiment Telescope), sensitive to γ-rays from about 1 MeV to 30 MeV and 30 MeV to 100 GeV respectively, performed the first all-sky surveys at these energies during the first 18 months of the mission and since then carried out a large number of pointed observations across the whole sky.

As with earlier measurements of the low-energy EGB, the SMM/GRS and COMPTEL analyses, covering the energy ranges 0.3–7 MeV and 0.8–30 MeV respectively, relied heavily on the variable nature of the instrumental background for its elimination. In the case of SMM/GRS even the signal itself exhibited a characteristic variation. Because of its wide field of view, the largest contributor to the measured SMM/GRS count rate, after internal and atmospheric background, was the EGB. To extract the spectrum of the EGB, the observed count rate in each energy channel was fitted with a model attributing the time-variation of the count rates to the following components: EGB and albedo γ radiation,[3] prompt and delayed cosmic-ray induced backgrounds, and decay of radioactivity within the instrument and the spacecraft (Watanabe 1997). One of the fit parameters in this preliminary analysis accounts for the Galactic diffuse emission (Watanabe, private communication). In the COMPTEL analysis the so-called prompt instrumental background, which follows closely the local instantaneous cosmic-ray intensity, and the long-lived instrumental background, which arises from the decay of radioactive isotopes produced in interactions of cosmic-ray particles and neutrons in the instrument material, are accounted for as follows (see, e.g., Kappadath 1998; Weidenspointner 1999; Weidenspointner et al. 2000a,b):[4] In each energy band the event rate is measured as a function of the trigger rate of the charged-particle or veto domes, which is used as a measure of the incident cosmic-ray intensity. The contributions from individual radioactive isotopes are determined based on their characteristic line structures in the energy spectra of different detectors. After subtraction of the activity from long-lived isotopes, the prompt background is accounted for in a linear extrapolation of the remaining event rate as a function of the veto dome trigger rate to conditions of zero incident cosmic-ray intensity. The Galactic diffuse emission is not explicitly subtracted in this analysis because its intensity is still uncertain (see Chap. 9). Its contribution was minimized, however, by using only observations at high Galactic latitudes. As can be seen in Fig. 9.15, the intensity of the high-latitude Galactic diffuse emission appears to be less

---

[3] The variation of these two components results from the variation of the position of the Earth relative to the instrument pointing direction, with the EGB term being the 4π integral of the SMM/GRS response minus about 4 sr blocked by the Earth.

[4] Recently, an alternative approach has been initiated by Bloemen et al. (1999), which exploits the temporal variation of the instrumental background and differences in the dataspace distribution of instrumental background and celestial signal to derive the EGB intensity in an iterative model-fitting procedure.

than the total (statistical and systematic) uncertainty in the low-energy EGB as measured by COMPTEL.

In contrast to measurements at low γ-ray energies, the EGRET instrumental background was a minor concern in the analysis of the high-energy EGB in the 30 MeV to 100 GeV range (Sreekumar et al. 1998). The EGRET instrumental background intensity was demonstrated to be more than an order of magnitude less than the EGB intensity derived from SAS-2 data. The extragalactic emission was derived by assuming that the observed emission at high Galactic latitudes is the sum of a galactic and an extragalactic component, with the former being described by a numerical model. Both components were separated by fitting the numerical model for the Galactic emission and an isotropic model for the EGB to the observed intensities as a function of energy and location in the sky.

The EGRET instrument was also used to examine the large-scale anisotropy of the EGB above 100 MeV. A comparison of 36 independent regions of the sky showed that the integrated sky intensity above 100 MeV is consistent with a uniform distribution in regions outside the inner Galaxy (Sreekumar et al. 1998). The average integrated intensity above 100 MeV in 36 independent sky regions is $(1.47 \pm 0.33) \times 10^{-5}$ photons cm^{-2} s^{-1} sr^{-1}; the corresponding result from the combined data is $(1.45 \pm 0.05) \times 10^{-5}$ photons cm^{-2} s^{-1} sr^{-1}. There is evidence that these observations contain Galactic diffuse emission that has not yet been accounted for. The intensity distribution is, however, again consistent with isotropy if a sample of 28 independent observations, excluding in addition the inner regions of the Galaxy which are not more than 60° from the Galactic Center, is studied. A first attempt to examine the fine-scale anisotropy of the EGB above 100 MeV was made in a fluctuation analysis on EGRET data (Willis 1996). The quality of the data turned out to be insufficient to rule out a significant, truly diffuse component in the EGB. The contribution of unresolved point sources was found to be 6–100% of the observed EGB intensity.

The isotropy of the low-energy EGB on scales of a few steradian was studied for the first time using the COMPTEL instrument (e.g. Weidenspointner 1999). No significant anisotropies in the diffuse emission at Galactic latitudes $|b| > 30°$ could be found in the 0.8–30 MeV energy range. This finding is consistent with the assumption of an extragalactic origin of the EGB, which implies isotropy on large angular scales. Upper limits on the relative deviations from isotropy were found to be about 25–45% at the 95% confidence level.

As can be seen in Fig. 14.3, the shape of the EGB spectrum appears to be considerably simplified as compared to the situation around 1982 (compare Fig. 14.2). The COMPTEL and preliminary SMM/GRS measurements provide no evidence for the existence of the putative MeV-bump, which therefore must have been due to instrumental background not accounted for in previous analyses. Instead, the overall spectrum seems to consist of a softer

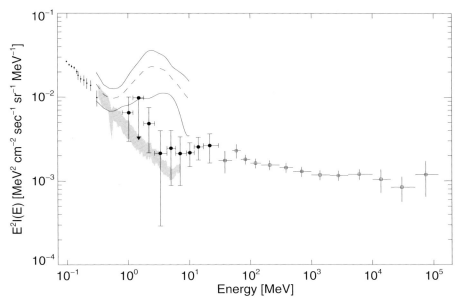

Fig. 14.3. A comparison of current results on the spectrum of the EGB. The hard X-ray data are from Kinzer et al. (1997) obtained with HEAO-A; the *grey-shaded band* depicts the preliminary GRS/SMM result of Watanabe et al. (1997). The historic MeV-bump is illustrated by the Apollo result of Trombka et al. (1977) (compare the illustration of the historic measurements of the EGB in Fig. 14.2). For simplicity, the COMPTEL measurements in the 0.8–30 MeV range are represented by the result of Weidenspointner (1999) (the *arrow* of the 2σ upper limit reaches down to the intensity value), since the results by Kappadath (1998) and Bloemen et al. (1999) are consistent within uncertainties, as can be seen in Fig. 14.1. The high-energy result has been derived by Sreekumar et al. (1998) using EGRET data. To date, no extragalactic diffuse γ radiation has been detected above 100 GeV

low-energy component and a harder high-energy component, with the transition occurring around a few MeV. The high-energy component has now been determined over a much larger energy range with greatly improved accuracy and precision by EGRET.

As discussed in the following, the current results suggest that it is more likely that the EGB up to 100 GeV results from the superposed emission of various classes of discrete, extragalactic sources than from a truly diffuse process. Above about 30 MeV a significant fraction of the EGB may be due to the blazar class of AGN; other possible contributors include starburst galaxies and normal galaxies. At MeV energies and below candidate source populations are radio-loud Seyfert galaxies, misaligned blazars and MeV-blazars, as well as cosmological Type Ia supernovae. The EGB contributions of the different source classes, however, are still quite uncertain, and it is possible that other, yet unidentified, source populations may be required to account for the observed intensity at these energies.

Above 100 GeV no significant extragalactic diffuse γ radiation has been detected so far. Atmospheric Cherenkov telescopes (see Chap. 3), however, have already measured significant emission from individual extragalactic point sources, namely BL Lac objects, in the ~ 300 GeV to ~ 30 TeV range (Weekes et al. 1997).[5] It is therefore expected that future instruments with improved sensitivity will be able to detect extragalactic diffuse γ radiation at energies above 100 GeV, which may include a contribution from interactions of extremely high-energy cosmic-rays with the CMB. The extension of the spectra of extragalactic sources well beyond 1 TeV opens a new, exciting aspect of γ-ray astronomy. The absorption features in the γ-ray spectra of distant extragalactic sources due to γ–γ pair production of primary γ-rays of very high energy with the diffuse EBR at lower energies, i.e. microwave through optical wavelengths, contain unique cosmological information about the intergalactic photon and magnetic fields, and their evolution in time (see, e.g., Stecker et al. 1992; Weekes et al. 1997, and references therein). In particular, γ–γ absorption may allow us to determine the cosmic IR background, a direct measurement of which is notoriously difficult because of strong foreground emission from interstellar and interplanetary dust. Early claims for attenuation by pair production on optical or near IR photons using TeV observations were inconclusive (see, e.g., Krennrich et al. 1997, and references therein), as it is very difficult to disentangle the effects of a high-energy cut-off in the primary source spectrum, intrinsic γ–γ absorption in the source, and γ–γ absorption during propagation to Earth. However, more recent observations at TeV energies, supplemented by simultaneous observations in other wavebands, increasingly constrain the physical parameters involved (see, e.g., Bednarek and Protheroe 1999).

14.4 Possible Origins

The EGB is of particular interest to cosmology because the universe is transparent to γ-rays back to very high redshifts of 100 or more (see, e.g., Stecker 1971). A large number of possible origins for the EGB have been proposed, all of which fall into one of the following two categories: the superposition of unresolved, discrete sources or a truly diffuse origin. None of the proposed models can, however, by itself explain the entire spectrum of the EGB, which spans at least six orders of magnitude in energy. Most likely, therefore the EGB is the sum of a number of different components, each dominating a specific energy range, which may be disentangled once the spectrum of the EGB is known with sufficient accuracy. In the future, additional leverage for the

---

[5] At even higher energies, air-shower arrays (see Chap. 3) are used to search for ultra-high energy photons and particles. To date, no verifiable γ-ray source has been detected at these energies.

separation of the supposedly different components contributing to the EGB may come from the enhanced ability to resolve ever fainter point sources and from observations of the large-scale and fine-scale anisotropy of the intensity of the EGB, which is beyond the capabilities of current instruments.

Historically, diffuse mechanisms were favoured over the superposition of discrete point sources until the launch of the CGRO in 1991, simply because only four extragalactic sources, all AGN, had been reported to be γ-ray sources until then (see Chap. 12):[6] Cen A, NGC 4151, 3C 273[7] and MCG 8-11-11. Following the detection of more than 80 active galaxies by CGRO (see, e.g., Johnson et al. 1997; Hartman et al. 1997), the current thinking on the origin of the EGB was reversed: now it seems more likely that the EGB originates from the superposition of various classes of discrete, extragalactic sources than from truly diffuse mechanisms.

14.4.1 Unresolved Point Sources

Normal Galaxies

Normal galaxies are an obvious candidate class of discrete sources that may contribute to the EGB as they are abundant in the universe and known to emit at all observable γ-ray energies. The diffuse emission from our Galaxy stands out clearly against the isotropic glow of the EGB as an intense band of emission along the Galactic Plane, dominating the γ-ray sky (see Chap. 9, in which the origin of the diffuse Galactic continuum emission is described in detail).

A reliable estimate of the contribution of normal galaxies to the EGB is currently hampered by our ignorance concerning their γ-ray luminosity function. A determination of the γ-ray luminosity function of normal galaxies, describing the evolution of space density and luminosity of these sources with cosmic epoch, is beyond the capabilities of current instruments, which lack the required sensitivity and spatial resolution. It is therefore necessary to resort to simplifying assumptions and models to predict the diffuse γ-ray intensity from normal galaxies. For example, it is often assumed that the γ-ray luminosity of a normal galaxy is proportional to its luminosity in another waveband, for example, the radio luminosity, at all cosmic epochs. Following different approaches typical estimates for the contribution of normal galaxies to the EGB above 100 MeV are less than about 10% (see e.g. Strong et al. 1976; Setti and Woltjer 1994). Provided the evolution of normal galaxies is stronger at γ-ray energies than at radio wavelengths, however, these sources may account

---

[6] In addition, γ-ray line emission had been detected from SN 1987A in the Small Magellanic Cloud (see Chap. 10 and Sect. 14.4.1).

[7] Interestingly, the power-law index of $-2.5^{+0.6}_{-0.5}$ derived for this quasar (Bignami et al. 1981) is consistent with the SAS-2 result for the EGB spectrum (see Sect. 14.2).

for a significant fraction of the EGB (see e.g. Lichti et al. 1978; Prantzos and Casse 1994).

Currently, it cannot be ruled out that normal galaxies can account for an appreciable fraction of the EGB intensity. The importance of their contribution depends upon evolutionary effects which are still uncertain.

Active Galaxies

Up to 10% of all galaxies in the observable universe harbour an active nucleus (Robson 1996), a very small central region of very high luminosity, which can outshine by far the remainder of the host galaxy. In contrast to the spectral energy distribution of normal galaxies, which peaks around the optical waveband as it is dominated by the thermal emission from stars, the spectral energy distribution of active galaxies or AGN is usually dominated by nonthermal emission and often has maxima in the radio, IR and UV wavebands, and at γ-ray energies. The origin of the γ-ray emission of AGN, as well as an overview of the different AGN classes, can be found in Chap. 12.

Recent observations of the high-energy emission of AGN revealed the existence of two classes of γ-ray-emitting AGN: firstly radio-quiet Seyfert galaxies and quasars (QSOs), and secondly jet-dominated, radio-loud quasars (FSRQs, SSRQs) and BL Lac objects commonly referred to as blazars (see Chap. 12). The intrinsic spectra of radio-quiet AGN, in particular Seyferts and QSOs, cut off around 100 keV (see Fig. 14.4) and therefore these objects are not expected to be important contributors to the EGB. Radio-quiet AGN do, however, provide a substantial contribution to, or may even account for, the extragalactic X-ray background (XRB) from ~ 0.1 keV to ~ 100 keV. In particular, the hard XRB, characterized by the pronounced peak in the spectral energy distribution at ~ 30 keV, can be accounted for by the combined emission from different types of Seyfert galaxies, the dominant contribution being from strongly absorbed Seyfert Type 2 galaxies, in accord with the unification paradigm (see Sect. 14.5).

In marked contrast, the γ-ray emission of blazars extends to GeV and even TeV energies (see Chap. 12). Above a few tens of MeV the blazar spectra are well described by a power law, the spectral index of the "average blazar" spectrum being $\alpha = -2.16 \pm 0.31$ (Hartman et al. 1997). Observations at lower energies show that a spectral break is required for blazars at energies between a few MeV and about 100 MeV, depending on the individual source (e.g. McNaron-Brown et al. 1995). The close agreement of the average spectrum of radio-loud blazars above about 100 MeV with the spectrum of the EGB, which was found to be consistent with a power law of spectral index -2.10 ± 0.03 (Sreekumar et al. 1998), immediately establishes γ-ray blazars as prime candidate sources for the EGB.[8] There is also evidence for the existence of

---

[8] A cosmological integration of a power law in energy yields the same functional form and slope.

a subclass of the blazar population, the so-called MeV blazars, which are characterized by a time-variable spectral bump, instead of a break, at MeV energies (see e.g. Bloemen et al. 1995). This renders them candidate sources for the low-energy EGB.

In addition to the two "standard" classes of γ-ray-emitting AGN briefly described above, namely radio-quiet Seyferts/quasars (QSOs) and radio-loud blazars, there is growing evidence for the existence of an intermediate class of γ-ray-emitting AGN, referred to as radio galaxies with Seyfert nuclei, radio-loud Seyfert galaxies or misdirected/misaligned blazars (see Chap. 12). The best-studied of these objects is the well-known FR I radio galaxy Centaurus A (Cen A). The γ-ray emission of these radio-loud Seyfert galaxies does not cut-off exponentially at ~ 100 keV, but breaks to another power law. Nevertheless, no emission at MeV and higher energies has yet been detected within the sensitivity limits of current instruments except for Cen A, the closest AGN (e.g. Steinle et al. 1998).

Most calculations of the blazar contribution to the EGB are based on the assumption that there is a linear correlation between the radio and the γ-ray emission of these objects (see, e.g., Stecker et al. 1993; Salamon and Stecker 1994; Setti and Woltjer 1994; Comastri et al. 1996; Salamon and Stecker 1998). The general conclusion from these calculations is that most of the EGB can be explained as originating from unresolved blazars, although the uncertainties are still quite large, even more so for the relative contributions from BL Lacs and FSRQs, two classes of radio-loud blazars (see Chap. 12). It has to be pointed out, however, that the degree and nature of the correlation between the radio and the γ-ray emission is still under debate (see, e.g., Mücke et al. 1996; Mattox et al. 1997). In addition, it is possible that blazars are detected in γ-rays only during brief flaring episodes, with their quiescent emission (if it exists at all) being below the current sensitivity limits. Another possibility is that the γ-ray emission of blazars is steady, but with a broad luminosity distribution, so that only the high-luminosity wing is observed. It was concluded by Stecker and Salamon (1996) and Kazanas and Pearlman (1997) that blazars have been preferentially detected in flaring states. Including a spectral hardening during flare emission, as compared to the softer quiescent emission, results in a curvature of the calculated blazar contribution to the EGB (Salamon and Stecker 1998), which is already constrained by the latest EGRET measurements (Sreekumar et al. 1997, 1998).

The variability of the blazar γ-ray emission is also a major source of uncertainty for the γ-ray luminosity function deduced from the γ-ray data itself by Chiang et al. (1995) and Chiang and Mukherjee (1998). They showed that there is indeed evolution of the γ-ray emitting blazars, and found the γ-ray evolution to be consistent with pure luminosity evolution and similar to that seen at other wavelengths. While the conclusion of the first analysis was that radio-loud AGN contribute a significant fraction of the EGB above 100 MeV, the improved calculation of the low-luminosity end of the luminosity function in the latter analysis reduces the emission of unresolved blazars to $\sim 25\%$ of the observed EGB intensity. Clearly, improved knowledge of the γ-ray luminosity function is needed before firm conclusions can be drawn.

A unique approach to estimating the contribution of unresolved radio-loud AGN to the EGB was taken by Mücke and Pohl (2000), who calculated the nonthermal emission of blazars in the framework of leptonic models and then proceeded to estimate the diffuse γ-ray background due to these sources in the context of the AGN unification paradigm. The model was constrained by observations of resolved BL Lacs and FSRQs: their $\log N$–$\log S$ distribution, their relative number, and their redshift distribution. It was found that the model is consistent with the observed FSRQ number density, but underpredicts the number of BL Lacs. This model, integrated back to $z_{\max} = 3$, can account for only 20–40% of the EGB above about 30 MeV, to which unresolved BL Lacs/FR I contribute 70–90%. However, this model considers only those AGN whose power-output peaks between about 100 MeV and 1 GeV. For members of the BL Lac/FR I class, this may only be true for a small fraction of the total population (see e.g. Fossati et al. 1998), possibly resolving the discrepancy concerning the BL Lac number density. In this case, unresolved BL Lacs/FR I could produce the observed EGB (Mücke and Pohl 2000).

To summarize, emission from radio-quiet AGN can account for the entire XRB up to about 100 keV. In the 100 keV to a few MeV range radio-loud Seyfert galaxies and misaligned blazars, as well as MeV blazars, could contribute significantly to the EGB at MeV energies. Finally, there are strong indications that radio-loud AGN may provide a substantial fraction of, or even account for, the EGB above a few MeV. Estimates of their contribution, however, are still uncertain.

Infrared-Luminous Galaxies

The IR emission of IR-luminous galaxies exceeds that of normal galaxies by 1–2 orders of magnitude. The origin of this strong IR emission is not yet fully understood, but may involve starburst activity, Galactic winds, or a buried AGN (see, e.g., Dermer 1996, and references therein). IR-luminous galaxies may also be stronger emitters at γ-ray energies than normal galaxies for two reasons. In scenarios of galaxy unification through evolution, galaxy mergers and interaction are supposed to lead to IR-luminous sources which harbour active nuclei that are fueled and uncovered as the dust settles (see, e.g., Dermer 1996, and references therein). In addition, the star formation rate is high in many IR-luminous galaxies,[9] resulting in a high density of cosmic rays, in particular of cosmic-ray electrons. The latter produce γ-ray photons in interactions with the interstellar medium through emission of

---

[9] These are then referred to as starburst galaxies.

bremsstrahlung and via the inverse-Compton process (see, e.g., Soltan and Juchniewicz 1999, and references therein).

A recent estimate of the EGB contribution of galaxies luminous at far-IR wavelengths was given by Soltan and Juchniewicz (1999), who estimated that these galaxies may account for a substantial fraction of the observed EGB intensity. Direct testing of the viability of this picture by observations at γ-ray energies will be difficult. Although the number density of far-IR-luminous galaxies is relatively high, individual members of this population are fainter than AGN by orders of magnitude.

Supernovae

The explosion of a star in a supernova (SN) is a truly cosmic event, as described in Chap. 10. SNe are copious sources of radioactivity, in particular of γ-ray line emission from iron production, and hence SNe could provide a significant fraction of the EGB around 1 MeV (Clayton and Silk 1969; Clayton et al. 1969). In addition, the observation of characteristic steps or edges (see below) in the low-energy EGB spectrum due to individual γ-ray lines could provide valuable information on the history of nucleosynthesis in the universe – provided that the γ-ray lines can emerge from the source without degradation. These basic considerations were put on solid grounds by observations of γ-ray lines from SN 1987A (e.g. Leising and Share 1990, and references therein).

Recently, improved calculations of the diffuse background arising from Type Ia and Type II SNe have indicated that SNe may account for most or all of the observed EGB intensity at MeV energies (see The et al. 1993; Watanabe et al. 1999). The SN contribution to the EGB is dominated by emission from Type Ia SNe. While the product of SN rate and γ-ray yield is comparable for Type Ia SNe and for core-collapse SNe, the massive envelopes of the latter inhibit γ-ray escape, making them much less luminous in γ-rays than SNe of Type Ia (see Chap. 10). The major source of high-energy radiation from Type Ia SNe is γ-ray line emission from the decay chain ^{56}Ni $\to$ ^{56}Co $\to$ ^{56}Fe (0.85 MeV, 1.24 MeV, 1.77 MeV, 2.03 MeV, 2.60 MeV and 3.25 MeV).[10] As these photons traverse the expanding SN debris, they may lose energy due to Compton scattering off electrons, photoelectric absorption, or pair production, resulting in a broad continuum spectrum underlying the γ-ray lines. The photon number spectrum of the diffuse background due to Type Ia SNe is characterized by a steep rise up to about 100 keV, a broad maximum at energies from 100–500 keV, and a steep decline towards higher energies with a cut-off near 3.5 MeV. On top of this continuum some structure is expected due to the integrated line emission from different redshifts: for each γ-ray line, an abrupt downward step above its rest energy, and a characteristic but smoother drop below its rest energy (The et al. 1993; Watanabe

---

[10] Long-lived radio-isotopes such as ^{26}Al, ^{44}Ti, and ^{60}Co are minor contributors.

et al. 1999). The shape of the steps and the corresponding drops give information on the space and time distribution of Fe nucleosynthesis at the current epoch ($z \approx 0$) and at high redshifts respectively. The SN contribution to the EGB is of particular interest because it offers an independent tool for studies of the star formation history (see, e.g., Watanabe et al. 1999), which are difficult at UV and optical wavelengths due to the substantial extinction in the gas clouds ambient in star forming regions.

SNe are promising contributors to the EGB around 1 MeV, although a definite assessment of the diffuse intensity due to SNe may have to wait until the detection of the predicted structures (features) in the low-energy EGB spectrum.

Primordial Black Holes

At the present cosmic epoch the only known process for the formation of black holes is self-gravitational collapse, which requires that the mass of the modern black hole must exceed a few solar masses. In the early universe, however, so-called primordial black holes of almost arbitrarily small masses may have formed. A variety of mechanisms for the formation of primordial black holes have been proposed, for example, the collapse of overdense regions in the presence of significant density fluctuations, cosmic phase transitions or the collapse of oscillating loops of cosmic strings (see Halzen et al. 1991; MacGibbon and Carr 1991, and references therein). As was first suggested by Hawking (1974), black holes thermally emit particles created by the strong gravitational fields with a temperature of about $1.2 \times 10^{26}\ M^{-1}$ K, where M is the black-hole mass in grams. For stellar-mass black holes this Hawking radiation is negligible, but for much smaller primordial black holes the effects of this quantum-mechanical decay through particle emission are significant. Primordial black holes with the critical mass $M_* \approx 10^{15}$ g evaporate at the current epoch, while primordial black holes with mass less than M_* have evaporated at earlier epochs.

The integrated emission of a uniform distribution of primordial black holes with $M \lesssim M_*$ over the lifetime of the universe results in an isotropic γ-ray background with a photon number spectrum $\sim E^{-3}$ for $E \gtrsim 100$ MeV and $\sim E^{-1}$ below (MacGibbon and Carr 1991). Thus a break at an energy of about 100 MeV is expected in the EGB spectrum if primordial black holes make a significant contribution.

Currently, a substantial contribution from evaporating primordial black holes to the EGB seems unlikely, since the spectral shape of the EGB is significantly different from what is predicted from primordial black holes.

14.4.2 Truly Diffuse Sources

Matter–Antimatter Annihilation

One of the fundamental questions of cosmology is the existence of antimatter on a cosmological scale. Unmistakable evidence for the presence of antimatter comes from the observation of matter–antimatter annihilation. The primary products of matter–antimatter annihilation are pions, which subsequently decay into muons which further decay into electrons, positrons, neutrinos, and γ-rays (from π^0 decay). Roughly one third of the energy released in a typical annihilation is deposited in γ-rays. The typical rest-frame γ-ray spectrum produced by π^0 decay from proton–antiproton annihilation has a broad maximum at $m_{\pi^0}c^2/2 \sim 68$ MeV and a minimum and maximum cutoff at ~ 5 MeV and ~ 1 GeV, respectively (Stecker 1971, see also Chap. 2.2). However, γ-rays with energies $\gtrsim 100$ MeV do not provide a unique signature of annihilation (see Steigman 1976, and references therein), since other processes such as π^0 production in cosmic-ray collisions, or the inverse-Compton scattering of low-energy photons and cosmic-ray electrons, or bremsstrahlung emitted by cosmic-ray electrons will also produce γ-rays in this energy range.

Two different approaches have been pursued to use existing γ-ray data to address the question of cosmic matter-antimatter symmetry. The first approach is to model a symmetric cosmology and explore whether its parameters can be adjusted so that it is consistent with observations.[11] Assuming baryon symmetry Stecker and collaborators calculated the cosmological γ-ray spectrum from annihilations occurring on the boundaries of colliding domains of matter and antimatter up to redshifts of ~ 100 (see, e.g., Stecker 1971; Stecker and Puget 1972; Stecker 1989). They obtained an annihilation spectrum with a flat peak near 1 MeV due to proton–antiproton annihilation at the highest redshifts. At higher energies, $\sim$5–200 MeV, the annihilation spectrum could be approximated by a power law ($\sim E^{-2.8}$). Above about 200 MeV the spectrum falls almost exponentially, finally cutting off at ~ 1 GeV. As can be seen in Fig. 14.2, this annihilation spectrum was in good agreement with observations of the EGB spectrum at that time, in particular with the observed excess intensity around 1 MeV, the so-called MeV-bump, which was claimed to be strong evidence for the validity of the matter–antimatter hypothesis. Recent observations of the spectrum of the EGB show no evidence for an excess intensity around 1 MeV (see Sect. 14.3).[12] In addition, at energies above 30 MeV the spectrum of the EGB is flatter than predicted from

---

[11] In addition to γ-ray observations, constraints imposed by the element abundances produced in primordial nucleosynthesis and from distortions of the spectrum and the isotropy of the CMB are of particular importance.

[12] The expected large-scale fluctuation patterns or anisotropies resulting from annihilation in symmetric universes and from AGN were investigated by Cline and Gao (1990); Gao et al. (1990a,b). They found that the fluctuations produced by annihilation on the boundaries of matter and antimatter domains, which resemble diffuse "ridges" in the 100 MeV range, are intrinsically different from EGB fluc-

matter–antimatter annihilations, with no indications for a break over the entire accessible energy range, which extends up to about 100 GeV. Thus it appears from the first approach that matter–antimatter annihilations do not contribute significantly to the EGB.

The second approach is to derive upper limits to the possible antimatter fraction from upper limits set to the annihilation rate by γ-ray observations. Following this line of thought it was shown conclusively that antimatter must remain separated from ordinary matter on scales at least as large as clusters ($\sim$10 Mpc) and even super-clusters ($\sim$100 Mpc) of galaxies, if present at all (see, e.g., Steigman 1976; Dudarewicz and Wolfendale 1994). In a more general approach Cohen et al. (1998) analyzed the dynamics of a patchwork universe consisting of distinct regions of matter and antimatter (as, e.g., suggested by Stecker 1985) together with the resulting contribution to the EGB and distortion of the CMB. The domains were characterized by their current size, which had to exceed $\sim$20 Mpc. To be conservative, only annihilations from the time of decoupling to the onset of structure formation ($1000 \gtrsim z \gtrsim 20$) were considered. They found that for empirically allowed Roberston–Walker universes the expected contribution to the EGB by far exceeds observational limits for all domain sizes $\lesssim 10^3$ Mpc, comparable in order of magnitude to the size of the visible universe, and arguably for even larger domains, thus excluding a matter-antimatter symmetric universe.[13]

To conclude, there is currently no evidence from either of the two approaches requiring the presence of antimatter in the universe.[14] Therefore a contribution from matter-antimatter annihilation to the EGB seems unlikely.

Exotic Elementary Particles

It has been suggested that the EGB may originate from the decay or the annihilation of some sort of exotic, relic elementary particle from the early universe not yet discovered. This speculative notion is motivated by astrophysical as well as particle physics considerations (see e.g. Silk and Srednicki 1984; Ellis et al. 1988; Rudaz and Stecker 1988; Kamionkowski 1995). Astrophysical observations strongly suggest the existence of dark matter, e.g. in galactic halos. This dark matter may be composed of non-baryonic, weakly interacting, massive particles (WIMPs) – which are natural consequences of many, in particular supersymmetric, extensions to the standard model of particle physics.

---

tuations produced by AGN, thus providing a probe for cosmological antimatter. These fluctuations are, however, unobservable with current instruments.

[13] The calculated distortions of the spectrum of the CMB were well below the observed limit and therefore provided no constraint at all.

[14] Neither of the two approaches, however, can exclude the existence of small and distant pockets of antimatter (Dolgov and Silk 1993).

Depending on the choice of particle theory, a large number of different exotic particles may exist, which may produce characteristic features in the EGB. Hence observations of the EGB can be used to constrain the parameter space of new particle theories (e.g. Daly 1988; Rudaz and Stecker 1988; Kamionkowski 1995) or, which would be more exciting, may lead to the discovery of new physics. None of the predicted spectral features, such as a cutoff at the rest energy of an annihilating particle (Rudaz and Stecker 1988), a series of γ-ray lines in the GeV range (Srednicki et al. 1986), or a broad distribution or a single line at GeV energies (Kamionkowski 1995), has been observed so far, and in many cases their observation is beyond the capabilities of current instruments. This is also true for the predicted angular features, namely anisotropies due to the annihilation or decay in a Galactic Halo or enhanced emission above the general EGB intensity from other galaxies or clusters of galaxies (e.g. Kamionkowski 1995).

Extended Galactic Halo

A major concern for measurements of diffuse extragalactic emission at any wavelength is the subtraction of the diffuse Galactic emission. In particular, in case an extended Galactic Halo exists, the diffuse Galactic emission may extend to the highest Galactic latitudes and may thus preclude the possibility of directly determining the EGB intensity in directions free of any Galactic "foreground" emission. Even worse, the isotropy of the emission from a very extended halo could mimic truly extragalactic diffuse emission to a level where the components can no longer be discerned.

In general, two diffuse components can only be separated on the basis of models for their energy spectra and/or the spatial distribution of their intensities on the sky. The Galactic diffuse emission including the extended Galactic Halo depends on the type and distribution of the cosmic-ray sources, on the propagation of cosmic rays, on the structure, distribution and composition of the interstellar medium, and on the interstellar radiation field (see Chap. 9). Not all of these quantities are well known, making the modelling of the Galactic diffuse emission a very complex task. Using astrophysical input data on the interstellar medium and radiation field it has been possible to construct models of the Galactic diffuse emission that agree well with the γ-ray observations at most, in particular high γ-ray, energies and most regions on the sky (see, e.g., Hunter et al. 1997a,b; Strong et al. 2000). Employing these models, the diffuse Galactic emission can be subtracted from the total observed intensity to derive the intensity of the EGB above about 100 MeV (e.g. Sreekumar et al. 1998).

It is clear that the subtraction of the diffuse Galactic emission, which is still far from being understood in detail, is a major concern for measurements of extragalactic diffuse emission. Current models for the diffuse Galactic emission appear to provide a reliable basis for the determination of the intensity of the EGB at high γ-ray energies; below about 100 MeV the diffuse Galactic emission is less well understood.

14.5 Conclusions and Prospects

The current status of extragalactic γ-ray astronomy is reminiscent of the status of extragalactic X-ray astronomy about 20 years ago. On the one hand the initially diffuse extragalactic emission begins to be resolved into discrete sources. On the other hand the properties of the extragalactic point sources observed thus far, almost all of which are AGN[15] (see Chap. 12 and Sect. 14.4.1), are only poorly understood owing to the small number observed. Currently available data firmly establish AGN as a viable source class contributing to the EGB, but considerable uncertainty remains concerning their "typical" emission characteristics. This uncertainty in the characteristics of individual source populations translates into corresponding uncertainties in the expected absolute intensity and spectral shape in any model of the integrated emission of AGN. It is therefore not surprising that different investigators arrived at quite different conclusions concerning the EGB contribution from these sources, as illustrated in Fig. 14.4. Consequently, the origin of the EGB is still far from being understood. Our fragmentary knowledge of the properties of identified sources indicates, however, that the XRB as well as the EGB are the sum of a number of different components, each dominating a specific portion of the high-energy extragalactic background spectrum. These components most likely arise from the superposed emission of various classes of point sources.

Before discussing the EGB, the following brief excursion into the X-ray regime is given as the results on the XRB strongly influence investigations of the EGB. Currently, the soft XRB is best understood. Based on the log N–log S function derived from deep ROSAT (Röntgen Satellit) observations in the so-called Lockman Hole region[16] 70–80% of the 0.5–2 keV XRB is resolved into discrete sources (Hasinger et al. 1998). Fluctuation analyses have shown that the remaining soft XRB intensity can be accounted for by source fluctuations at the faintest fluxes (see e.g. Hasinger et al. 1993). Identification of the optical counterparts of the X-ray sources in the Lockman Hole survey revealed that they are mostly QSOs and Seyfert galaxies with broad emission lines (Schmidt et al. 1998). A large fraction of the faint X-ray sources are optically resolved low-luminosity AGN (mostly Seyfert galaxies), some of which show clear evidence for gas and dust obscuration in the soft X-ray and optical bands (Hasinger et al. 1998). At harder X-ray energies, in the 2–10 keV band, medium-deep surveys such as that carried out with ASCA (Advanced

---

[15] In addition, γ-ray line emission has been detected from two extragalactic supernovae: SN 1987A and SN 1991T (see Chap. 10).

[16] The direction of the absolute minimum in the Galactic neutral-hydrogen column density.

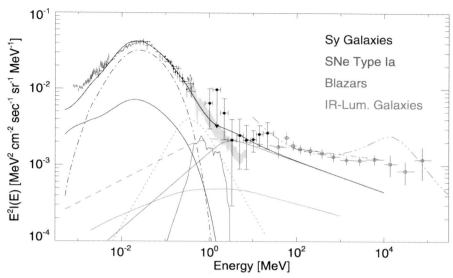

Fig. 14.4. Possible contributions of various classes of point sources to the XRB and EGB compared to recent experimental results (the data points are those from Fig. 14.3, extended to lower, X-ray energies with the data of Geandreau et al. (1995) and Gruber (1992); the historic matter–antimatter annihilation model of Stecker and collaborators is depicted in Fig. 14.2). *Blue lines* illustrate the expected intensity arising from Seyfert Type 1 (*solid*) and Seyfert Type 2 (*dash-dotted*) galaxies according to Zdziarski (1996). The *green line* represents the integrated emission from Type Ia SNe as calculated by Watanabe et al. (1999). The *cyan line* gives an estimate of the contribution of IR-luminous galaxies (Soltan and Juchniewicz 1999). *Red lines* represent various estimates of the emission of blazars: a possible contribution from MeV blazars is given by the *dotted line* (Comastri et al. 1996); *dashed* and *solid lines* represent the expected intensity due to FSRQs according to Zdziarski (1996) and Comastri et al. (1996), respectively; the *dash-triple-dotted line* depicts a model for the emission of blazars by Salamon and Stecker (1998). Finally, the *black line* gives the combined model for the XRB and EGB by Comastri et al. (1996), which includes the emission from Seyfert galaxies, MeV blazars, and FSRQs

Satellite for Cosmology and Astrophysics) have already resolved about 30% of the XRB (Ueda et al. 1998). Based on different assumptions on the generic spectral shapes and the luminosity function of specific source classes, various models have been proposed for the XRB. Up to about 100 keV, the spectrum of the XRB can be described well by the integrated emission of radio-quiet AGN (Seyfert galaxies and QSOs), as illustrated in Fig. 14.4. In particular, the models attribute the bulk of the emission to heavily absorbed Seyfert Type 2 galaxies, in accordance with current AGN unification schemes (e.g. Madau et al. 1993; Zdziarski and Zycki 1993; Comastri et al. 1995; Zdziarski et al. 1995; Zdziarski 1996). As the data on source counts and redshift as well as absorption distributions for individual source classes improve, these models become more and more constrained (Comastri 1999).

The least explored portion of the extragalactic high-energy background is that of the EGB around 1 MeV. This energy range is particularly challenging not only for experimentalists, but also for theorists. In the 100 keV to 10 MeV region a transition seems to occur in the physical processes governing the emission of various source populations. Below this region, thermal processes seem to dominate, as in Seyfert galaxies; above, nonthermal processes seem to dominate, as in the jets of blazars (see Chap. 12 and Sect. 14.4.1). In addition, the low-energy EGB covers the region of nuclear line emission from, for example, SNe. A fraction of the EGB intensity at MeV energies is expected to be accounted for by the emission "tails" of source populations dominating at lower and higher energies. The emission from classical Seyfert galaxies cuts off exponentially above about 100 keV, but the emission from so-called radio-loud Seyfert galaxies or misaligned blazars – such as Cen A – may typically exhibit high-energy tails extending to MeV and even GeV energies. Similarly, the spectra of blazars, which break at energies between about 1 MeV and 100 MeV, exhibit low-energy tails extending down to MeV energies and below. In addition, significant contributions to the low-energy EGB intensity may be due to source classes which emit the bulk of their high-energy radiation around 1 MeV, such as MeV blazars and Type Ia SNe. So far, few representatives of these candidate source populations have been observed around 1 MeV, and thus the absolute intensity and spectral shape of their integrated emission at these energies is still uncertain.

At higher energies, above about 100 MeV, the only extragalactic sources observed thus far are two types of radio-loud blazars, namely FSRQs and BL Lacs, whose γ-ray emission properties are similar but not necessarily identical (see Chap. 12 and Sect. 14.4.1). As indicated in Fig. 14.4, these AGN may account for the bulk of the EGB above about 100 MeV (see, e.g., Setti and Woltjer 1994; Comastri et al. 1996; Salamon and Stecker 1998). Despite the detection of more than 80 AGN their high-energy emission properties are only poorly constrained, and therefore the magnitude and spectral shape of the contributions from individual AGN classes are still under debate (see e.g. Chiang and Mukherjee 1998; Mücke and Pohl 2000; Salamon and Stecker 1998). Other source populations such as IR-luminous galaxies or normal galaxies may also contribute to the EGB.

The next two γ-ray missions scheduled or planned to be launched in the near and intermediate future should both be able to make important contributions to the study of the EGB: INTEGRAL (International Gamma-Ray Astrophysics Laboratory), covering the hard X-ray and soft γ-ray bands, and GLAST (Gamma-ray Large-Area Space Telescope), sensitive to high-energy γ-rays above about 50 MeV. The unprecedented angular resolution of IBIS/INTEGRAL, combined with a continuum sensitivity slightly exceeding that of OSSE (Oriented Scintillation-Spectrometer Experiment) and COMP-

TEL, makes it well suited for searching for extragalactic point sources with continuum emission such as AGN, and for studying their emission properties in the hard X-ray/soft γ-ray regime, which is least understood at present. These observations will supplement deep surveys at medium X-ray energies (2–20 keV) performed by current missions such as Newton XMM (X-Ray Multi-Mirror Mission) and Chandra. Together, these data are expected to greatly improve our models for the medium and hard XRB and to extend these models possibly up to a few MeV. In addition, SPI/INTEGRAL may determine the spectrum of the low-energy EGB with unprecedented energy resolution, allowing us to search for spectral features such as edges due to the redshifted line emission from cosmological Type Ia SNe. Also, SPI/INTEGRAL may detect γ-ray line emission from individual extragalactic SNe of Type Ia. At high γ-ray energies above about 50 MeV, GLAST will be an order of magnitude more sensitive than EGRET, in addition to providing improved angular resolution. With GLAST it will therefore be possible to study a much larger number of extragalactic sources in much greater detail, providing us with the database needed to investigate "typical" properties and the luminosity function of various source classes, and ultimately to better assess their contributions to the high-energy EGB. Combined with COMPTEL and INTEGRAL observations at MeV energies, it should also be possible to derive improved estimates of the blazar contribution to the low-energy EGB. Also, an improved determination of the Galactic diffuse emission with GLAST will allow us to determine the EGB spectrum above about 100 MeV with greater accuracy. On account of the increased effective area, it will be feasible to probe the EGB at energies above 100 GeV, where γ–γ absorption may become significant. Although direct measurements of the EGB in the TeV range may not be feasible in the near future, upcoming ground-based TeV telescopes, such as MAGIC (Major Atmospheric Gamma Imaging Cherenkov Telescope) or VERITAS (Very Energetic Radiation Imaging Telescope Array System), will increase our knowledge on the emission properties of individual sources at the highest observable γ-ray energies.

To conclude, recent and current γ-ray experiments have greatly improved our knowledge of the high-energy EBR. The overall shape of the EGB spectrum has been determined over a much larger energy range and appears to be considerably simplified as compared to the situation in the early 1980s. No evidence for the existence of the putative MeV-bump has been found. Instead, the results are consistent with a spectral transition from a softer to a harder component, which occurs around a few MeV. It appears likely that the EGB arises from the superposed emission of various classes of point sources. Significant progress in understanding the origin of the medium and hard XRB and the high-energy EGB is to be expected from future missions. These advances will also improve our knowledge of the origin of the low-energy EGB, which spans the transition region between thermal and nonthermal astrophysical processes, by improving extrapolations from lower and higher energies into the MeV regime. Knowledge of the emission characteristics of individual source classes at MeV energies, particularly γ-ray line emission from SNe, will also improve, although progress will be slower in this experimentally most difficult and challenging γ-ray energy band.

References

Bednarek, W., Protheroe, R.J., 1999, Mon. Not. R. Astron. Soc. **310**, 577
Bignami, G.F. et al., 1981, Astron. Astrophys. **93**, 71
Bloemen, H. et al., 1995, Astron. Astrophys. Lett. **293**, 1
Bloemen, H. et al., 1999, Astro. Lett. and Communications **39**, 681
Bowyer, S., Leinert, Ch., eds., 1990, *The Galactic and Extragalactic Background Radiation*, Cambridge University Press, Cambridge
Calzetti, D., Livio, M., Madau, P., eds., 1995, *Extragalactic Background Radiation*, Cambridge University Press, Cambridge
Chiang, J., Mukherjee, R., 1998, Astrophys. J. **496**, 752
Chiang, J. et al., 1995, Astrophys. J. **452**, 156
Clayton, D.D., Silk, J., 1969, Astrophys. J. Lett. **158**, 43
Clayton, D.D. et al., 1969, Astrophys. J. **155**, 75
Cline, B.C., Gao, Y.-T., 1990, Astrophys. J. **348**, 33
Cohen, A.G. et al., 1998, Astrophys. J. **495**, 539
Comastri, A., 1999, Astro. Lett. and Communications **39**, 181
Comastri, A. et al., 1995, Astron. Astrophys. **296**, 1
Comastri, A. et al., 1996, Astron. Astrophys. Suppl. Ser. **120**, C627
Daly, R.A., 1988, Astrophys. J. Lett. **324**, 47
Daniel, R.R., Lavakare, P.J., 1975, Proc. of the 14th Internat. Conf. on Cosmic Rays, Vol. 1, 23
Daniel, R.R. et al., 1972, Astrophys. Space. Sci. **18**, 462
Dermer, C.D., 1996, Proc. of the 2nd INTEGRAL workshop, 405
Dolgov, A.D., Silk, J., 1993, Phys. Rev. D **47**, 4244
Dudarewicz, A., Wolfendale, A.W., 1994, Mon. Not. R. Astron. Soc. **268**, 609
Ellis, J. et al., 1988, Phys. Lett. B **214/3**, 403
Fixsen, D.J. et al., 1996, Astrophys. J. **473**, 576
Fossati, G. et al., 1998, Mon. Not. R. Astron. Soc. **299**, 433
Fukada, Y. et al., 1975, Nature **254**, 398
Gao, Y.-T. et al., 1990a, Astrophys. J. Lett. **357**, 1
Gao, Y.-T. et al., 1990b, Astrophys. J. Lett. **361**, 37
Geandreau, K.C., et al., 1995, Publ. Astron. Soc. Japan Lett. **47**, 5
Gruber, D.E., 1992, in: *The X-Ray Background*, Cambridge University Press, Cambridge, p. 44
Halzen, F. et al., 1991, Nature **353**, 807
Hartman, R.C. et al., 1997, Proc. of the Fourth Compton Symposium (AIP 410), 307
Hasinger, G. et al., 1993, Astron. Astrophys. **275**, 1
Hasinger, G. et al., 1998, Astron. Astrophys. **329**, 482
Hawking, S.W., 1974, Nature **248**, 30
Hunter, S.D. et al., 1997a, Proc. of the Fourth Compton Symposium (AIP 410), 192

Hunter, S.D. et al., 1997b, Astrophys. J. **481**, 205
Johnson, W.N. et al., 1997, Proc. of the Fourth Compton Symposium (AIP 410), 283
Kamionkowski, M., 1995, in *The Gamma Ray Sky with CGRO and SIGMA*, Kluwer Academic Publishers, Dordrecht, p. 113
Kappadath, S.C., 1998, Ph.D. thesis, University of New Hampshire
Kazanas, D., Pearlman, E., 1997, Astrophys. J. **476**, 7
Kinzer, R.L. et al., 1997, Astrophys. J. **475**, 361
Kraushaar, W.L., et al., 1965, Astrophys. J. **141**, 845
Kraushaar, W.L., et al., 1972, Astrophys. J. **177**, 341
Krennrich, F. et al., 1997, Astrophys. J. **481**, 758
Leising, M.D., Share, G.H., 1990, Astrophys. J. **357**, 638
Lichti, G.G. et al., 1978, Astrophys. Sp. Sci. **56**, 403
Lockwood, J.A. et al., 1981, Astrophys. J. **248**, 1194
Longair, M.S., 1994 in *The Deep Universe*, Saas Fee Advanced Course 23, Lecture Notes 1993, Springer-Verlag Berlin Heidelberg
MacGibbon, J.H., Carr, B.J., 1991, Astrophys. J. **371**, 447
Madau, P. et al., 1993, Astrophys. J. Lett. **410**, 7
Mather, J.C. et al., 1994, Astrophys. J. **420**, 439
Mattox, J.R. et al., 1997, Astrophys. J. **481**, 95
Mazets, E.P. et al., 1975, Astrophys. Sp. Sci. **33**, 347
McNaron-Brown, K. et al., 1995, Astrophys. J. **451**, 575
Metzger, A.E. et al., 1964, Nature **204**, 766
Mücke, A., Pohl, M., 2000, Mon. Not. R. Astron. Soc. **312**, 177
Mücke, A. et al., 1996, Astron. Astrophys. Suppl. Ser. **120**, C541
Prantzos, N., Casse, M., 1994, Astrophys. J. Suppl. Ser. **95**, 575
Robson, I., 1996, *Active Galactic Nuclei*, John Wiley & Sons
Rudaz, S., Stecker, F.W., 1988, Astrophys. J. **325**, 16
Salamon, M.H., Stecker, F.W., 1994, Astrophys. J. Lett. **430**, 21
Salamon, M.H., Stecker. F.W., 1998, Astrophys. J. **493**, 547
Sandage, A.R., Kron, R.G., Longair, M.S., 1994, *The Deep Universe*, Springer-Verlag Berlin Heidelberg
Schmidt, M. et al., 1998, Astron. Astrophys. **329**, 495
Schönfelder, V., Lichti, G., 1974, Astrophys. J. Lett. **191**, 1
Schönfelder, V. et al., 1977, Astrophys. J. **217**, 306
Schönfelder, V. et al., 1980, Astrophys. J. **240**, 350
Setti, G., Woltjer, L., 1994, Astrophys. J. Suppl. Ser. **92**, 629
Silk, J., Srednicki, M., 1984, Phys. Rev. Lett. **3/6**, 624
Soltan, A.M., Juchniewicz, J., 1999, Astro. Lett. and Communications **39**, 665
Srednicki, M. et al., 1986, Phys. Rev. Lett. **56/3**, 263
Sreekumar, P. et al., 1997, Proc. of the Fourth Compton Symposium (AIP 410), 344
Sreekumar, P. et al., 1998, Astrophys. J. **494**, 523
Stecker, F.W., 1971, Cosmic Gamma Rays (NASA SP-249)
Stecker, F.W., 1985, Nucl. Phys. B **252**, 25
Stecker, F.W., 1989, Proc. of Gamma Ray Observatory Science Workshop, 4-73
Stecker, F.W., Puget, J.L., 1972, Astrophys. J. **178**, 57
Stecker, F.W., Salamon, M.H., 1996, Astrophys. J. **464**, 600
Stecker, F.W. et al., 1992, Astrophys. J. Lett. **390**, 49

Stecker, F.W. et al., 1993, Astrophys. J. Lett. **410**, 71
Steigman, G., 1976, Ann. Rev. Astron. Astrophys. **14**, 339
Steinle, H. et al., 1998, Astron. Astrophys. **330**, 97
Strong, A.W. et al., 1976, Mon. Not. R. Astron. Soc. **175**, 23P
Strong, A.W. et al., 2000, Astrophys. J. **537**, 763
The, L.-S. et al., 1993, Astrophys. J. **403**, 32
Thompson, D.J., Fichtel, C.E., 1982, Astron. Astrophys. **109**, 352
Trombka, J.I. et al., 1977, Astrophys. J. **212**, 925
Ueda, Y. et al., 1998, Nature **391**, 866
Vedrenne, G. et al., 1971, Astron. Astrophys. **15**, 50
Watanabe, K., 1997, Ph.D. thesis, Clemson University
Watanabe, K. et al., 1997, Proc. of the Fourth Compton Symposium (AIP 410), 1223
Watanabe, K. et al., 1999, Astrophys. J. **516**, 285
Weekes, T.C. et al., 1997, Proc. of the Fourth Compton Symposium (AIP 410), 361
Weidenspointner, G., 1999, Dissertation, Technical University Munich
Weidenspointner, G. et al., 2000a, Proc. of the Fifth Compton Symposium (AIP 510), 467
Weidenspointner, G. et al., 2000b, Proc. of the Fifth Compton Symposium (AIP 510), 581
White, R.S. et al., 1977, Astrophys. J. **218**, 920
Willis, T.D., 1996, Ph.D. thesis, Stanford University
Zdziarski, A.A., 1996, Mon. Not. R. Astron. Soc. Lett. **281**, 9
Zdziarski, A.A., Zycki, P.T., 1993, Astrophys. J. Lett. **414**, 81
Zdziarski, A.A., et al., 1995, Astrophys. J. Lett. **483**, 63

15 Gamma-Ray Bursts

Martin Varendorff

15.1 Introduction

15.1.1 The History of Gamma-Ray Burst Measurements

First Measurements

The puzzle of γ-ray bursts began in the year 1967 with the detection of inexplicable increases in the count rate of γ-ray detectors on board the Vela series of satellites meant to monitor violations of the nuclear explosion test ban treaty (Strong et al. 1973). The discovery of the then so-called γ-ray bursts was published in the year 1973 by Klebesadel et al. (1973), after several years of confidence building through the consistent accumulation of data [a summary of historic remarks can be found in Ramanamurthy and Wolfendale (1984)].

The Vela satellites were equally spaced in a circular orbit (1.2×10^5 km) around the earth and measured γ-rays in the energy range from 0.2 to 1.5 MeV with a nearly isotropic sensitivity. From the arrival times of the γ-ray bursts, which were recorded with an accuracy of about 0.05 s, a solar and terrestrial origin could be excluded (see explanation of the triangulation method in Fig. 15.3). Inverse square law considerations produced a lower limit to the distance of the burst sources of several million kilometers. Time profiles of one of the first bursts (GRB 700822) are shown in Fig. 15.1. Measurements of the energy spectra in the hard X-ray and low γ-ray range showed exponential spectra with characteristic energies clustering near 150 keV (Cline et al. 1973).

The discovery of γ-ray burst events with their high energy output and short time scales raised the general interest in the difficult-to-observe, lower γ-ray energy range. As a consequence several space instruments in the planning phase or under construction were modified to add burst-detection capabilities.

In the year 1976 the first dedicated γ-ray burst instrument on board the Helios 2 satellite was launched. Surprisingly, the burst location information it obtained (in collaboration with the Vela instruments) was inconsistent with the locations of all obvious candidate sources like pulsars, supernova remnants, the Galactic Center and other sources.

At the end of the 1970s and in the beginning of the 1980s several satellites equipped with instruments capable of measuring γ-ray bursts were launched.

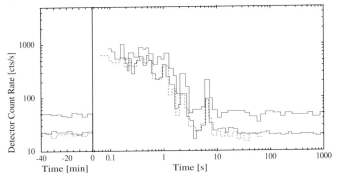

Fig. 15.1. Time profile of GRB 700822 measured by three Vela spacecraft (Vela 5A, *upper line*; Vela 6A, *dashed line*; Vela 6B, *lower line*). Left is the background count rate immediately preceding the burst event (time in minutes). Several structures in the time profile of the burst are about the same in all three measurements (time in seconds). The small but significant difference in the arrival times of the signal is not consistent with an origin from earth. (Data from Klebesadel et al. 1973)

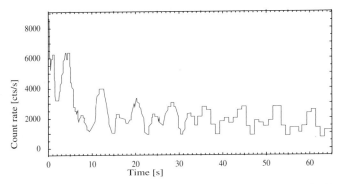

Fig. 15.2. Time profile of GRB 790305 with a resolution of 1/4 s in the energy range 50–150 keV measured by Venera 12. After the initial first pulse, which reaches far above the plotted range, the signal from the burst shows a clear periodicity of 8.1 s. (Data from Mazets et al. 1979)

The satellites Helios 2, Pioneer Venus Orbiter (PVO), International Sun-Earth Explorer (ISEE 3), Venera 11 and 12, and the Earth orbiter Prognoz 7 contributed to the newly formed Interplanetary Network [IPN, 1978, see Laros81]. Locations of several strong γ-ray bursts with errors of less than 1 arc min were achieved each year of observation by means of triangulation. This improvement in location accuracy by more than two orders of magnitude became possible with two noncollinear extremely long baselines of up to 500 light seconds (Helios 2 and PVO).

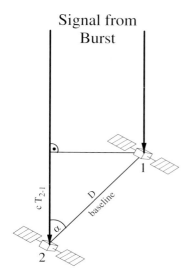

Fig. 15.3. Triangulation method. The early γ-ray detectors had only very limited imaging capabilities (if any) and therefore could not accurately determine the incoming direction of a detected γ-ray signal (see Chap. 3). However, from the measured difference in the arrival times T_{2-1} of a signal in 2 detectors at a distance D (baseline), one can derive the angle α between the direction from the first satellite to the source and to the second satellite: $\cos\alpha = \frac{T_{2-1} \cdot c}{D}$. This reduces the possible source locations to an annulus in the sky with radius α. The width of the annulus $\Delta\alpha$ is proportional to the accuracy of the time measurement ΔT relative to the baseline and indirectly proportional to $\sin\alpha$

The Gamma-Ray Burst of March 5, 1979 (GRB 790305[1])

A very important result of the IPN collaboration was the location of a very strong and very special γ-ray burst event on March 5, 1979 (Evans et al. 1980; Mazets et al. 1979, 1980). This was the first and for 18 yr the only burst which showed a periodicity (8.1 ± 0.1 s period) in the time profile (see Fig. 15.2). Using the data from the IPN and several other satellites, it was possible to derive a fairly accurate location (2 arc min²) coincident with the supernova remnant (SNR) N49 in the Large Magellanic Cloud (LMC).

Subsequently more bursts were detected from this direction which showed similiar temporal and spectral behaviour, but no periodicity (Norris et al. 1991). The very likely correlation of GRB 790503 with N49 led rather to difficulties in understanding the origin of γ-ray bursts. This was due to the fact that this source belongs to a special group, the so-called Soft Gamma-Ray Repeaters (SGRs) (Sect. 15.1.2).

The Solar-Maximum Mission satellite, built for the study of properties of solar flares, extended with its large NaI and CsI crystals the spectral range of burst investigations up to 100 MeV. The high-energy part of the spectrum of many bursts showed no deviation from a power law (no lines, no break).

In the data of the Konus experiment on the Wind spacecraft, line features were found in 7% of its detected bursts, which were interpreted as cyclotron lines from moving particles in strong magnetic fields. Similiar line features were found in the spectra of two bursts (GRB 870303 and GRB 880205),

---

[1] GRB 790305 = γ-ray burst, recorded on March 5, 1979. The term γ-ray burst is used for all transient sources in the γ-ray energy range, irrespective of the nature of the burst source.

detected by the Ginga satellite, raising the credibility of the Konus detections. However, the reality of these line features is still under debate as no other instruments have detected any lines so far.

The general accepted theory at this time, inspired partly by the identification of the SGRs and the possible detection of cyclotron lines, was that γ-ray bursts originate from neutron stars in our Galaxy. The resulting concentration of the bursts along the Galactic Plane was not in contradiction with the distribution of the few bursts localised up to then.

In the last decade several instruments had been launched to detect γ-ray bursts. The Russian–French experiment Phebus operated successfully aboard the GRANAT spacecraft. The WATCH experiment (wide angle telescope for cosmic hard x-rays) installed on two spacecrafts (GRANAT and EURECA) could rapidly locate γ-ray bursts to an accuracy of $1°$ using a rotating collimation modulator. In addition the IPN received a new outpost in the solar system with a burst detector aboard the Ulysses spacecraft providing triangulation arcs with a width of arc minutes.

A new era of burst measurements began with the launch of the Compton Gamma-Ray Observatory (CGRO) in April 1991 (see Chap. 3). All four instruments on board this observatory detected γ-ray bursts. The Burst and Transient Source Experiment (BATSE), dedicated mainly to burst investigations, measured the location of more than two thousand bursts with errors of only $1°$ for strong bursts (by the end of 1998).

The first image of a single γ-ray burst with an accuracy of about $1°$ was obtained from COMPTEL measurements (Varendorff et al. 1992). The image is shown in Fig. 15.4 together with a triangulation arc using Ulysses and COMPTEL data and six high-energy γ-ray events detected by EGRET (Energetic Gamma-Ray-Experiment Telescope). Gamma-ray photons up to the GeV range have been registered by EGRET in several other bursts.

In spite of expectations, the derived spatial distribution of the bursts detected by BATSE is isotropic, but limited in distance, invalidating the contemporary models of the Galactic origin of γ-ray bursts. This brought up again the basic question: How far away are γ-ray burst sources? A "great debate" on this topic was held in April 1995 between Lamb and Paczynski in commemoration of the great debate between Shapley and Curtis in April 1920 on the Scale of the Universe. Both debates led to no conclusion, but brought their topics to public attention. Analogous to the first problem, which was solved by new optical measurements a couple of years later, the latter question was also answered only two years later by optical observations: in 1997 the optical afterglow of a few bursts, localized by the X-ray satellite Beppo-SAX (see Sect. 15.3.2), was seen. The redshift of line features in some optical spectra of afterglows clearly positioned γ-ray burst sources at extragalactic distances.

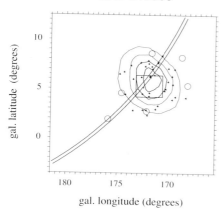

Fig. 15.4. Image of GRB 910503 as derived from COMPTEL data (contour lines). The triangulation arcs ($\pm 1\sigma$) were derived from COMPTEL and Ulysses timing measurements. The *rectangle* indicates the error box calculated from the six high-energy γ-quanta recorded by EGRET (*circles*). The *dots* mark the position of X-ray sources from the ROSAT (Röntgen Satellit) sky survey, emphasizing the ambiguity in the search for counterparts

15.1.2 Soft Gamma-Ray Repeater

In spite of the detection of several SGRs (SGR 0526-66, SGR 1900+14, SGR 1806-20) up to 1979, it took 8 more years before SGRs were recognized to be of a clearly different origin. SGRs are distinct from the classical γ-ray bursts in their rather uniform characteristics: soft spectra, with no evolution, of the form of an optically thin thermal bremsstrahlung spectrum ($T \approx 30\text{--}40$ keV), short durations (0.1 s), short rise times (down to 5 ms) and simple light curves (triangular or flat-topped). For more details see Kouveliotou et al. (1993).

The year 1998 brought a boost in the research of SGRs. The scarcely recurring burster SGR 1900+14 emitted over 50 bursts in the last week of May 1998, with one event showing a record-breaking high fluence comparable to GRB 790305. Additionally in August of that year a pulsation (5.16 s) in the steady X-ray radiation from this object was discovered. Later on SGR 1806-20 was found to show a pulsation (7.5 s) too. The pulse period of both SGRs is gradually slowing down at a rate far beyond typical slow-down rates of pulsars (see Chap. 6). To achieve such large braking, a magnetic field of about 5×10^{10} T is required. The magnetar model (Duncan and Thomson 1992) of SGRs predicted such a large magnetic field and could explain the other observed phenomena of SGRs as well. A magnetar is a neutron star which is created in a supernova explosion with a rotation period fast enough ($\ll 20$ ms) to allow the creation of a very strong magnetic field by the dynamo effect. Ordinary pulsars fail to build such a strong magnetic field in the short available time interval (~ 20 s) after birth until the neutron star has cooled down, and the convection, and with it the dynamo action, has ceased. The large spin-down rate slows down the rotation to a period of a few seconds in several thousand years only. This explains why observed magnetars usually show no lighthouse beams like ordinary pulsars, which require a faster rotation. The energy dissipated by the spin down heats the interior of the SGR and causes the steady X-ray emission, which can show some pulsation due to the rotation of the neutron star.

The enormous magnetic field puts a strong stress on the crust of the neutron star, and once in a while it cracks. The resulting seismic waves create magnetic waves which energize particle clouds above the surface of the neutron star. The frequency distribution of the amplitude of bursts, as a function of the energy of the burst event, follows the Gutenberg–Richter law for earthquakes, giving evidence for this model. The exorbitantly strong events (such as GRB 790305) are produced by a different mechanism analogous to solar flares. The magnetic field occasionally becomes unstable and abruptly changes to a state of lower energy, rearranging the magnetic field via reconnection, which releases magnetic energy in an enormous γ-ray burst.

All of the SGRs lie in or near young (<10 000 yr) SNRs; three are in our Galaxy and one is in the LMC (Kouveliotou et al. 1994; Kulkarni et al. 1994). The displacement of SGR 1900+14 from the possibly associated SNR requires a recoil velocity of 1500–2000 km s^{-1}, which the object could have obtained during a asymmetric supernova explosion.

15.1.3 Bursts from Earth

A by-product of the search for cosmic γ-ray bursts was the discovery of bursts from the earth – terrestrial gamma-flashes (TGFs) – in BATSE data by Fishman et al. (1994). TGFs last typically only a few ms with spectral changes during the event. A size scale for the emitting region of about 15 km could be derived from the minimum time scales in the range of about 50 µs. The time profiles show no significant time asymmetry. TGFs are weak BATSE events, and therefore only coarse spectral data are available with power-law indices from -0.6 to -1.5.

TGFs are hypothesized to be related to thunderstorms, but might also be related to recently discovered upper atmospheric lightning phenomena: red sprites, blue jets, and trans-ionospheric pulse pairs. Red sprites are favoured because both phenomena have event durations of ms and are typically composed of a few dominant episodes of emission (Sentman et al. 1995). Additionally red sprites rise to higher atmospheric regions (90 km), where the escape of the γ-rays from the atmosphere is easier.[2]

15.2 Properties of Gamma-Ray Bursts

The following section, which focusses on the properties of γ-ray bursts, is mostly based on measurements obtained by the CGRO.

---

[2] At 90 km the rest mass of the atmosphere is about 5 mg cm^{-3}, resulting in an opacity of about 1/1000 for γ-rays at an energy of 1 MeV.

15.2.1 Time Characteristics of Gamma-Ray Bursts

One of the greatest problems γ-ray bursts pose to theorists is the multitude of their time profiles. Bursts show time structures in the range from sub-ms spikes up to 1000 s long events. The burst morphology can be subdivided into four groups (Fishman and Meegan 1995):

- single-pulsed or spiked events;
- smooth events, which are either singly pulsed or multi-peaked, but with well-defined peaks;
- bursts with distinct, well-separated episodes of emission; and
- very erratic, chaotic, and spiky bursts (see Fig. 15.5 for examples).

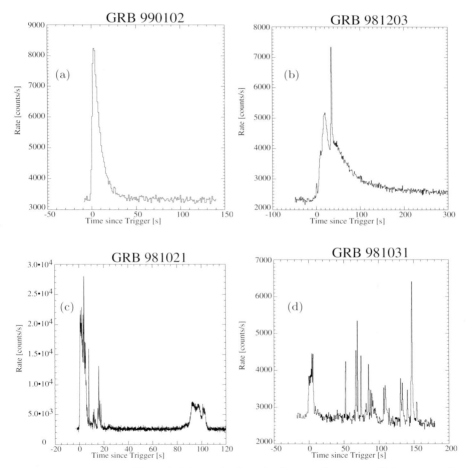

Fig. 15.5. Time profiles of (**a**) a single-pulsed smooth event, (**b**) a smooth event with well-defined peaks, (**c**) a burst with distinct episodes of emission, and (**d**) a chaotic and spiky burst

A comparison of the structure of time profiles (complexity, smoothness) with scale-independent methods (such as wavelet analysis) showed that the pulses in short bursts seem to be identical to the pulses in long bursts compressed by a factor of 20. However, long bursts tend to have many pulses, whereas short bursts only have a few major pulse structures.

It is not easy to derive the duration of a γ-ray burst, especially for weak bursts with a high background. Precursors, trailing pulses or low continuous emission might be hidden by background noise, thus biasing the measurement of the burst duration. Therefore instruments with different sensitivities give different burst durations for the same event. Fairly independent of the intensity of the burst are duration definitions based on the integral over time. The BATSE team used T_{90}, the time enclosing 90% of the burst fluence (from 5% to 95% above background) as a compromise between the bias from the intensity (instrument sensitivity) and missing separated leading and trailing emissions. The distribution of the T_{90} burst durations is shown in Fig. 15.6. A clear separation in short bursts with a peak near 0.33 ± 0.21 s and long bursts with a peak around 26 ± 1.7 s can be seen. The two broad peaks overlap with a minimum around 2 s. The two duration classes show no difference in their spatial distribution on the sky.

A detailed multivariate analysis revealed that bursts can be classified into three groups (instead of two) with the following duration/fluence/spectrum bulk properties: Class I with long/bright/intermediate bursts, Class II with short/faint/hard bursts and Class III with intermediate/intermediate/soft bursts (Mukherjee et al. 1998).

In general short bursts are harder than long bursts, and bursts soften within the pulses and from beginning to end. The softer spectra of long

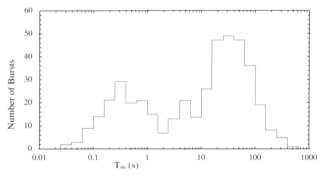

Fig. 15.6. Distribution of the T_{90} duration of γ-ray bursts. The distribution is separated into at least two subsets which overlap. Both subsets are isotropic and spatially inhomogeneous. The longer bursts are in general softer than the shorter bursts, which might be related to different source classes, or the fact the longer bursts are farther away

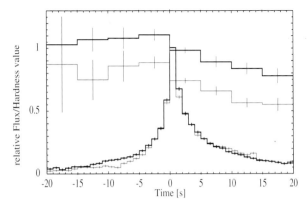

Fig. 15.7. Average of the temporal evolution of the intensity (*lower two curves*) and hardness (*upper two curves*) of 209 bright (*thick lines*) and 204 dim bursts (*thin lines*) of the third BATSE catalog are compared. Bursts are hardest just before the maximum flux is reached. Dim (far) bursts are generally softer than bright (near) bursts. (Data from Mitrovanov 1995)

bursts may be caused by radiation transfer from previous pulses. But there are other bursts which behave differently.

The average temporal evolution of the intensity and the hardness[3] of bursts is asymmetric (see Fig. 15.7) with an average decay time (4.2 s) nearly twice as long as the rise time (2.3 s). The hardness of γ-ray bursts generally rises within a burst event with the amplitude of the flux and drops with time. The main peak is usually more symmetric than the overall profile.

Cosmological Effects on Time Profiles

If γ-ray burst sources are at cosmological distances, then the time profile of "far distant" bursts is stretched relative to "near" bursts due to relativistic effects (expansion of the universe).

Several methods have used to find such a time dilation. One way is to correlate the burst duration T_{90} with the burst fluence or peak flux as a measure of the distance (see Fig. 15.8). Another way is to investigate time scales within bursts, such as the average time between peaks or the average width of peaks, and correlate them with a distance indicator. A third method is to compare the distribution of the peak energy[4] E_p of weak bursts with that of strong bursts. The different methods lead to dilation factors of the order of two, supporting the idea of a cosmological origin. However, the relativistic time dilation might be partly compensated by an intrinsic narrowing of the light

---

[3] The hardness is defined as the ratio of a high-energy interval to a low-energy interval.

[4] E_p is the maximum of the νF_ν energy spectrum $[= E^2 \times \text{particles}/(\text{area} \times \text{time})]$.

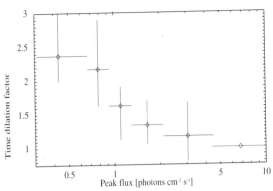

Fig. 15.8. The bursts of the third BATSE catalog were divided into 6 brightness groups according to their peak flux. For each burst the median interval between pulses or peaks was derived. For each group the stretching factor for the time interval distribution (due to a possible time dilation) was determined relative to the time interval distribution of the group containing the brightest bursts. The significance of the time dilation is about 4 σ. (Data from Norris et al. 1995)

curve: the observed photons can be redshifted from an energetically higher part of the spectrum where time scales were found to be shorter. However, the methods of data reduction and the results obtained are not universally accepted because of systematic selection effects in the data analysis. Detailed studies of these effects using Monte Carlo methods and sophisticated analysis tools are necessary; such studies are just beginning.

15.2.2 Spectral Characteristics of Gamma-Ray Bursts

Much knowledge in astronomy is gained by the analysis of energy spectra. The global spectra, as well as structures like emission and absorption lines or edges, very often reveal the production process of the radiation. In this respect the spectra of γ-ray bursts were disappointing. No characteristic energies are evident in the spectra. In general the measured spectra are simple and can be described by a function with four independent parameters only, the so-called Band function (Band 1995):

$$N(E) = \begin{cases} A \times E^\alpha e^{-E/E_0} \\ A' \times E^\beta \end{cases} \text{ for } \begin{matrix} E \leq E_b \\ E \geq E_b \end{matrix}, \quad \begin{matrix} E_b = (\alpha - \beta) \times E_0, \text{ and } A' \\ \text{is such that } N(E) \text{ is continuous.} \end{matrix}$$

This function results in a curvature in the low-energy part and a power law in the high-energy part of the spectrum. Typical values for the parameters are: $\alpha = -1$, $\beta = -2$; the characteristic energy $E_0 = 150$ keV. E_b is the "break" energy, where the power-law index changes, leading to a steepening of the spectrum.

Most analyzed νF_ν ($= N(E) \times E^2$) spectra peak in the energy range from 0.1 MeV to 1 MeV, with a general decrease of the peak energy from stronger

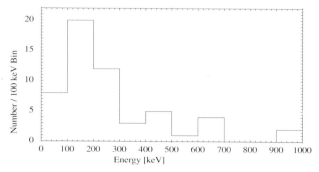

Fig. 15.9. Distribution of the peak energy E_p of $N(E) \times E^2$ energy spectra of 54 intense bursts. The peak of a $N(E) \times E^2$ spectrum indicates the region in the spectrum where most energy is radiated. Most burst spectra peak around 150 keV. (Data from Band 1995)

to weaker bursts (see Fig. 15.9). Combining the data from all four CGRO instruments, spectra could be compiled which span more than four decades of energy (see Fig. 15.10).

Interesting but not easy to explain is the deficit of soft X-ray fluences (below 20 keV) compared to γ-ray fluences (L_X/L_γ). The X-ray fluence is usually in the range of a few percent of the γ-ray fluence. But on the other hand about 10% to 15% of the bursts show a significant excess up to a factor of ~ 5 (but still far below the γ-ray fluence) in the spectrum below 20 keV compared to a generic burst spectrum, providing evidence for an additional low-energy spectral component or an upturn of the spectrum. A relatively stronger soft component has also been found in the X-ray precursors and tails of some bursts (Laros et al. 1984).

The continuation of the power law has been measured up to the GeV range for some bursts (Dingus et al. 1998). No softening or break at the highest energies has been found. In one burst event (GRB 940217) several γ-ray photons above 2 GeV (one at 18 GeV) were detected. The high-energy emission lasted up to 90 min after the onset of the burst, by far outlasting the medium-energy emission, suggesting a different energy production mechanism.

An extensive search has been undertaken for γ-ray lines. Only in data from the Konus experiment and the Ginga satellite have line features been found, and these were interpreted as cyclotron absorption lines from electrons moving in the intense magnetic field of a neutron star. However, the detected line features could be instrumental artifacts produced by the superposition of spectra with different shapes in adjacent energy ranges. In the burst spectra measured by BATSE no significant line was found.

Burst spectra have been observed to change on very short time scales. GRB 930131, which was the strongest burst in the first years of CGRO/BATSE observations, showed hardness variations within 2 ms. This burst

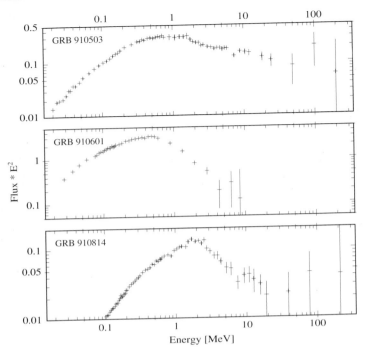

Fig. 15.10. Flux × E^2 spectra in (photons cm^{-2} keV^{-1}) × $(E/100\,\text{keV})^2$ of GRB 910503, GRB 910601 and GRB 910814 measured by all CGRO instruments. The bursts were observed by COMPTEL and EGRET since they happened to be in their field of view, and the spectra were exceptionally hard with peak energies between 500 keV and 2 MeV. While the peak energies are similar, the steepness of the power-law spectra is quite different (Data from Schaefer et al. 1994)

is also a typical example of the temporal evolution of γ-ray spectra. The spectrum hardens when the count rate increases, especially spikes after the onset of the burst are hard, and there is a general evolution from hard to soft within a spike and from beginning to end. The spectral hardness of γ-ray bursts is best described by E_p and not the characteristic energy E_0. The measured distribution of E_p is broad and peaks at 150 keV with only a few bursts having $E_p > 500$ keV (see Fig. 15.9). Variations of E_p within the duration of a burst are intrinsic to the emission process, and very likely not due to a cosmological redshift effect.

15.2.3 Spatial Distribution of Gamma-Ray Bursts

The burst distributions measured prior to BATSE already suggested an isotropic, but inhomogeneous distribution[5] in space, however, most Galactic source distributions could not be ruled out because of the low number of localized bursts.

Isotropy is tested by comparing statistical measures of the data, such as the dipole moment $\langle \cos\theta \rangle$, with expected values from the hypothesis that the measured burst sample is drawn from an isotropic distribution. The hypothesis cannot be excluded (but is not proved) if the two values agree within the statistical errors. Dipole moments are sensitive to concentrations towards one direction in space (i.e. Galactic Center, Sun), while quadrupole moments test concentrations toward poles or a plane. Using the first 1005 BATSE bursts the Galactic dipole moment $\langle \cos\theta \rangle$ differs from the value predicted for isotropy by 0.9 σ, and the observed quadrupole moment $\langle \sin^2 b - \frac{1}{3} \rangle$ by 0.3 σ. Also coordinate-system independent methods did not reveal any significant dipole or quadrupole moment. With these results on a very firm statistical basis it can be said that γ-ray bursts are much more isotropic than any other observed Galactic population, strongly favoring, but not requiring, an origin at cosmological distances (Briggs et al. 1996).

Homogeneity is tested by comparing the radial distribution of bursts with the distribution of a homogeneous sample. Fig. 15.11 shows the number of bursts with a peak flux larger than a flux P as a function of P for 772 bursts. If the bursts are homogeneously distributed in space, then the number of bursts within a sphere depends on the cube of the radius. The measured flux of a burst diminishes with the square of the distance to the burst. Combining these two effects results in a dependence of the integral number of bursts versus the peak flux according to a power law with index $-3/2$. The function $N(>P) \sim P^{-3/2}$ for a homogeneous distribution is also plotted in Fig. 15.11. The obvious deficiency of weak bursts indicates that the end of the distribution is seen (or that bursts are not standard candles).

A general measure to test the deviation from homogeneity is the average of $V/V_{\max}$ (Schmidt et al. 1988). $V_{\max}$ is the total volume in which a given source can be detected, depending on the current trigger threshold $C_{\lim}$. The idea of this test is: the peak count rate C_p of a source in Eucidean space decreases with the square of its distance, so that $C_\mathrm{p}/C_{\lim} = (r/r_{\max})^{-2}$. The relative radial location of the source is thus related to the volume contained within r. The ratio of the volumes is then $V/V_{\max} = (C_\mathrm{p}/C_{\lim})^{-3/2}$. This makes $V/V_{\max}$ independent of variations of the trigger threshold due to sensitivity variations of the detector, but the observed values in an inhomogeneous distribution depend on the average sensitivity of the instrument. Comparisons of the results of different detectors are therefore difficult. The value of $V/V_{\max}$ varies from 0 to 1. A homogeneous parent distribution of

---

[5] Isotropic means that the distribution is the same in all directions, whereas homogeneous means the same density of sources at all distances.

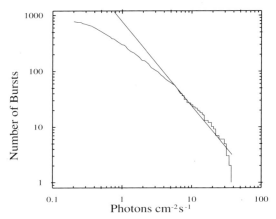

Fig. 15.11. Integral distribution of 772 bursts versus the peak flux. The *curve* shows the measured number of bursts observed, the *straight line* represents the $-3/2$ power law expected for a homogeneous distribution of bursts. The deviation from the power law for faint bursts indicates the deficiency of sources towards the lower end of the distribution. (Data from Meegan et al. 1995)

the sources yields a value of 0.5. Smaller values are obtained if the sources are concentrated towards the center, larger values if the sources tend to be at the outer rim of the observed space.

With the BATSE instrument a value of $\langle V/V_{\max}\rangle = 0.33 \pm 0.01$ was obtained, which clearly deviates from the homogeneous distribution.

Because of the isotropy and inhomogeneity, three classes of burst models have been discussed: an origin of the bursts within the Oort cloud of the solar system, within a Galactic Halo or at cosmological distances. A discussion of the different models is given in Sect. 15.4.

15.3 Counterparts of Gamma-Ray Bursts

15.3.1 Search for Counterparts in Other Wavelengths

The difficulty in the search for counterparts was the large search regions (error boxes) combined with the long time delay before the burst-location data were available. The precision of most burst triangulations was in the range of a few degrees for satellites in earth orbit and arc minutes if interplanetary space probes could be used. The afterglow of γ-ray burst events in other bands of the electromagnetic spectrum was expected to last only from hours to weeks. Since acquiring and analyzing of the data took at least several weeks at the beginning of γ-ray burst research, rapid follow-up observations were impossible. The probability for the simultaneous, successful observation of a burst region with telescopes in other wavelength bands by chance is very small because either the field of view of a telescope/detector is very small or the sensitivity is low. The search for counterparts on photographic plates in the archives of large sky surveys in the form of transient objects in the error boxes was very ambiguous. Several candidate objects were found, but most of them could be explained by plate defects and artifacts caused by airplanes and satellites in the field of view (Greiner and Moskalenko 1994; Schaefer 1994). The remaining candidates were doubtful, because the probability of finding some variable object in such large regions was high.

Another resource in the search for optical counterparts were the archives of patrol cameras for the observation of meteors in the atmosphere. The advantage is the wide field of view of those instruments, but the sensitivity of the detectors is limited to bright objects in the sky (10–15 mag). Several candidates for counterparts were found, but dismissed later after a more thorough analysis (Zytow 1990).

The general agreement in the astronomical community was that rapid follow-up observations with sensitive detectors are needed. Several projects were set up with quick analysis of satellite data and follow-up observations by small automatic telescopes (within seconds of the burst), larger observatories (half an hour) and satellites (few hours).

15.3.2 A New Approach with the Beppo-Sax Satellite and First Success

Beppo-Sax, an Italian–Dutch satellite experiment (see Fig. 15.12) had the ability to detect γ-ray bursts simultaneously at low and hard X-ray energies. The combination of an all-sky X-ray detector (40–700 keV) with two wide-field (2–28 keV) and four narrow-field X-ray cameras (0.1–10 keV and 1–10 keV) allowed the Beppo-SAX team to detect γ-ray bursts and search for X-ray counterparts with the same instrument platform. For each burst trigger in the all-sky monitor there is a search in the wide-field cameras (WFCs) for a simultaneous signal. On a positive detection the location of the signal is obtained and the narrow-field instruments are pointed in this direction. Combining the fields of view of the two WFCs ($40° \times 40°$), the probability of observing the source direction of a γ-ray burst with one of the wide-field cameras was about 5%. Beppo-SAX was launched on April 30, 1996, and started its search for γ-ray bursts at the end of the year.

On January 11, 1997, a burst was detected by Beppo-SAX and the source region was observed by an optical telescope only 16 h after the burst. No counterpart at any other wavelength was found in subsequent observations.

However, on February 28, 1997, at 02:58 h another burst triggered the all-sky monitor. A quick-look analysis of the data of the WFC revealed an excess in the count rate. A preliminary image analysis showed a point-like X-ray signal which occurred simultaneously with the γ-ray burst. The location error of about 10 arc min was small enough to point the narrow-field instruments (NFIs) to this position at the next opportunity (satellite manoeuvers need very detailed planning and therefore usually take several hours at least).

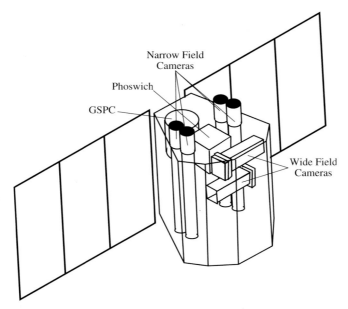

Fig. 15.12. The Beppo-SAX satellite carries an all-sky γ-ray detector (40–700 keV), 2 wide-field X-ray cameras (WFC, 2–28 keV) and 4 narrow-field instruments (NFI), 3 for the medium-energy range (1–10 keV) and one for the low-energy range (0.1–1 keV). The GSPC (high precision gas szintillation proportional counter) is also a narrow field instrument. Since the WFCs are oriented perpendicularly to the NFIs, the satellite has to be reoriented to locate a WFC source with the NFIs

Meanwhile a more detailed analysis of the attitude yielded an error box with a radius of 3 arc min only. Then, 8 h after the burst trigger, the observation with the NFIs started at 11:14 h and lasted more than 8 h. In the resulting image only one source was found inside the error box[6] with a 2σ location error of 50 arc sec.

The first optical observations prior to the publication of the more detailed NFI position was obtained 20.8 h after the burst with the 4.2 m William Herschel Telescope on La Palma (van Paradijs et al. 1998). This image contained a galaxy, which was considered to contain the counterpart of the burst. However, a follow-up observation on March 8 with the 2.5 m Isaac Newton Telescope on La Palma revealed the true counterpart (see Fig. 15.16): a transient source that was clearly detected in the first observation, but was not visible in the images of the follow-up observation.[7] The image was consistent with a point source. The magnitude of this object dropped in the V-Band (visible)

---

[6] At the position 5 h 1 m 44 s (right ascension) and 11° 46.7′ (declination), equinox 2000.

[7] The position of this object was 5 h 1 m 46.66 s right ascension and 11° 46′ 53.9″ declination (equinox 2000) at an accuracy of 0.2″.

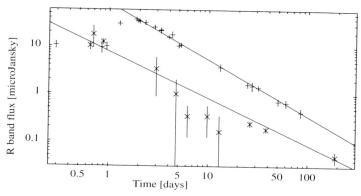

Fig. 15.13. Temporal evolution of the optical emission in the R-band of the counterparts to GRB 970228 (*lower plot*) and GRB 970508 (*upper plot*). A power law is fitted to both data sets (Garcia et al. 1997). The power law indices are −0.96 for GRB 970228 and −1.18 for GRB 970508

from 21.3 mag in the first observation to >23.6 mag in the second observation. The temporal evolution of this optical transient and the transient of a later burst is shown in Fig. 15.13. The decay of the emission nicely follows a power law as predicted by the fireball model (see Sect. 15.4).

The reliability of the identification was further enhanced by a ROSAT observation of the region, which fixed the position of the X-ray transient within a radius of 10″. Its centroid coincides within 2″ with the position of the optical transient.

Later images by the Hubble Space Telescope and several Earth-bound observatories revealed a faint Galaxy at the position of this source. A correlation of these objects is fairly likely. The redshift of this Galaxy was estimated from the observed optical intensity to lie between 0.2 and 2.

Further confirmation that γ-ray bursts are extragalactic objects was provided by the optical spectrum of the counterpart to GRB 970508 (also found using Beppo-SAX data), which showed absorption lines of Fe and Mg at a redshift of 0.835. This minimum distance requires a luminosity of 7×10^{51} erg, assuming isotropic emission. Also, for the first time, a transient radio source was detected at the burst position.

Radio observations of another burst (GRB 970508) at 8.46 GHz, 4.86 GHz and 1.43 GHz (Frail et al. 1997; Taylor et al. 1997) showed a flickering in the light curve, which was consistent with the diffractive radio-scintillation effect. Due to the interaction with the local interstellar medium several images of the source (or source regions) are produced. The interference of these images produces a temporal modulation of the signal, if the angular size of the source is smaller than about 1 μarc sec (Goodman 1997). The diffractive scintillation is measured by the variation of the relation of the intensities at different wavelengths. After about a month the diffractive scintillation ceased, because the source became too large, and only a smaller modulation remained due to varying refraction (focussing and defocussing of the source). This gave a first independent measure of the size of the object ($\sim 10^{17}$ cm) one month after the burst event, which is consistent with the cosmic fireball model.

At the end of this fruitful year a burst (GRB 971214) was detected with an optical counterpart coinciding with a galaxy at a redshift of 3.42. If this association holds, then some γ-ray bursts may originate from the edge of the observable universe at a distance of several Gigaparsecs, leading to a luminosity of 4×10^{53} erg. For several seconds the source was as bright as the entire universe (again assuming isotropic emission).

But that was not the end of the superlatives. At 09:47 h UTC on January 23, 1999, a very strong γ-ray burst (among the brightest 1%) was registered by BATSE (Kippen et al. 1994, GCN 224) and its position (accurate to a few degrees) was immediately distributed through the GRB Coordinates Network (GCN). The Robotic Optical Transients Search Experiment (ROTSE) automatically turned its CCD cameras towards the broadcasted position. The first image was captured just 22.18 s after the onset of the burst event, still well within the ongoing γ-ray emission, which lasted more than 100 s. The image revealed a source with a visual brightness of 11.82 mag. The signal peaked 25 s later at 8.95 mag. Another 25 s later it dimmed to 10.08 mag, and finally 10 min later to 14.53 mag, close to ROTSE's detection limit ($\sim$15 mag). A plot of the light curve including later observations by ground-based telescopes is depicted in Fig. 15.14. It shows nicely the decaying nature of the afterglow emission. This burst was detected by Ulysses, Beppo-SAX, and COMPTEL as well.

The time profile in γ-rays consists of two hard peaks, followed by some softer irregular emission (see Fig. 15.14). The X-ray signal from the WFC shows a complex light curve with only one clear peak about 40 s after the γ-ray peak and a structured high plateau thereafter. The X-ray fluence is a few percent, and the optical fluence is $< 0.1\%$ of the γ-ray fluence.

Further follow-up observations in several energy bands consistently showed a signal decaying according to a power law (index ≈ -1.1), with some variation in the index. An exception to this behaviour was the detection of a radio signal of 260 µJy 1 day after the burst, with no signal before and after (upper limit $\sim$ 64 µJy).

The optical spectrum of the afterglow revealed several strong absorption lines at a redshift of $z = 1.6$ and some weaker lines at a redshift of 0.20 and 0.28. The origin of the burst is very likely in the Galaxy producing the larger redshift or even farther away, but not beyond $z = 2.2$, because then the redshifted Lyman-α lines should have been seen in the optical part of the spectrum. The exceptional brightness of the optical flash and afterglow suggested an enhancement of the signal by gravitational lensing in one of the foreground systems. Assuming the source lies at $z = 1.6$, an isotropic emission of 3.4×10^{54} erg, equivalent to nearly $2\,M_\odot$, can be inferred.

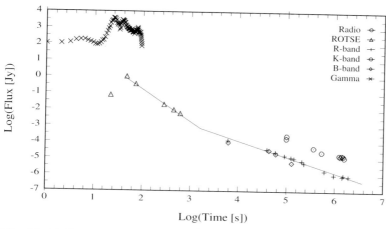

Fig. 15.14. Temporal evolution of the γ-ray, optical (V, R, K, B) and radio emission of GRB 990123. Separate power laws are fitted to the afterglow emission. The power-law indices for the temporal evolution are −2 for the first and −1 for the later part of the afterglow emission

Among the latest news was the discovery of polarization in the optical afterglow of GRB 990510 (Covino et al. 1999), giving evidence for nonthermal production of the radiation (synchrotron).

The wealth of new data becoming available during this exciting time resulted in a major step forward in the 25 year search for a solution of the γ-ray burst enigma. It is now safe to accept a cosmic origin of γ-ray bursts with all its consequences, especially the extreme physical properties of the source objects.

15.4 Models for Gamma-Ray Bursts

15.4.1 The Distance to the Gamma-Ray Burst Sources

There are probably as many different burst models and proposed possible origins of γ-ray bursts as there have been theorists working in this field. Most of the models become obsolete with the next generation of measurements. The discovery of the isotropic but inhomogeneous (distance-limited) distribution of the γ-ray bursts ruled out the previously favoured Galactic Disk models. The discovery of optical counterparts with red-shifted absorption lines in their spectrum pushed the origin of the γ-ray bursts to cosmological distances, ruling out the local heliocentric and Galactic Halo models. These older models are explained briefly in the following for historic and educational reasons only.

Local Heliocentric Models

such as comet–comet or comet–black hole collisions (Bickert and Greiner 1993) releasing an energy of about 10^{28} erg are consistent with the measured isotropy of the bursts, but the predicted concentration of the strongest bursts in the sample towards the Sun was not found.

Galactic Halo Models

are mostly relics from earlier Galactic models favouring neutron stars as the origin of γ-ray bursts. The general idea is that neutron stars are produced in the Galactic Disk in supernova events. Due to asymmetries in the supernova explosion, the neutron stars receive a strong impulse, which results in a high peculiar velocity of up to $600\,\mathrm{km\,s^{-1}}$. The distribution of these neutron stars becomes nearly isotropic after more than 100 million years. Bursts can only be produced after this early phase and no low-velocity neutron stars may produce bursts without disturbing the isotropy. The typical amount of energy released in a burst event is about 10^{43} erg. The Sun's location in the outer part of the Galaxy would require deviations from isotropy in the halo model. Since no deviations are observed, a lower limit of about 100 kpc for the inner edge of this halo can be inferred. An outer limit is derived with a similar argument. If the average distance to a γ-ray burst were larger than about 300 kpc, an excess towards the Andromeda Galaxy (M31), our Galactic neighbour should be found. This already drove the parameters of Galactic Halo models to nearly impossible limits.

Cosmological Models

naturally explain the observed isotropy and inhomogeneity: the distribution of galaxies is isotropic on angular scales larger than a few degrees (about the BATSE location error). The temporal evolution of the source distribution (creation of the source objects in the course of cosmic evolution) and cosmological effects due to the expansion of the universe (especially redshift and number frequency of bursts) produces the observed deficiency of weak bursts.

The main problem for cosmological models is to generate the enormous amount of energy (10^{50}–10^{54} erg) and to get γ-rays out of the small region (about 100 km from variability arguments) of highly concentrated energy production. γ–γ pair production causes an optical depth of 10^{12}, severely limiting the energy density for γ-rays above several MeV, which would lead to an observable break in the spectrum at this energy. However, γ-ray bursts are observed with unbroken power laws up to 100 MeV. Thus either the production region is large enough that it is optically thin for γ-rays, or the energy is produced in a highly relativistic frame blueshifting the pair-production threshold to the range of a few 100 MeV, or the emission is focussed in a narrow beam, reducing the total amount of energy required by the factor (beam solid angle)/4π. A combination of these mechanisms is of course possible too.

In general the γ-ray burst process can be separated into two steps: the initial production of energy and the subsequent dissipation of the energy creating the γ-rays of the burst event and the afterglow emission in other wavebands. The entire phenomenon is often referred to as a hypernova.[8]

I first concentrate on the second step, since its general mechanism, the cosmic fireball, seems to be widely agreed upon.

15.4.2 Cosmic Fireball: The Evolution of the Blast Wave

A simple model of the dynamics in the blast wave, independent of the initial explosion, is the relativistic generalization of the method used to understand supernova remnants. A comprehensive summary of the models currently discussed can be found in Meszaros (1999).

The initially accelerated blast wave or beams produce the γ-rays by interactions within outflowing material, the interstellar medium (ISM), or the stellar wind or outer shell of the companion in a binary system. In the subsequent cooling phase, the deceleration of the blast wave, lower-energy radiation may be released, causing an afterglow. Independent of the specific model, the broken power-law spectral shapes and the rapid variability of γ-ray bursts are almost certainly produced by nonthermal particles in a synchrotron process or to some extent in inverse-Compton interactions.

The typical numbers characterizing the luminosity L, duration T, total energy E per burst and event rate R are

$$L \approx 10^{52}(\Omega/4\pi)\,\mathrm{erg\,s^{-1}},\ T \approx 10\,\mathrm{s},$$
$$E \approx 10^{53}(\Omega/4\pi)\,\mathrm{erg},\qquad R \approx 10^{-6}(\Omega/4\pi)^{-1}\,\mathrm{galaxy^{-1}\,yr^{-1}},$$

where Ω is the solid angle into which the energy is channeled (Meszaros 1999). The durations have a large spread from 10^{-3} s up to 10^3 s.

From causality considerations the initial dimensions of the source are constrained by the shortest observed time variations of the bursts ($\sim 10^{-3}$ s), which leads to an upper limit of about 100 km. The explosion of the compact source creates an optically thick plasma that expands and accelerates to a relativistic velocity (Waxman 1997). The fireball energy is then converted to proton kinetic energy. A cold shell of thickness cT is formed and continues to expand with a time-dependent Lorentz factor $\Gamma \sim 200$. Such a high Lorentz factor is required to avoid attenuation of the high-energy γ-rays due to γ–γ pair production.[9] The dissipation of the kinetic energy generates the

---

[8] The scope of the term "hypernova" is not yet well defined. Some authors use it for a model of the afterglow, others as a name for the entire phenomenon/object.

[9] The maximum collision angle of the protons is restricted by the relativistic kinematics to values lower than $\sim \Gamma^{-1}$. The pair production is then effectively diminished, since the cross section for this process is small at small angles.

main γ-ray burst event at a radius of $R > 10^8$ km, and the expanding cooling fireball produces the delayed emission at lower energies. The dissipation may occur because of internal collisions within the expanding shell and/or collisions with the ISM. Internal shocks develop in the relativistic wind itself, when faster portions of the flow catch up with slower portions. External shocks with the surrounding medium occur when the lab-frame energy of the swept-up matter equals the initial energy of the fireball at about 10^{12} km.

The time scales involved in the external shock are of the order of the deceleration time (minutes). Therefore, external shocks cannot produce highly variable time profiles, but might be the source for long smooth bursts with a few peaks. Internal shocks can occur at much smaller radii, reflecting to some degree the time variability of the internal source. Such internal shocks have been shown explicitly to reproduce some of the more complicated light curves (Panaitescu and Meszaros 1998; Sari and Piran 1998). The time profile of the arriving γ-rays is not smoothed out by the simultaneous emission from different points on the expanding shell, because at $\Gamma \approx 100$ the γ-rays are emitted in a cone of opening angle of only ∼1°. Therefore only radiation from a small sphere/annulus of the expansion shell is observed. It is therefore not easy to distinguish between a spherical fireball and an emission in a jet directed towards the observer. At later stages, when the Lorentz factor decreases to about one, emission from a wider range of angles and deviations from spherical symmetry can be detected. The recent detection of a steepening in the decay of the afterglow of GRB 990510 provides evidence for a jet geometry.

The recent discovery of an optical flash within seconds of the γ-ray burst (GRB 990123) can be explained by the emission from a reverse shock propagating in backward direction (see Fig. 15.15). The ratio between the typical observed energies from these shocks is Γ^2 due to the relativistic movement in opposite directions. The forward-propagating shock, producing the γ-rays, is caused by the internal collision of blast waves with different velocities. The backward propagating shock is the reverse shock of the collision of the blast wave with the external medium. This is the site of the emission in the optical range and also possibly of the 8.46 GHz radio signal observed one day later. The later afterglow in X-rays and γ-rays is then produced by the forward-propagating shock interacting with the external medium. This model can account for the uncorrelated, but concurrent signals in the optical and γ-ray range and for the afterglow emission. This was proposed by Sari and Piran (1999), about a month prior to the above-mentioned discovery.

The spectrum resulting from such an afterglow model consists of a power-law spectrum with three breaks. The first part of the spectrum is a steeply rising, self-absorbed synchrotron spectrum, which above the self-absorption break is followed by a slowly rising power-law up to the synchrotron break (corresponding to the minimal energy of the accelerated electrons). The third part corresponds to the electrons in the adiabatic energy regime. The last

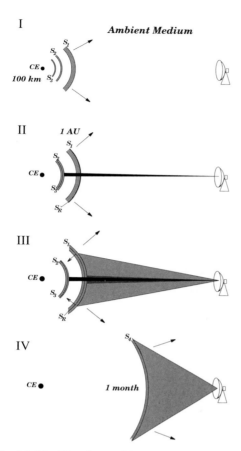

Fig. 15.15. The forward/reverse shock model for GRB 990123 : (I) Several seconds after the onset of the explosion. Several shells are ejected by instabilities in the emission process of the central source (size ≈ 100 km). S3 was ejected at a larger initial velocity than S2. (II) The onset of the γ-ray emission of the burst. S3 overtakes S2 and produces by this internal shock the γ radiation. Since the emission happens with ultrarelativistic motion, only radiation within a cone of about 1° is visible to the observer. At about the same time S1 builds up a shock-front with the surrounding medium, causing a reverse shock. (III) Some seconds later the reverse shock traverses S1 and starts to emit radiation. Since it travels backwards, the mean energy of the photons from this process is lower by a factor of Γ^2, which shifts the radiation into the optical range. At about the same time the forward shock starts to produce the X-ray emission by its interaction with the surrounding medium. This explains the rough correlation between the optical and X-ray emission with no correlation to the γ radiation. The reverse shock might also produce the observed radio emission when it hits another shell a day later. (IV) The scenario about a month later. Now most emission is produced by the first shock sweeping up the ambient matter. The velocity is now slowed down to $\Gamma \approx 1$

break is expected at energies where the electron cooling time becomes short compared to the expansion time.

In the further expansion of the fireball, the main emission moves to lower frequencies, so that in a given band the flux decays as a power law in time, with changes in the index when a spectral break moves through the observed band.

The production of the afterglow emission is similiar to nonthermal synchrotron models for the radio-through-optical continua of blazars, but with larger Lorentz factors. Hence certain blazar characteristics like superluminal motion and high polarization should be seen.

Although the delayed X-ray/optical emission must result from synchrotron emission of accelerated electrons of the ISM, inverse-Compton emission may dominate during the first hours/days, suppressing the lower-energy emission.

15.4.3 The Central Engine

The general scheme of most models is the gravitational collapse or rapid accretion of one or more objects resulting in the production of neutrinos and antineutrinos which later annihilate and lead to the production of a rapidly expanding shell of electron–positron pairs.

The different proposed sources of γ-ray bursts have to be consistent with the number of source objects in comparison to the observed burst rate. If the weakest events are at $z = 1$ and the γ-ray emission is not beamed, then about 10^{-6} events galaxy^{-1} year^{-1} are required.

All plausible progenitors so far (mergers of compact objects or of stars with compact objects, or gravitational collapses)[10] are expected to lead to a black-hole-plus-debris torus system, and they are all capable of producing relativistic outflows through the same mechanism (Meszaros 1999). Energy can then be drawn from the spin energy of the black hole or from the potential energy of the orbiting debris. In the merger scenarios, the black hole can contain more energy than the torus, because it is rapidly spinning and usually has more mass than the orbiting debris. The rotational energy of the black hole can be extracted through magnetohydrodynamic coupling by the Blandford and Znajek (1977) effect.

The rotating torus can dissipate its energy via two processes. First, the differential rotation of the torus generates magnetic fields, which then dissipate their energy. Second, the torus heats up by the friction of the differential rotation and produces neutrinos and antineutrinos directly. A maximum of 30% to 40% of the total energy of the rotating system can be extracted, resulting in an energy release of up to nearly $\epsilon \times 10^{54}$ ergs, where ϵ is the efficiency for the conversion of gravitational energy into energy of the blast

---

[10] An origin in quasars requires the burst to coincide with the center of the host Galaxy, but some γ-ray bursts, localized by the optical afterglow, are found offset from the likely host Galaxy.

wave or jet. The different progenitors corresponding to different central engines vary in their energy injection into the blast wave/jet by up to a factor of 20.

In the following some scenarios for the central engine and/or the subsequent evolution are presented in some more detail. The scenarios in general do not exclude each other. A multitude of events in the cosmos might lead to γ-ray bursts:

Coalescing compact objects, such as neutron stars (NS), black holes (BH), He cores or white dwarfs, were one of the first models for the energy production in γ-ray bursts. A close binary system of two such objects is not stable due to energy loss by the radiation of gravitational waves. Due to this the two objects will approach each other and finally coalesce. In most cases a disk is produced, from which matter is quickly accreted. The rapid accretion of H-rich matter is difficult since the energy production by fusion due to the extreme high temperatures slows down the accretion (He, C and O accrete faster). An accretion of 2 to 10 $M_\odot$ at a rate of at least $0.1\,M_\odot\,s^{-1}$ is necessary. Such a merger event can release up to 10^{53} erg in 10 ms. A significant fraction is radiated in neutrinos and antineutrinos, which annihilate and produce electrons and positrons. The final blast results in an expanding sphere or beams in opposite directions along the rotation/revolution axis. The shock-heated matter has a temperature of $k_b T \approx 100$ MeV.

The *collapsar model* assumes the gravitational collapse of a star, as in a *failed supernova*, for the progenitor of the γ-ray burst. The model starts either with the progenitor of a Type Ib supernova (a close binary system) or a Type II supernova (single massive star).

In the first scenario a huge amount of kinetic energy (10^{54} erg) is deposited into the expanding envelope of the supernova via the extremely strong magnetic field (10^{11} T) of the nascent rapidly rotating NS[11] (Paczynski 1998). The mechanism of a Type Ib supernova fails to create a shock above a certain He core mass ($>6\,M_\odot$). After the initial collapse with a duration around 1 s, the outer He core is drawn inwards too. The infall stops at 100 km to 1000 km because of the angular momentum and an accretion disk is formed.

The second scenario starts with a single massive star, very likely a Wolf–Rayet star with low metallicity, which causes the initial star to be more compact and therefore favours the formation of a BH with an accretion disk instead of the collapse to a NS. The value of the angular momentum is critical in this model (Woosley and MacFadyen 1999). If the angular momentum is too low, then most of the disk is just accreted by the BH with no explosion-like phenomenon, and if it is too large, then the temperature in the disk is not sufficient for the production of neutrinos, a very efficient cooling mechanism which allows a rapid outflow of energy because of the small neutrino cross section for interaction with matter. The accretion time for the disk is about

---

[11] 10% of the rapid rotating massive Wolf–Rayet stars can lose their envelope if their initial mass is larger than 35 $M_\odot$, or if they are in a binary system.

1 min and at least larger than 10 s. A plasma jet (opening angle roughly 30°) of neutrinos and electron–positron pairs is formed which plows through the remaining shell of the star along the axis of the accretion disk. The radius of the outer layer of hydrogen must be small enough not to sufficiently slow down the outflowing jet. The break-through of this jet might explain the occurrence of soft γ-ray/hard X-ray precursors of γ-ray bursts. If bursts are produced according to this model, then each burst should be accompanied by a supernova explosion, whereas for $\Omega/4\pi = 10^{-2}$ one event per 10 000 yr per Galaxy would be needed. Hence only a small percentage of all supernovae would cause a burst event, assuming about 1 supernova per 100 yr per Galaxy.

This model can account for the class of longer bursts. Short bursts, below fractions of a second, which are weaker as well, are more likely produced by the merger of a NS with another NS or a BH. The recent discovery of the unusual supernova SN 1998bw and its apparent correlation with the γ-ray burst GRB 980425 (Wheeler et al. 1999) together with observations of some γ-ray burst afterglows over longer time periods indicating a supernova light component gave new emphasis to this model.

Stellar collisions in globular clusters is another possiblity for the progenitor event. The probability of the collisions of compact objects with stars or other compact objects occurring in globular clusters is strongly enhanced by the locally high density of stars. The large number of millisecond pulsars in globular clusters, relative to the Galactic Disk, is evidence for a significant population of compact objects there. The difference from most other models is the involvement of an old stellar population, which allows the occurrence of burst events in all galaxies, not only those with high star-formation rates.

He star/BH mergers, in contrast to the last model, tend to occur in regions of star formation. The rate of He star mergers is generally greater than the rate of NS/BH mergers. The scenario starts with a close binary system of two massive stars ($>8\,M_\odot$). The evolution of the system leads to a supernova explosion (see Chap. 8) leaving a NS and a massive main-sequence star. The second star evolves, finally overflowing its Roche lobe and builds a common envelope with the NS. This causes the NS to spiral towards its companion within 1 to 1000 yr, depending on the original distance between the stars. When the system merges, up to $1\,M_\odot$ may get accreted within 1 s leading to an energy release of 10^{52} erg. The strong magnetic field produced in the disk might focus the energy enough to account for even the stronger burst events.

The collapse of a supermassive object is far more energetic than the collapse or merger of stars with an average mass only. The progenitor object is either a single supermassive star or a relativistic star cluster. The resulting object is a highly massive BH, which could be the origin of AGN (see Chap. 12). The problem of this model is the large creation rate of baryons, which damps the burst event.

A mini-supernova model for the optical afterglow was proposed by Blinnikov and Postanov (1997). In a close binary system a γ-ray burst originates from one object in the system and impacts on the other, a red supergiant, producing the afterglow emission. The precise nature of the γ-ray burst source is not relevant in this model, only an initial energy release of 10^{50}–10^{51} erg is required. During 10 s the energy of the γ-rays is deposited in the outer layer ($10^{-3}\,M_\odot$) of a supergiant ($M = 15\,M_\odot$, $R = 4000\,R_\odot$), which is heated up to $\sim 5\times 10^5$ K. The layers outside of $10^{-4}\,M_\odot$ acquire a speed of $2\times 10^4\,\mathrm{km\,s^{-1}}$ during a few hours. The inward moving heat wave dies out during 10 days reaching around $1\,M_\odot$. By that time the kinetic energy is about 2% of the initially deposited energy. For the first 10 days the light curve is consistent with observations; after 30 days the flux is two orders of magnitude below the observed values. However, this model is very flexible, since different objects in the binary system might produce quite different results. Especially in the combination of a collapsing NS with an OB supergiant the speed of the heated matter reaches the velocity of light and the model becomes similar to the relativistic fireball model.

15.5 Prospects

New insight into the enigma of γ-ray bursts is expected from a detailed study of the counterparts, especially from an investigation of the distribution of the properties of the counterparts and correlations with astronomical objects like galaxies, Galaxy clusters or AGN. For "near" γ-ray bursts the progenitor object might have been previously detected by chance or on a deep sky survey, depending on its nature. Therefore a much higher number of counterparts need to be discovered.

HETE-2 (high-energy timing explorer) is one of the instruments which will help in this respect. It will be launched in the near future and is dedicated to the precise localization and rapid follow-up observation of γ-ray bursts. With its coded mask X-ray proportional counters, bursts can be localized with a precision of about $0.1°$. The soft X-ray CCD cameras determine the position of possible counterparts in the field of view with an accuracy of 3 arc sec for strong events.

The γ-ray burst Coordinates Network GCN (former name: BACODINE, see Barthelemy et al. 1994) demonstrated its capabilities in the recent discovery of the optical flash of GRB 990123. This strongly depended on the proper function of the BATSE instrument aboard CGRO. BATSE will hopefully be in operation until a new satellite instrument can take over its role as the provider of burst positions in near real time (within seconds).

It is still necessary to get independent accurate localizations of γ-ray bursts by triangulation, to verify the correlation of a counterpart with the burst. Therefore the space probes next launched should still carry small γ-ray detectors. The IPN should remain in operation until the angular resolution of γ-ray detectors reaches a comparable accuracy.

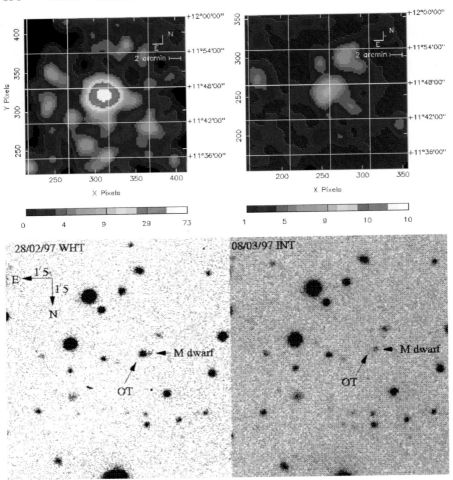

Fig. 15.16. Discovery images of the X-ray (http://www.sdc.asi.it/first/grb970228.html) and optical (http://www.ing.iac.es/PR/AR1997/high_97.htm) counterpart of GRB 970228. *Upper left*: X-ray image of the sky around the counterpart about 8 h after the burst trigger. The bright source in the center was not found in any previous X-ray observation. *Upper right*: The same sky region 2 days later in X-rays. The bright source has dropped by a factor of 20. *Lower left*: Optical image of the burst region captured by the William Herschel Telescope (4.2 m) in La Palma 20.8 h after the onset of the burst. *Lower right*: The same region of the sky a few days later as seen by the Isaac Newton Telescope (2.5 m). The source has faded beyond the sensitivity limit of this telescope

The last two years from the first discovery of an afterglow to the detection of an optical flash within seconds of the γ-ray burst onset have brought a tremendous amount of new information bearing on the mystery of γ-ray bursts, which will keep the scientific work in this field interesting for many more years.

References

Band D.L., 1995, AIP Conv. Proc. **384**, 123
Barthelemy S., Cline T., Gehrels N., 1994, AIP Conv. Proc. **307**, 643
Bickert K., Greiner J., 1993, III. Compton Symposium St. Louis, 1059
Blandford R.D., Znajek R.L., 1977, MNRAS 179, 433
Blinnikov S.I., Postanov K.A., 1997 , MNRAS, **293**, L29
Briggs M.S., Paciesas W.S., Pendleton G.N. et al., 1996, Astrophys. J. **459**, 40
Cline T.L. et al., 1973, Astrophys. J. Lett. **185**, L1
Covino S. et al., 1999, IAU Circular 7172
Dingus B.L., Catelli J.R., Schneid E.J., 1998, API Conf. Proc. **428**, 349
Duncan R.C., Thomson C., 1992, Astrophys. J. **392**, L9
Evans W.D., Klebesadel R.W., Laros J.G. et al., 1980, Astrophys. J. **237**, L7
Garcia M.R., Callanan P., Moraru D. et al., 1997, Astrophys. J. **500**, L105
Fishman G., Barthelemy S., 1995, Proc. IAU Symp. No. **151**
Fishman G.J., Meegan C.A., 1995, Ann. Rev. Astron. Astrophys. **33**, 415
Fishman G.J. et al. 1994, Science **264**, 1313
Frail D.A., Kulkarni S.R., Nicastro L. et al., 1997, Nature **389**, 261
Goodman J., 1997, New Astronomy 2, 449
Greiner J., Moskalenko E.I., 1994, Astron. Astrophys. **283**, 693
Kippen R., Connors A., Macri J. et al., 1994, AIP Conv. Proc. **307**, 418
Klebesadel, R.W., Strong, I.B., Olson, R.A., 1973, Astrophys. J. Lett. **182**, L85
Kouveliotou C., Fishman G.J., Meegan C.A., 1993, Nature **362**, 728
Kouveliotou C., Fishman G., Meegan C. et al., 1994, Nature **368**, 125
Kulkarni S., Frail, D., Kassim N. et al., 1994, Nature **368**, 129
Laros J.G. et al. 1981, Astrophys. J. Lett. **245**, L63
Laros J.G. et al., 1984, Astrophys. J. **286**, 681
Mazets, E.P., Golenitskii, S.V., Ilyinski, V.N., 1979, Nature **282**, 587
Mazets E.P., Golenitskii S.V., Ilyinski V.N. et al., 1980, Sov. Astron. Lett. **5**, 163
Meegan C.A., Pendleton G.N., Briggs M.S. et al., 1995, AIP Conv. Proc. **384**, 291
Meszaros P., 1999, Astron. Astrophys. Suppl. Ser. **138**, 533
Mitrovanov I. G., 1995, AIP Conv. Proc. **384**
Mukherjee S., Feigelson E.D., Babu G.J. et al., 1998, Astrophys. J., **508**, 314
Norris J., Hertz P., Wood K. et al., 1991, Astrophys. J. **366**, 240
Norris J.P., Bonnell J.T., Nemiroff R.J., 1995, AIP Conv. Proc. **384**, 80
Paczynski B., 1998, Astrophys. J. Lett. **494**, L45
Panaitescu B., Meszaros P., 1998, Astrophys. J. **492**, 683
Ramanamurthy P.V., Wolfendale A.W., 1984, Cambr. Astroph. Series **22**, 43
Sari R., Piran T., 1998, Astrophys. J. **485**, 270
Sari R., Piran T., 1999, Astron. Astrophys. Suppl. **138**, 537
Schaefer B., 1994, AIP Conv. Proc. **307**, 382
Schaefer B., Teegarden B., Cline T., 1994, AIP Conv. Proc. **307**, 280

Schmidt M., Higdon J.C., Hueter G., 1988, Astrophys. J. **329**, L85
Sentman D.D. et al., 1995, Geophysical Res. Lett. **22**, 1205
Strong, I.B. et al., 1973, Los Alamos Conf. on Transient Cosmic Gamma and X-ray sources, unpublished
Taylor G.B., Frail D.A., Beasley A.J., 1997, Nature **389**, 263
van Paradijs J. et al., 1998, Nature **386**, 686
Varendorff M. et al., 1992, AIP Conf. Proc. **265**, 77
Waxman E., 1997, Astrophys. J. **485**, L5
Wheeler J.C., Hoeflich P., Wang L, AAS Meeting 193, Austin, Texas, January 1999, http://hires.gsfc.nasa.gov/stis/science/aas/aas193/abs_193.html
Woosley S.E., MacFadyen A.I., 1999, Astron. Astrophys. Suppl. Ser. **138**, 499
Zytow A., 1990, Astrophys. J. **359**, 138

Index

α–α line complex, 97, 101, 110, 122
α-rich freeze-out, 239, 246
γ-ray, 191
– blazars, 78, 86
– bursters, 78
– bursts, *see* bursts
– horizon, 24
– sources, 82
γ^2 Velorum, 262
π^0-decay, 207, 208, 219
H_α emission, 94, 105, 108
^{3}He abundance, 98
^{10}Be, *see* beryllium
^{26}Al, *see* aluminium-26
^{44}Ti, *see* titanium-44
^{56}Co, *see* cobalt
^{57}Co, *see* cobalt
^{60}Fe, *see* iron
1ES 2344+514, 298
1E 1740.7-2942, 170
2.223 MeV line, 98, 106
3C 120, 308
3C 273, 286, 291, 292, 296, 297
3C 279, 292, 296, 297

A 0535+26, 176
absorption of γ-rays
– in materials, 50
abundances, 271
– primordial, 233
– standard, 233
acceleration, 186
accelerator, 22, 23
accreting neutron stars, 82, 279
accreting pulsars, 82
accreting X-ray binaries, 77, 82
accretion, 21, 250, 259, 268, 279

accretion disks, 82, 160, 163, 173, 175, 180, 289, 296, 307, 311, 312
accretion–disk corona, 172, 176, 311
ACE, 252
activators in scintillators, 33
active galactic nuclei, 21, 22, 78, 86, 320, 327, 341, 352
– 1ES 2344+514, 298
– 3C 120, 308
– 3C 273, 286, 291, 292, 296, 297
– 3C 279, 292, 296, 297
– Centaurus A, 286, 291, 312, 314
– Cygnus A, 286
– GRO J0516-609, 295
– MCG-8-11-11, 291
– Mkn 421, 298, 320
– Mkn 501, 298, 320
– NGC 1068, 288, 290
– NGC 4151, 291, 308
– PKS 0528+134, 292–294
– PKS 0208-512, 295
– PKS 1622-297, 293
– PKS 2155-304, 298
active galaxies
– classification, 285
– unification, 287
AE Aquarii, 179, 180
AGB star, 242, 249
air showers, 60, 62
air-shower arrays, 64, 179, 314
Alfvén speed, 191
all-sky maps, 77, 78, 80
aluminium-26, 5, 243, 250, 253, 259, 260, 262, 264, 267, 268
AM Hercules, 179
Anger-camera principle, 48, 55
annihilation, 18, 86, 233, 251, 268, 269, 306

annihilation line, 5, 91, 99, 101, 311
anticoincidence method, 28
Apollo 15, 3
ASCA, 172
astration, 233, 252
atmosphere, 10, 12, 25
Atmospheric-Cherenkov Technique, 60
– IACT, 62
– wave-front sampling, 62
atomic hydrogen, 213
atomic nucleus, 236
atomic transitions, 233
attenuation coefficient, 50
auroral processes, 120
avalanche models, 121

B/C ratio, 217
background
– active background suppression, 30
– charged-particle background, 28
– cosmic diffuse, 22
– external, 29, 40
– from neutral particles, 29
– instrumental, 25
– internal, 29, 40
– passive background suppression, 30
BACODINE, 393
Barium ion experiments, 120
BATSE, 2, 65, 77, 86, 87, 102, 114, 164–166, 168, 170, 175, 307, 312, 370
beamed emission, 301
Bell, Jocelyn, 127
Beppo-Sax, 6, 88, 162, 381
beryllium, 252
Bessel function spectrum, 106, 118, 119
Big Bang, 21, 186, 233, 253
Big Bang nucleosynthesis, 233, 239
binary systems, 244, 248, 260
BL Lac objects, 292, 294, 298, 305, 306, 314
black hole binaries, 160, 163, 165, 169, 173, 181
black holes, 21, 22, 186, 246, 247, 289
blackbody radiation, 12, 194
blazars, 22, 285, 291, 292, 298, 306, 307, 312, 314, 352
– γ-ray luminosity function, 353
blue bump, 296
blue jets, 372

boron, 252
bremsstrahlung, 15, 22, 193, 207, 208, 269
broad-line regions, 290
bubble, 186
bulge, 269
burst models, 21, 385
– black hole, 391
– central engine, 390
– collapsar, 391
– collapse of supermassive object, 392
– Cosmic Fireball, 387
– Cosmologic, 386
– failed supernova, 391
– fireball, 21, 253
– Galactic Halo, 386
– heliocentric, 386
– mergers, 390, 392
– mini-supernova, 392
– neutron star, 391
– progenitors, 390
– stellar collisions, 392
bursts
– afterglow, 380, 384
– cosmologic effects, 375
– counterparts, 380
– duration, 374
– energy spectra, 376, 378
– first optical counterpart, 381
– from Earth, 372
– hardness, 375, 378
– high-energy emission (GeV), 377
– individuals
– – GRB 790305, 368, 369
– – GRB 970228, 381, 383, 394
– – GRB 970508, 383, 388
– – GRB 990123, 389, 393
– – GRB 990510, 385, 388
– intensity, 375
– lines, 377
– origin, 370, 385
– polarization, 385
– properties, 372
– radio observations, 383
– relativistic time dilation, 375, 376
– soft X-ray fluence, 377
– spatial distribution, 370, 374, 379, 380, 385

- temporal evolution, 375, 378, 385
- time profile, 373

calcium, 246, 255
CANGAROO, 194–196
Carina, 80, 263, 265
Cas A, 186, 199, 201, 202, 255, 257
cataclysmic variables, 160, 177, 179
Centaurus A, 286, 291, 312, 314
Centaurus X-3, 178–180, 182
central bulge component, 86
Chandra, 257
Chandrasekhar limit, 129
Chandrasekhar mass, 248, 249
chemical evolution, 234, 253, 267
Cherenkov, 86
- light, 2
- telescopes, 193, 194, 298, 314, 316, 320
- threshold, 60
chimney, 267
chopper technique, 41
CO molecule, 213
coalescence of compact objects, 88
cobalt, 254, 255, 268
COBE, 263
collapse, 185
collimation
- active, 42
- modulator, 370
- passive, 30, 42
COMPTEL, 2, 55, 77–79, 93, 114, 164, 168, 170, 178, 204, 215, 218, 219, 254, 255, 257, 259, 262, 263, 265, 267, 292, 293, 307, 312, 370, 371
Compton Gamma-Ray Observatory, 2, 5, 44, 55, 77, 91, 93, 106, 114, 291, 370, 372
Compton reflection, 172, 310, 312
Compton scattering, 12, 20, 24, 250
- inverse, 16, 22, 24
Comptonization, 302
continuous spectrum, 95
convection, 243
conversion layer, 37
core, 185
- carbon oxygen, 185
- nickel, 185
coronal mass ejection, 94

COS-B, 3, 5, 56, 58, 177, 191, 291
COS-B source catalog, 5
cosmic diffuse gamma-ray background, 22, 78, 343
cosmic microwave background, 193
cosmic-ray
- diffusion, 211
- electrons, 80, 192, 207–210, 212, 222
- gradients, 217, 223
- halo, 217, 228
- in external galaxies, 229, 230
- nucleons, 207, 208, 210, 221
- propagation, 210, 253
- protons, 211
- reacceleration, 211
cosmic-rays, 9, 18, 21, 22, 25, 77, 93, 186, 207–210, 217, 221, 222, 233, 252, 267
cosmology, 253
Coulomb barrier, 236
counterparts, 88
Crab, 78, 86
Crab Nebula, 128, 133, 186, 188–190
Crab pulsar, PSR B0531+21, 91, 134
crystal diffraction, 72
- Bragg reflection, 72
- von-Laue diffraction, 72
curvature radiation, 15
cyclotron line, 370
cyclotron radiation, 19
cyclotron resonance, 86
CYGNUS, 64
Cygnus, 78, 80, 264
Cygnus A, 286
Cygnus region, 263
Cygnus X-1, 159, 161, 164–166, 168, 170–172, 174, 175
Cygnus X-3, 161, 177–179

dark matter, 358
Darwin width, 72
data analysis
- GEANT, 67
- imaging equation, 66
- maximum-entropy method, 67
- maximum-likelihood method, 67
- Monte-Carlo techniques, 66
- response function, 66
- response matrix, 66

400   Index

decay, 17
- β^+, 18
- β^-, 18
- pion, 18
- radioactive, 18
detector
- gamma-ray, 23
deuterium, 233
diffuse background, 331
diffuse continuum gamma rays, 207
- observations, 215
diffuse Galactic γ-ray emission, 80, 322, 359
- models, 219
diffusive shock acceleration, 197, 201
DIRBE, 263
DIRBE-infrared map, 81
disk component, 86, 269
Doppler broadening, 96, 99
Doppler effect, 25
Doppler factor, 301
dredge-up, 242, 243, 251, 259
dust, 214, 243, 251, 267
- presolar grains, 243
dynodes, 34

Eddington limit, 287
EGRET, 2, 56, 58, 77–79, 91, 114, 164, 178, 179, 182, 191–193, 196, 204, 215, 218, 219, 225, 291, 293, 307, 312, 316, 370, 371
electric field acceleration, 116, 118, 120
electromagnetic radiation, 9, 10, 12
electromagnetic shower, 25
electromagnetic spectrum, 1
electron bremsstrahlung, 91, 95, 101
electron induced flare emission, 95
electron–positron pairs, 37, 300, 301, 304
electron-dominated events, 104
electron-rich events, 104
electronic leakage-rate, 29
elemental abundances, 110, 122
elemental composition, 106
emission lines, 285, 286
entropy, 238
equilibrium
- hydrostatic, 242
- nuclear freeze out, 239

- nuclear quasi e., 239
- nuclear statistical e., 238
EURECA, 370
event circle, 52
evolution, 185
excitation cross sections, 275
Explorer 11, 3
explosive nucleosynthesis, 233, 244
extended γ-ray emission, 114, 120
extragalactic γ-ray background, 339, 343, 362
extragalactic X-ray background, 352, 360

fan beam distribution, 109
Fe Kα line, 310, 312
FIP effect, 110, 111, 113
fireball, see burst models
first order Fermi acceleration, 116
flare
- solar, 21, 23
flare size distribution, 102
flat-spectrum radio quasars, 293
fluorine, 268
fountain, 267
Fourier-transform telescopes, 47
free–free emission, 17, 263

Galaxy, 185
- Bulge region, 81, 269
- Carina, 263, 265
- Center region, 86, 278
- chimney, 267
- Cygnus, 263, 264
- disk, 263, 269
- fountain, 267
- gamma-ray emission, 230
- global luminosity spectrum, 228
- Gould Belt, 264
- Halo, 267, 359
- magnetic field, 213
- microquasars, 5
- Plane, 263
- rotation curve, 267
- Sco–Cen association, 264
- spiral arm, 263, 265
- structure, 213
- Vela, 263
GAMMA-1, 93, 106, 114

gamma-ray bursts, *see* bursts
Gamow peak, 236
GCN, 384
Ge-detectors
– high-purity, 36
– planar strip detectors, 70
Geminga, 5, 78, 91, 292
GINGA, 220, 370
GLAST, 71, 182, 204, 316
Goldreich–Julian density, 133
Gould Belt, 264
gradual flare emission, 112
grains, 243
GRANAR, 370
GRANAT, 2, 5, 48, 65, 77, 93, 106, 370
gravitational collapse, 244
gravitational radiation, 249
GRB 790305, 368, 369
GRB 970228, 381, 383, 394
GRB 970508, 383
GRB 990123, 388, 389, 393
GRB 990510, 385, 388
GRB Coordinates Network, GCN, 393
Great Annihilator, 269
GRIS, 44, 264, 266, 268
GRO J0422+32, 164, 166–168, 170, 174, 175
GRO J0516-609, 295
GRO J1655-40, 165, 169
ground-based γ-ray astronomy, 1
GRS 1009-45, 169
GRS 1716-249, 169
GRS 1915+105, 165
GS 2023+338, 176
GX 339-4, 165
gyrosynchrotron radiation, 104

hadronic models, 304
halo, 267
Hawking radiation, 356
HEAO-C, 5, 262, 267
heating and ionization of the interstellar medium, 275
heating of the solar corona, 103
HEGRA, 62, 196, 298
Helios-2, 65
– satellite, 367
heliosphere, 221
helium, 253

Hercules X-1, 179
HESS, 204
HESSI, 123
HEXAGONE, 31
HII region, 186, 197, 214, 263, 266
Hinotori, 93, 106, 113
history of γ-ray astronomy, 2
homogeneity, 379
horizon
– γ-ray, 24
hot-bottom burning, 243
Hubble Space Telescope, 189
hydrogen, 80, 81
hydrostatic nucleosynthesis, 233
hyperaccreting black hole, 89

IBIS, 49
impulsive flare emission, 112
in-flight calibration, 44, 46, 55
instrumental background, 25
instruments
– ASCA, 172
– BATSE, 2, 65, 77, 86, 87, 164–166, 168, 170, 175
– Beppo-SAX, 162
– Cherenkov telescopes, 178–180, 320
– COMPTEL, 2, 55, 77–79, 164, 168, 170, 174, 178
– COS-B, 3, 56, 58, 134, 177
– CYGNUS, 64
– EGRET, 2, 56, 58, 77–79, 164, 178, 179, 182, 321
– Explorer 11, 3
– GLAST, 71, 182
– GRIS, 44
– HEAO-C, 5
– HEGRA, 62
– Helios-2, 65
– HEXAGONE, 31
– Hubble Space Telescope, 88
– IBIS, 49
– INTEGRAL, 182
– Konus, 65, 369
– LXeGRIT, 68
– MAGIC, 183
– MEGA, 70
– OSO-3, 3, 42
– OSO-7, 3

- OSSE, 2, 44, 77, 86, 166–170, 172, 175, 178
- Phebus, 65
- PVO, 65
- RXTE, 162
- SAS-2, 3, 56, 134, 177
- SIGMA, 2, 5, 48, 77, 81, 83, 85, 170, 175
- Solar Maximum Mission, 31, 43
- SPI, 50
- TD-1, 56
- TIGRE, 70
- Ulysses, 65
- Vela, 65
- Venera, 65
- VERITAS, 182
- Whipple, 62

INTEGRAL, 49, 182, 255, 316
Interplanetary Network, 368, 393
interstellar γ-ray emission, 80
interstellar gas, 207, 213, 215, 217, 223, 225
interstellar matter, 77
interstellar medium, 185, 244, 262, 264, 268, 275
interstellar radiation fields, 207, 209, 214
inverse bremsstrahlung, 16, 282
inverse component, 96
inverse Compton emission, 207, 209, 219
inverse Compton scattering, 16, 22, 24
inverse-Compton process, 193
inverse-Comptonization, 306
ion induced flare emission, 95
ionized hydrogen, 214
IR-luminous galaxies, 354
IR radiation, 215, 264
iron, 254
- iron-60, 250, 267
isotropic γ-ray background, 227
isotropy, 379

jets, 22, 269, 286, 288, 290, 301, 306, 313

Klein–Nishina formula, 52, 70
knock-on electrons, 282
Konus, 65, 369

Large Magellanic Cloud, 185, 229
leptonic models, 302, 305
light cylinder, 133
light elements, 275
lithium, 252, 253
Loop I, 264
Lorentz factor, 16
low energy catastrophy, 104
low-energy cosmic-ray component, 275
low-frequency noise, 166
low-mass X-ray binaries, 175, 176, 179, 279

Mach number, 191
Magellanic Clouds, 77, 185, 229
MAGIC, 183
magnesium, 251, 253, 259
magnetar, 371
magnetic bremsstrahlung, 152
magnetic fields, 229
magnetic particle trapping, 114
magnetic reconnection, 95, 118
Markarian 421, 86, 298
Markarian 501, 86, 298
mass accretion rate, 287
mass cut, 246
mass function, 159, 162, 163
massive black holes, 78
matter–antimatter annihilation, 357
MCG-8-11-11, 291
MEGA, 70
metallicity, 233
meteorites, 243, 253
microflares, 93, 94, 102
microquasar, 21, 86, 161
microwave background, 214, 341
Milky Way, 185, 295
molecular clouds, 227
molecular hydrogen, 213, 225
molecular tori, 289, 296
Moon, 78
multi-frequency monitoring, 296
multi-wavelenght spectra, 296
MeV-blazars, 295, 306, 353
MeV-bump, 346, 348, 357, 363

nanoflares, 93, 102
narrow-line regions, 290

neon, 251, 259
network flares, 103
neutrino process, 241, 252
neutrinos, 19, 186, 238, 245, 252, 304, 305
neutron
− capture, 98, 240
− capture line, 91, 98, 101, 106, 280
− production, 97
neutron star, 21, 77, 89, 246
− angular momentum, 130
− binaries, 160, 175, 182
− history, 128
− magnetic dipole radiation, 130
− magnetic flux, 130
− mass, 130
− rotational energy, 130
NGC 1068, 288, 290
NGC 4151, 291, 308
nickel, 247, 250, 254–256
Nikischow effect, 20
nitrogen, 268
normal galaxies, 341, 351
Nova Muscae 1991, 85, 164
novae, 21, 233, 244, 249, 250, 259, 264
nuclear astrophysics, 236
nuclear binding energy, 236
nuclear deexcitation lines, 91, 96, 101, 106, 110
nuclear excitation, 18, 20, 233
nuclear force, 236
nuclear fusion, 287
nuclear interaction lines, 275
nuclear reaction, 236, 238
− network, 237
nuclear transition, 17, 18, 233
nuclei, 186
nucleonic flare emission, 95
nucleosynthesis, 77, 233
− α-rich freeze-out, 239
− explosive, 233, 244
− history, 236
− hot-bottom burning, 243
− hydrostatic, 233, 242, 244
− neutrino process, 241
− primordial, 233, 239, 253
− r-process, 240
− rp-process, 239

− s-process, 240
− shell burning, 242
− solar, 252
− spallation, 252
− yield, 235

OB associations, 197, 265, 278
occultation technique, 47
on/off technique, 41
Orion complex, 278, 280
OSO-3, 3, 42
OSO-7, 3, 91, 92
OSSE, 2, 44, 77, 86, 101, 110, 114, 166–170, 172, 175, 178, 217, 220, 255, 256, 269, 293, 307, 312
oxygen, 259

pair creation, 20, 24, 300
parent particles, 95
particle acceleration, 93–95, 108, 111, 113, 122
particle transport, 94, 95, 115
passive shielding, 30
pencil beam distribution, 109
periodicity, 369, 371
PHEBUS, 65, 93, 106, 370
Phobos, 105
phoswich detector, 45
photodiodes
− avalanche diodes, 35
− drift diodes, 35
− normal, 35
− PIN-diodes, 34
photoelectric effect, 20
photomultiplier tubes, 34
pions, 18, 305
− decay, 22
− emission, 91, 99, 101, 108
PKS 0528+134, 91, 292–294
PKS 0208-512, 295
PKS 1622-297, 293
PKS 2155-304, 298
planar shock, 117
plasma, 22, 24
positron, 18, 251, 268
positronium, 99
positronium decay, 5
power-density spectra, 167

power-law spectra, 16, 104, 106, 117, 122
primary bremsstrahlung, 99
primordial black holes, 356
primordial nucleosynthesis, 233, 253
PSR B0531+21, Crab pulsar, 138, 142, 144
PSR B0633+17, Geminga pulsar, 136, 138, 143, 146, 149
PSR B0656+14, 138, 144
PSR B0833-45, Vela pulsar, 134, 138, 143, 145
PSR B1509-58, 138, 144
PSR B1706-44, 86, 136, 138, 144, 146
PSR B1951+32, 138, 144, 146
pulsar light curve tests
– H-test, 137
– Z_m^2-test, 137
– χ^2-test, 137
pulsars, 188, 227, 247, 256, 263, 320, 328
– γ-ray beaming factor, 135
– γ-ray efficiency, 135
– γ-ray limits, 138
– electric fields, 131
– frame dragging, 133
– Geminga, 5, 292
– Goldreich–Julian density, 133
– magnetosphere, 150
– outer gap, 133, 150
– outer gap model, 329
– pair-formation front, 151
– phase of rotation, 137
– photon splitting, 151
– polar cap, 133, 150
– polar cap current, 133
– polar cap model, 329
– PSR B0355+54, 329
– PSR B0531+31, Crab pulsar, 319
– PSR B0633+17, Geminga pulsar, 329, 335
– PSR B0656+14, 329
– PSR B1046-58, 329
– radio-load, 329
– radio-quiet, 329, 336
– rotational age, 128, 131
– solar system barycenter, 136
– unipolar inductor, 131

pulse-shape discrimination, 31, 32, 52
PVO, 65
PeV energies, 87

quantum efficiency, 34
quasars, 285, 288, 290, 291, 305
quasiperiodic oscillations, 166
quiet Sun γ-ray emission, 93

r-process, 240
radiation
– blackbody distribution, 12
– bremsstrahlung, 15, 22
– curvature, 15
– cyclotron, 19
– free–free emission, 17, 263
– gravitational, 249
– inverse bremsstrahlung, 16
– inverse Compton, 16
– IR, 264
– nonthermal, 13, 15
– synchrotron emission, 14, 263
– thermal, 12, 15
radio bursts, 113
radio emission, 112
radio galaxies, 78, 285, 290, 307, 312, 353
radio-loud Seyfert galaxies, 308, 310, 315
radio-synchrotron emission, 81
radioactive decay, 18, 233, 234
radioactive isotopes, 93, 99
radioactivity, 21, 250, 268
– β decay, 251, 268
– decay chains, 234
– lifetime, 235
Ranger 3 and 5, 3
red sprites, 372
redshift, 88, 286, 301
rest-mass energy of the Sun, 88
Roche lobe, 249
Roche-lobe overflow, 159
rocking curve, 72
ROSAT, 189, 190, 371
rotation curve, 267
ROTSE experiment, 384
rp-process, 239
RXJ0852, 259
RXTE, 162, 220

s-process, 240, 243
sandwiched collimator, 31
SAS-2, 3, 56, 177
satellites
– Apollo 15, 3
– Beppo-Sax, 6, 88
– Compton Gamma-Ray Observatory, 2, 5, 44, 55, 77
– COS-B, 3
– Explorer 11, 3
– GRANAT, 2, 5, 48, 65, 77
– HEAO-C, 5
– INTEGRAL, 49
– OSO-3, 3
– Ranger 3 and 5, 3
– SAS-2, 3
– Solar-Maximum Mission (SMM), 5, 264, 267, 269
– Vela satellites, 3
– XMM, 204
Schwarzschild radius, 289, 302, 307
scintillators
– BGO, 30, 33
– CsI, 30, 33
– doped CsI, 33
– NaI, 30, 33
– plastic scintillators, 30
– pure CsI, 33
Sco–Cen association, 264
Scorpius X-1, 159, 175, 176
second-order Fermi acceleration, 118
secondary bremsstrahlung, 99
Sedov phase, 193, 198, 199
Sedov–Taylor solutions, 187
sensitivity, 39
Seyfert galaxies, 78, 86, 285, 288, 291, 307, 308, 310, 313, 314, 352
shadowgram, 48
shock, 245
– acceleration, 116, 119
– wave, 186
SIGMA, 2, 5, 48, 77, 81, 83, 170, 175, 271
SIGMA sources, 84
silicon, 254
silicon-strip detectors, 69, 71
Small Magellanic Cloud, 229

SMM, 93, 104, 105, 108–110, 113, 114, 119
SN 1987A, 199, 256
SN 1991T, 254
SN 1998bu, 255
sodium, 260
soft γ-ray repeater, 89, 369, 371
solar flares, 3, 21, 23, 78
Solar Maximum Mission, 5, 31, 43, 93, 369
solar system, 287
sound speed, 191
sources
– unidentified, 196
spaceborne γ-ray astronomy, 1
spallation, 196, 252
spectral states, 166, 168, 169, 171, 181
spectrum, 186
SPI, 50, 255
spin-down pulsars, 77
spiral arms, 80, 263, 265
star formation, 264
starlight, 24
stars, 185, 214
– massive, 185
– neutron, 186
statistical flare models, 121
stellar black hole, 5, 77, 82
stellar evolution
– burning stage, 242
– convection, 243
– giants, 242
– time scales, 242
stellar wind, 243, 244
stochastic acceleration, 115, 116, 118–120
strong interaction, 236
sub-MeV domain, 80, 86
sulfur, 251
Sun, 78
superhot component, 104
superluminal jet sources, 77
superluminal motion, 301, 308
supermassive black holes, 302, 307, 314
supernova remnants, 22, 77, 89, 185, 212, 220, 222, 225, 227, 247, 265
– Cas A, 255, 257, 258
– composite-type, 189

- Loop I, 264
- plerion-type, 188
- RX J0852.0-4622, 259
- shell-type supernova remnant, 188
- Vela, 259
supernovae, 21, 22, 77, 185, 191, 233, 244, 285, 355
- core-collapse, 185
- delayed detonation, 254
- gravitational collapse, 244, 267, 271
- shock wave, 245
- SN 1987A, 256
- SN 1991T, 254
- SN 1998bu, 255
- thermonuclear, 248, 268
- Type I, 185
- Type Ia, 248, 254
- Type Ib/c, 246, 255
- Type II, 246, 255
supersoft, 249
synchrotron radiation, 14, 80, 152, 188, 191, 263
synchrotron-self Comptonization, 303, 305

TAU A, *see* Crab Nebula
TD-1, 56
terrestrial gamma-flash (TGF), 372
TeV blazars, 305
TeV emission, 305
TeV energies, 298
TeV sources, 86, 87
thermal bremsstrahlung, 371
thermal Comptonization, 311
thermal radiation, 12
Thompson scattering, 17
TIGRE, 70
time-projection chamber, 68
titanium-44, 246, 250, 255, 257, 259, 268
transients, 86
triangulation, 367, 369, 370, 380
trigger telescope, 37
turnover kinetic energy, 117
Tycho SNR, 188, 189

UHE band, 298
Ulysses, 65, 252, 370, 371
unidentified sources

- 2CG135+01, 325, 334
- 3EG J0010+7309, 335
- 3EG J1418-6038, 336
- 3EG J1744-3039, Galactic Center, 333
- 3EG J1835+5918, 334
- 3EG J2020+4017, 335
- associations
-- black holes, 331
-- Gould Belt, 327, 332
-- OB stars, 328, 330
-- SNOBs, 328
-- SNRs, 328, 330
-- Wolf Rayet stars, 331
- correlations, 328
-- blazars, 332
- distributions, 322
- general, 319, 321
- GRO J1838-04, 325, 334
- GRO J2227+61, 336
- high-latitude, 327
- low-latitude, 330
- source catalogs, 321, 322
- spectra, 326, 328
- statistics, 322
- variability, 324
Universe, 185
unresolved γ-ray sources, 227

Vela, 3, 65, 78, 80, 263, 367, 368
Vela SNR, 189, 259
Vela X-1, 180
Venera, 65
VERITAS, 182
VHE band, 298

Watch, 370
wave–particle dualism, 9
weak interaction, 236, 238
weakly-interacting massive particles (WIMPs), 358
Whipple, 62, 196, 298
white dwarfs, 185, 248, 250, 259
WIMPs, 358
Wind, 369
Wolf Rayet star, 244, 263, 265

X-ray binaries, 159, 287, 320
- 1E 1740.7-2942, 170

Index   407

- A 0535+26, 176
- AE Aquarii, 179, 180
- AM Hercules, 179
- Centaurus X-3, 178–180, 182
- Cygnus X-1, 159, 161, 164–166, 168, 170–172, 174, 175
- Cygnus X-3, 161, 177–179, 335
- GRO J0422+32, 164, 166–168, 170, 174, 175
- GRO J1655-40, 165, 169
- GRS 1009-45, 169
- GRS 1716-249, 169
- GRS 1915+105, 165
- GS 2023+338, 176
- GX 339-4, 165
- Hercules X-1, 179
- Nova Muscae 91, 164
- Scorpius X-1, 159, 175, 176
- Vela X-1, 180

X-ray bursters, 82
X-ray nova, 5, 84
X-ray pulsars, 161, 178, 180
X-ray source
– supersoft, 249
X-rays, 191
XMM, 204

Yohkoh, 93, 104, 106, 110

Printing (Computer to Film): Saladruck Berlin
Binding: Stürtz AG, Würzburg